Wilted

CRITICAL ENVIRONMENTS: NATURE,
SCIENCE, AND POLITICS

Edited by Julie Guthman, Jake Kosek, and Rebecca Lave

The Critical Environments series publishes books that explore the political forms of life and the ecologies that emerge from histories of capitalism, militarism, racism, colonialism, and more.

Wilted

PATHOGENS, CHEMICALS, AND
THE FRAGILE FUTURE OF
THE STRAWBERRY INDUSTRY

Julie Guthman

UNIVERSITY OF CALIFORNIA PRESS

University of California Press, one of the most distinguished university presses in the United States, enriches lives around the world by advancing scholarship in the humanities, social sciences, and natural sciences. Its activities are supported by the UC Press Foundation and by philanthropic contributions from individuals and institutions. For more information, visit www.ucpress.edu.

University of California Press
Oakland, California

Chapter 6 is derived in part from Julie Guthman, "Paradoxes of the Border: Labor Shortages and Farmworker Minor Agency in Reworking California's Strawberry Fields," *Economic Geography* 93, no. 1 (2017): https://www.tandfonline.com/doi/abs/10.1080/00130095.2016.1180241. Copyright Clark University.

Library of Congress Cataloging-in-Publication Data

Names: Guthman, Julie, author.
Title: Wilted : pathogens, chemicals, and the fragile future of the strawberry industry / Julie Guthman.
Description: Oakland, California : University of California Press, [2019] | Series: Critical environments: nature, science, and politics ; 6 | Identifiers: LCCN 2018058268 (print) | LCCN 2019000194 (ebook) | ISBN 9780520973343 (ebook) | ISBN 9780520305274 (cloth : alk. paper) | ISBN 9780520305281 (pbk. : alk. paper)
Subjects: LCSH: Strawberry industry—California. | Strawberries—Diseases and pests—Control. | Soilborne plant pathogens—Control.
Classification: LCC HD9259.S83 (ebook) | LCC HD9259.S83 U625 2019 (print) | DDC 338.1/747509794—dc23LC record available at https://lccn.loc.gov/2018058268

Manufactured in the United States of America

28 27 26 25 24 23 22 21 20 19
10 9 8 7 6 5 4 3 2 1

CONTENTS

ILLUSTRATIONS

MAPS

FIGURES

ACKNOWLEDGMENTS

Much of the data contained in this book was gathered in the context of two research projects supported by the National Science Foundation (award nos. 1228478 and 1262064). I was delighted to receive these grants and hope that the program officers and other representatives of the NSF continue to successfully impress upon Congress the societal importance of research that does not necessarily lead to direct applications and technology transfer.

Strawberry fruit may be beloved by many, but not so the California strawberry industry, which has been subject to a great deal of critique—and is nearing research fatigue. It wasn't easy cold calling growers and some others in the industry, much less securing interviews. I am especially grateful to the seventy-five growers, fifty workers, and dozens of other industry representatives who took the time to meet with me or members of my research team and share their knowledge. Regardless of whether they did so out of a sense of duty, generosity, or the opportunity to have their perspectives made public, I commend them for engaging with me and other scholars, and hope they feel that I've been faithful to their truths. Researchers can only say as much as they learn.

Due to promises of confidentiality embedded in my human subjects' protocol, I cannot reveal the names of the most generous research subjects, but I trust they know who they are. I can, however, reveal names of those who provided important informational interviews (those devoid of personal attributes or opinions). These include Brian Leahy of the California Department of Pesticide Regulation, Jenny Broome of Driscoll's, Henry Gonzales of the Ventura County Agricultural Commission, Karen Stahlman of the Monterey County Agricultural Commission, Mary Lou Nicoletti of the Santa Cruz Agricultural Commission, and Carolyn O'Donnell of the California Strawberry Commission.

I want to recognize Sandy Brown, without whom this project would never have come into being, and express my regret that other demands took her away from it. Thankfully, the gaping lacuna she left was filled by a splendid array of students, who completed tasks large and small to bring the research to completion. These include Yajaira Chavez, Zoe Chertov, Savannah Coker, Rachel Cypher, Hanan Farhan, Jean Larsen, Sierra McCormick, Sarah Palmer, and Alex Sauerwein. I am most grateful to Madison Barbour, whose intelligence and curiosity so wowed me in her frosh year that she became my primary research assistant. While competent and willing to complete the most mundane tasks, she stepped up to tasks befitting an advanced graduate student in both conducting some field research and developing publications.

This book is a different beast than the journal articles that reported on my original research questions, requiring ecological as well as social and political-economic explanation. As someone who hasn't taken a biology class since high school and never took chemistry, I was challenged by many questions I wanted to ask and explanations I wanted to give. I owe a great deal of thanks to the scientists I met along the way at the University of California at Davis and Santa Cruz (UCSC), and later at Harvard University, who took the time to help with my undoubtedly naive-sounding queries. These included Glenn Cole, Louise Ferguson, Greg Gilbert, Tom Gordon, David Hibbett, Louise Jackson, Steve Knapp, Jane Lipson, Margaret Lloyd, Don Pfister, and Kate Scow. Special thanks to Carol Shennan at UCSC, with whom I met multiple times. Those working at university extension and the US Department of Agriculture were also helpful interpreters. Many thanks to Eric Brennan, Steve Fennimore, and Steve Koike.

I have been extremely fortunate to have received outstanding support in the form of sabbatical leave, fellowships, and writing residencies to complete the book. During a 2015–16 sabbatical courtesy of UCSC, my home institution, I was able to spend two weeks at the rustically beautiful and contemplative Mesa Refuge in Point Reyes, where I saw that there was a compelling book in this material; three weeks at the Sydney Environment Institute, where I was able to clear off some other nagging work; and four weeks at the absolutely breathtaking Rockefeller Bellagio Center, where I was able to churn out an outline and rough drafts of three chapters. In 2017, I hit the proverbial jackpot to receive 2017–18 fellowships from both the John Simon Guggenheim Foundation and the Radcliffe Institute for Advanced Study at Harvard University, the latter endowed by Frances B. Cashin. As a residential fellowship shared with fifty other artists and academics, the Radcliffe

Institute provided not just outstanding writing and library facilities, but also learning and friendship opportunities. Thank you to those whose support helped me get to these places: Susanne Freidberg, Don Mitchell, Nancy Peluso, Elspeth Probyn, Rachel Schurman, Richard Walker, and Michael Watts. In addition, I wish to thank just some of the others with whom I have shared conversation and laughter over the years, and whose imprint on my thinking is profound: Charlotte Biltekoff, Aaron Bobrow-Strain, Joe Bryan, Melissa Caldwell, Hugh Campbell, Melanie DuPuis, Susanne Freidberg, Ben Gardner, Jill Harrison, Jake Kosek, Rebecca Lave, Geoff Mann, Becky Mansfield, James McCarthy, Scott Prudham, Paul Robbins, Amy Ross, and Wendy Wolford. Mary Beth Pudup and Andrea Steiner have been important colleagues in the everydayness of university life.

I am grateful for the feedback I obtained when presenting slices of this book in talks at the Royal Geographical Society / Institute for British Geography conference; a workshop in Oslo, "Food's Entanglements with Life," sponsored by the European Research Council Overheating grant (295843); the University of Utah; Yale Agrarian Studies; the University of Buffalo; Boston University; Brown University; and Harvard STS Circle. Previous to the these talks, I received useful comments when presenting on other aspects of the research at the University of Hawai'i at Manoa, the University of Sydney, the Agri-food Research Network meeting in New Zealand, Dartmouth College, the University of Minnesota at Minneapolis, the University of Washington at Seattle, the University of North Carolina at Charlotte, the University of California at Davis, the University of California at Irvine, Colorado University at Boulder, and the University of Iowa, as well as various disciplinary conferences. At several of these events I collected comments or suggestions that really stuck with me and went on to inform my arguments in a significant way. I wish to thank the following scholars for such nuggets: Filippo Bertoni, Sarah Besky, Mark Bomford, Marion Dixon, Lisa Heldke, Hannah Landecker, Jamie Lorimer, Mara Miele, and Alex Nading. During fellowships, Patrick Keefe, Shireen Hassim, and Janina Wellman likewise inspired some useful directions. As she has done previously, Becky Mansfield helped me gain clarity about my arguments at multiple junctures. And longtime friend Debora Pinkas has given me advice on just about anything whenever I asked.

While at the Radcliffe Institute at Harvard, I organized a dream team to workshop works in progress. I clearly got the long end of the stick, as members of the group were so generous as to read and discuss several draft chapters.

Abundant thanks to Alex Blanchette, Susanne Freidberg, Lisa Haushofer, Allison Loconto, and Wythe Marschall for their incisive comments. I also want to acknowledge Evan Hepler-Smith, Jane Lipson, and Adam Romero for both instruction and comments on the chapter addressing chemistry. I owe a huge debt of gratitude to Adam, who sent me pages of unpublished research to use. The chapter would not have been possible without him.

Several kind souls did the truly generous task of reading the manuscript in its entirety. A big thank you to Alex Blanchette, Christopher Henke, Rebecca Lave, and Emily Reisman for their comments. My old friend Jerry Kohn also read the manuscript just for the hell of it. Very kind indeed.

At the University of California Press, Kate Marshall has been a stalwart colleague and keen editor. I appreciate all of her advice. Enrique Ochoa-Kaup, Jessica Moll, Lindsey Westbrook, and Ellen Sherron magnificently helped usher this manuscript to publication. Tom Sullivan did his usual magic with publicity.

Finally, and of course not least, I wish to thank the two people who most closely and lovingly inhabit my life: Michael and Sierra. Both have put up with my nonstop work habits, kept me sane, and helped make our household a pleasant and nourishing environment. Michael continues to be my giant supporter (he'll get the joke), and Sierra makes it all worthwhile. And then there's the dog, Bernie, who has more energy than the sun and is one piece of work. At least she keeps it on the funny side during dark times.

This book is dedicated to my mom, whose lifetime of support did not go unnoticed. Unfortunately, by the time this went to press, she could no longer appreciate my thanks. I wish her last moments had been as she had wished.

MAP 1. California, with key locations indicated. Map by Bill Nelson.

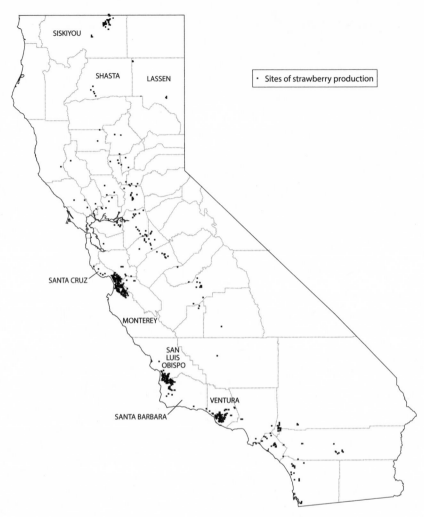

MAP 2. Strawberry production regions in California. Map by Bill Nelson.

Prologue

THE BATTLE AGAINST METHYL IODIDE

IN APRIL 2010, THE DIRECTOR OF THE CALIFORNIA Department of Pesticide Regulation (DPR), Mary-Ann Warmerdam, announced her intent to register the chemical compound methyl iodide as a replacement for methyl bromide. Methyl bromide had long been used as a soil fumigant in California strawberry production, injected into soils in advance of planting in order to kill weeds, nematodes, and soilborne pathogens, especially. Under certain air pressures, gaseous fumigants like methyl bromide can propel active ingredients throughout infested fields. Aboveground, they dissipate into the air, spreading these same materials into places where their effects are not wanted—for instance nearby communities or the upper atmosphere. As it happens, methyl bromide easily dissipates into the atmosphere, enough to have been categorized as a class-one stratospheric ozone–depleting chemical. It is for that reason that methyl bromide was seeing its last years of use, destined for phaseout in compliance with the internationally approved Montreal Protocol on Substances That Deplete the Ozone Layer.

An unprecedented wave of protest followed Warmerdam's announcement to register methyl iodide.[1] Unusually, an array of activist organizations, including anti-pesticide, environmental, public health, farmworker, and "foodie" groups, joined together to mount a major campaign to prevent the chemical from coming into use. They argued that it was even more acutely toxic and environmentally degrading than methyl bromide. They cited reports showing it to be a known neurotoxin and carcinogen, associated with suppression of thyroid hormone synthesis, respiratory illness, and lung tumors, and a probable cause of miscarriages and birth defects.[2] They noted that it had even been used to induce cancer in laboratory rats. And they made special efforts to communicate that those who would be put at risk were farmworkers, neighbors, and others in the vicinity of strawberry fields. For, unlike methyl bromide, methyl iodide would stay close to the ground and not dissipate into the upper atmosphere. Indeed, the chemical was so earthbound that the US Environmental Protection Agency had bestowed the maker of the chemical, Arysta LifeScience, with an Ozone Protection Layer Award!

To battle the chemical, the activists first launched an internet "clicktivism" campaign. Fifty-three thousand people responded with public comments to DPR opposing the chemical's registration. Notably, an appreciable number objected to the chemical with comments suggesting that their own health as consumers was at stake. "If you move forward on this proposal, I will ensure that I either purchase berries that were not turned into cancer carriers or omit the fruit from my daughter's diet," wrote one of many parents.[3] Warmerdam not only dismissed comments like these that misunderstood the nature of fumigation; she also opted to discount those that didn't deploy scientific research, which was virtually all. When she went on to approve methyl iodide for use in an emergency registration in December 2010, activists rebounded with petitions, picket lines, mock fumigations, public hearings, and more. Many of these actions were designed to shed light on the specious ways in which Warmerdam had approved the chemical, including her dismissal of public comments. In hopes of having the registration revoked, activists also filed a lawsuit against the DPR and director Warmerdam, as well as Arysta LifeScience. Most of the counts were about the failure of those responsible for the registration process to abide by California environmental laws for transparency in decision making and robust assessment of potential health and environmental effects. Warmerdam had even neglected the recommendations of her own staff and scientific review panels.

Just in advance of a court ruling on the lawsuit, Arysta voluntarily revoked its request for registration of the chemical in California and announced its plans to suspend operations in the US market altogether. In its press release announcing the withdrawal, the company publicly stated that the chemical was no longer economically viable. As it happens, very few growers had registered to use it. Some were unsure of its efficacy; many more feared further protest. The California Strawberry Commission (the industry trade organization) and several large shippers had been reticent to recommend adoption as well, concerned that retailers would no longer carry strawberries. Based on these events, the campaign to end methyl iodide was heralded as a major anti-pesticide and food movement achievement.[4]

My interest in the California strawberry industry was piqued during the battle over methyl iodide. After witnessing how much the food movement had, over the years, come to focus on consumer health and desires, I was relieved to see a turn to farmworker and community health. After participating in the growing critique that the movement had too readily adopted market-based approaches to change, I was intrigued to see it pursue a more oppositional and overtly political approach.[5] I was also interested in what kinds of knowledge were being brought to bear in the debate over the chemical. During this period, my Ohio State University colleague Becky Mansfield and I were writing on environmental epigenetics—the science that examines how environmental toxins alter gene expression and hence biological development—and I was curious about whether knowledge of the potential epigenetic effects of methyl iodide was entering into the regulatory discussions regarding farmworker lives and health. I specifically wondered whether the potential that the children of farmworkers might be adversely affected was changing the public conversation about agrochemicals. The answer to these questions turned out to be "no," as I learned in my subsequent research. But because methyl iodide was withdrawn from the market, I embarked on further research to investigate how growers would cope without methyl bromide or methyl iodide, and with tighter restrictions on the remaining fumigants.

In the course of visiting farms and talking to growers to learn about these questions, I began seeing things that far exceeded the scope of my research questions. What I found was a complex and deeply entrenched web of connections in the strawberry production system. These connections reflect the multiple ways that the California industry has coevolved with the plants, soils, chemicals, climate, and other nonhumans that roughly constitute the nature-based conditions of strawberry production, as well as the regulations, labor and land markets, scientific institutions, and marketing arrangements that roughly constitute the social conditions of production.

As I thought about it further, I came to realize how important fumigants are to this assemblage of soils, plants, biophysical conditions, and labor and land markets—they are in many ways the glue that holds much of it together. Economy, ecology, and science all developed with the presumption that fumigants were here to stay. Expectations of annual fumigations are built into strawberry land values and leasing arrangements. Workers' wages are calibrated based on the high yields that fumigation allows, and fumigation plays a role in labor recruitment, too. Expectations of fumigation are even bred into the modern strawberry itself, not least because breeding trials often take place in fumigated soil. Fumigation allows long periods of growth so that strawberries can be in the market year-round, and along with other innovations has boosted yields to the point that strawberries are widely affordable. With an entire strawberry industry hinging on something so contentious, I realized that the business itself may be as fragile as the fruit it produces. *Wilted* tells how this situation came to be, and what it means for the future of strawberry production in the Golden State.

ONE

California Strawberry Assemblages

Effective soil fumigation has been the forerunner of dramatic changes in the California strawberry industry. Instead of growing the crop 4–6 years, it is now grown as an annual or biennial crop, and planting is timed for each variety to achieve high first-year yields. First-year berries are superior to those of later years in fruit size and quality, and are the most economical to harvest. Also, through the research of Driscoll Strawberry Associates, Inc., a California corporation, it has become possible to grow a considerable acreage of the large-fruiting everbearing class of strawberries (the French remontant class). These exceptionally fruitful strawberries could not be grown on non-fumigated land because of extreme susceptibility to root diseases. Commercial breeding for *Verticillium* wilt resistance in strawberries has now been discontinued in California, and the breeding, thus, has been greatly *simplified*. Most importantly, soil fumigation has made lands available for strawberries which were previously avoided. These were the rich, fertile, alluvial lands with long crop histories.

Agricultural Scientists Stephen Wilhelm, Richard C. Storkan, and John M. Wilhelm, "Preplant Soil Fumigation with Methyl Bromide-Chloropicrin Mixtures for Control of Soil-Borne Diseases of Strawberries: A Summary of Fifteen Years of Development," 1974[1]

The *simplifications* of industrial farming multiply beyond the original target species. The multispecies modifications create ever more monsters—exploding numbers of parasites, drug-resistant bacteria, and more virulent diseases—by disrupting and torqueing the species that sustain life. The ecological simplifications of the modern world—products of the abhorrence of monsters—have turned monstrosity back against us, conjuring new threats to livability.

Anthropologists Heather Swanson, Anna Tsing, Nils Bubandt, and Elaine Gan, "Introduction: Bodies Tumbled into Bodies," 2017[2]

IN 2015, I WAS INVITED TO SACRAMENTO, California's state capital, to discuss my research with the director of the Department of Pesticide Regulation (DPR), Brian Leahy. Leahy is a former organic farmer who once presided over California Certified Organic Farmers, one of the premier organic farming organizations in the United States. Leahy was appointed director in 2012 by Democratic governor Jerry Brown following that tumultuous period when the previous DPR director, appointed by Republican governor Arnold Schwarzenegger, had all but ignored California environmental laws and her own agency staff in registering the highly toxic soil fumigant methyl iodide for use. It was expected that Leahy would take a more balanced approach to pesticide regulation, using science to weigh growers' needs against increasing concerns about the human and environmental health effects of agrochemicals. Having held many leadership roles in agriculture, Leahy had a reputation for working collaboratively with environmental organizations, agricultural groups, trade associations, and local government officials.[3]

I had just completed the interview phase of a project designed to understand how strawberry growers were faring with tighter regulations on soil fumigants. These regulatory changes included not only the international phaseout of methyl bromide and the abrupt withdrawal of methyl iodide from commercial use, but also tighter use restrictions on the remaining allowable chemicals. For fifty years growers had been using fumigants to control soilborne pests, most notably the fungal pathogen *Verticillium dahliae*, which can make plants wilt and die. At every regulatory juncture, the industry claimed that without these chemicals, it itself would wilt and die, and consumers would no longer see the luscious berries stacked on supermarket displays year-round. Director Leahy summoned me specifically to ask where strawberry growers now stood with soil fumigants. Just two years before, under his leadership, DPR had published an action plan for the development of practical and cost-effective ways to grow strawberries without soil fumigants. Along with laying out several lines of research for industry investment, the report suggested that fumigants were not long for this (California) world.[4] So Leahy was genuinely curious to know whether strawberry growers were undigging their heels, as it were.

I told him that fumigation restrictions were just one of the concerns irking growers. They were also complaining, mightily, of labor shortages, drought, high land values, low crop prices, and ... bad press. "Yeah," he said, "the strawberry production system is insanely complicated." He was not the first or last to make such a comment. How can a crop, for many imagined as an inconsequential spring delight, garner so much adversity?

In many regions of the world, strawberries are a minor crop, available for a few short weeks in the late spring. But in California, specialty crops, grown for a national market, are big business. As early as the 1870s California farmers were abandoning wheat and barley production to produce oranges, stone fruit, and grapes—crops that were highly desirable if not always essential, according to the nutrition canons of the day. Dried, canned, or refrigerated, these crops were shipped in railway cars so consumers in colder climes could have a taste of summer year-round. Intensive vegetable production began some three decades later, when iceberg lettuce gained ascendance.[5] Strawberries were late in taking their place among California's pantheon of specialty crops. But by 2017, they were the sixth most important crop in terms of sales. In that same year, California was growing 88 percent of the nation's strawberries, while Driscoll's, a California company albeit with operations elsewhere, was selling 29 percent of the world's.[6] Only in recent years have other berries become economically important as well, as many of the major strawberry shippers have diversified into blueberries, blackberries, and raspberries. But strawberries remain the undisputed leader in the field, even as they are reportedly the most challenging to grow.[7]

California strawberries became big business because of the extraordinary gains in productivity that fumigation and other technologies propelled. With such productivity the strawberry industry needed equally robust markets and, thus, it needed consumers who would see strawberries as a near necessity. Luckily, changing ideas in nutrition came to its aid. Nutritionists rarely see eye to eye on anything these days, but one thing they do agree on is that fresh fruits and vegetables should be the cornerstone of diets. Among recommended fruits, berries rate as particularly virtuous. Not only are they not too sweet—a problem for the glucose-concerned crowd—they are supposedly chock-full of essential vitamins, minerals, fiber, and antioxidants.[8] Parents love them because their kids will eat them—one of the few fruits and vegetables that don't require too much cajoling. As it happens, much public knowledge of the health benefits of strawberries came as a result of the vigorous public-relations efforts of the California strawberry industry, whose gluts in production compelled attention to marketing.[9] These efforts apparently paid off. Per capita consumption of fresh strawberries in the United States almost doubled between 1994 and 2014, and berries as a group became the number one produce category for US grocery retailers.[10]

Despite these successes, the California strawberry industry is undoubtedly beleaguered. And it *has* had a lot of bad press. Take the report "California's Strawberry Industry Is Hooked on Dangerous Pesticides," published by the

FIGURE 1. Cheap, abundant strawberries at Haymarket Boston in February. Photo by author.

Center for Investigative Reporting in its *Reveal News* in 2014. In that report, reporters blasted the industry for its use of highly toxic soil fumigants and called out regulators for failing to adequately control them.[11] Or consider that strawberries continue to rank first in the Environmental Working Group's "Dirty Dozen," the list of fresh fruits and vegetables tested to have the highest amounts of pesticide residues.[12] This ranking does not even include soil fumigants, which are applied to the soil before plants go in the ground and therefore do not directly contact the fruit. With the highest pounds per acre of active ingredients applied, strawberry production entails the most intensive agrochemical regime of all California crops.[13]

Pesticides are not the only arena in which the strawberry industry has been the object of journalists' derision. In 1995, Eric Schlosser, future author of the muckraking *Fast Food Nation* (2001), published an exposé in the *Atlantic Monthly* about the California industry. The piece condemned a sharecropping system in which farmworkers are enticed into becoming farmers, incurring mounds of debt along the way.[14] Nor have labor pay and conditions escaped the eyes of the press. Beginning in 2015, Driscoll's became subject to a highly publicized boycott when strawberry workers on both sides of the US-Mexico border called for union negotiations to address the poor pay of strawberry workers at certain berry farms. Although Driscoll's wasn't the chief offender—and Driscoll's itself doesn't even have farming operations—the idea of the boycott was to pressure Driscoll's, as the largest berry shipper in the world, to exert leverage on its contract growers to recognize a union.[15]

Denunciations of labor conditions and pesticide use have been standard fare for specialty crop industries—those that produce high-value fruits and vegetables. This is because the delicateness and perishability of many specialty crops require abundant, cheap labor at harvest time and chemical treatments to make the produce both affordable and attractive.[16] But, unusually, the strawberry industry has received a spate of unflattering press about its plant breeding arrangements, too. This occurred when two University of California plant breeders announced their intentions to leave the university and join a private plant breeding company where they could make a lot more money. A series of lawsuits ensued, precipitating much bad faith among institutions that were once allied and accusations that at least some in the industry are driven by naked greed.[17]

No wonder the strawberry industry has become so defensive—and elusive. Websites of industry organizations and shippers increasingly emphasize the industry's contributions to sustainability, grower and farmworker livelihoods, and economic stability in strawberry farming communities, while the ever-enlarging group of interested researchers and journalists find that actually talking to people in the industry is pretty challenging. It was no small matter for me as a researcher to get in the door to speak with growers and other industry representatives. Some of those who generously agreed to be interviewed did so based on the implied understanding that I would tell their side of the story. In certain respects that is what I am going to do in this book, although perhaps not always to their liking.

Wilted is not a muckraking account, and I'm not interested in shaming the strawberry industry just because. My goal, instead, is to show how the very features that once made strawberry production so lucrative in the Golden State now pose grave threats to that very industry. It is not only that chemical fumigation is under the gun because of its toxicity to humans. It is that the entire production system has been built on the presumption of fumigation, rendering it resistant to change—at the same time that several other once-advantageous conditions have evaporated, leaving a suite of problems that are all the more intractable because of their interconnections.[18] Moreover, years of managing pests with chemical solutions amid dynamic environments has unleashed organisms that defy control. These heterogeneous and interactive threats make it nearly impossible to continue to produce what was once a luxury crop, available for a short time and at high prices, for the mass market. In that way, the solution of fumigation, once lauded for its efficacy and the "dramatic changes" it brought to the rest of the production system (as noted in the opening epigraph) has become the problem. Fumigation, I suggest, is the source of iatrogenic harm, referring to the problem of a cure causing illness.

The uncertain fate of the California strawberry industry certainly makes for a cautionary tale about industrial agriculture, referring to scaled-up, *simplified* monoculture accompanied by forms of exploited and often spatially transported labor.[19] It also exemplifies more generally the frailties of the so-called plantationocene, a term coined with the "ocene" suffix to denote plantation agriculture's imbrications with human-induced planetary crisis.[20] Scaled-up agriculture—with its dependence on environment-changing fossil fuels and pesticides, that is—has both contributed to the crisis of the so-called Anthropocene, but is also highly vulnerable to the pests, pathogens, and other environmental problems (for instance saltwater intrusions) that have come with climate change.[21] But unlike some who have deployed the arguably apocalyptic language of the plantationocene in oddly optimistic terms, I'm less certain that ruination is an assured outcome, or sanguine that it presents a way forward. The social, economic, and environmental conditions in which strawberry plantations are embedded, not least of which are the high-octane real estate markets of California, are unlikely to create the space for more heterogeneous and de-scaled kinds of food production anytime soon.[22]

Unfortunately, my conclusions are unlikely to satisfy either activists or the industry. Activists imagine an agro-ecological ideal that can be achieved with the right kind of experimentation. They imagine that the problem lies with the intransigence of farmers. I will show that it's the intransigence of the entire edifice that has been created through 150 years of strawberry growing in California. For their part, the industry sees a public out of touch with the realities of growing food that is affordable, appetizing, and widely available. The industry wants to stop being shamed and gain public acceptance of its practices. Those in the industry imagine that the problem lies with public misperceptions of the possible. Both parties, in other words, see the problem as one of opposing worldviews that need to be altered.[23] While it cannot be denied that activists and growers see the challenges differently, neither party wants to admit how political-economic limits have interfaced with ecological dynamics to make sustainable and just strawberry production highly elusive except in rare and not readily replicable cases.

EXPLAINING INTRANSIGENCE

Wilted traces how California strawberry production, so ripe with possibility in the early years, became so challenging. Much hinges on the emergence of soil-based plant pathogens and the solution of chemical fumigation as a way to address them. Once widely adopted, fumigation reverberated throughout the rest of the production system—in plant breeding, land access, labor practices, marketing, and more—locking in a particular way of doing things, at the same time that the social and ecological conditions of strawberry production were themselves changing to make fumigation less effective. Elaborating this explanation requires attention to three different kinds of actors and, in two cases, their guiding rationales. The first is *growers*, whose embeddedness in political-economic dynamics typical of agro-industry has made fumigation seem to them a necessity; the second is *agricultural scientists*, whose role has been to support growers through practices of repair; and the third has been the *multifarious nonhuman entities, materials, and forces* that have collaborated with the industry at some moments and thwarted it at others. Together, these actors have formed what I will refer to as a *more-than-human assemblage* that has increasingly come up against the *limits of repair*. In discussing the scholarship that has brought attention to the roles of these three groups of actors as well as to the fragility of agricultural assemblages, I provide a methodological framework for understanding the fate of the strawberry industry in California.

Growers and Political-Economic Dynamics

Although romantics like to see farmers as pursuing the virtuous vocations of tending land and feeding people, modern growers are businesspeople, imbricated in the dynamics of capitalism. They grow food to make a profit, and therefore they worry about accessing capital and having crop yields and sales adequate to pay their debts, wages, and land rents. This mindset has been especially true in California. Early orange growers in California, for example, saw themselves as businesspeople, not "dirt farmers," and approached the work of fruit production with the same zeal as their corporate brethren, embracing industrialization at every turn.[24] Today, as geographer Richard Walker has detailed, California agriculture has been saturated with capital through and through—capital, he writes, is the "invisible thread" "that weaves together all of the elements of the agribusiness system."[25] It is hardly a stretch, therefore, to draw on capitalist exigencies to explain the strawberry industry's heavy reliance on chemical fumigants. Indeed, social scientists of agriculture have typically employed the tools of agrarian political economy precisely to explain how farmers meet numerous challenges in crop (and animal) production.

At the core of explanations in agrarian political economy are questions of how agricultural industries have formed, and how they have come to both serve farmers and constrain them. As it happens, many of these explanations also revolve around the role of nonhumans in farm production. Indeed, agrarian political economy's central departure from classical political economy is its attention to the difference nature makes in agricultural production, distinct from in manufacturing, and how those differences create particular challenges for growers.[26] For one, unpredictable weather, perishability, seasonality, and various pests are major sources of risk. Yields may falter from disease, for example, or crops may rot before they are sold, diminishing farmers' chances to earn revenue.[27] Second, land in agrarian production is not just a site of production, as it is in manufacturing. Land itself is a condition of production, making soil fertility and quality something of value, and diseased or degraded soil something of concern. Since the quality of land affects yield, it also affects land costs.[28] Third, agricultural labor processes are different than in manufacturing. Workers do not actually apply their labor directly to produce crops, which grow through biological processes. Therefore, the role of agricultural laborers is to tend to these biological processes to enable yields, and then of course to harvest the crops once they are ready.[29] Given the seasonality of cropping, this labor is rarely needed year-round, which in theory has created challenges in recruitment.

Much of technological development in agriculture has been spurred by these challenges, with the aim of mitigating crop risk, making land more fertile, and easing and smoothing the work of farmers and farm laborers. Agrochemicals, fertilizers, breeding-enhanced cultivars (plant varieties), and farm machinery are all in these ways supposed to assist farmers—to make their incomes less volatile. But the production of these technologies has largely been hived off by businesses that manufacture the inputs and sell them back to farmers, in a process that agrarian political economist David Goodman and colleagues coined "appropriationism."[30] Examples of how appropriationism tends to reduce the centrality and risk of on-farm processes, yet forces farmers to pay for inputs, include the replacement of animal power with farm machinery, saved seeds with hybrid or transgenic seeds, cover crops and manure with synthetic fertilizers, and in situ biodiversity as pest control with manufactured chemicals.

Meanwhile, growers have generally not been well positioned to market their crops. Not only is marketing a distinct task from farming, but moving fruit to distant markets is more economical with dedicated cooling, packing, and shipping facilities to address issues of fragility and perishability. Although many of the original shippers of California fruit were organized as growers' cooperatives, including those for strawberries, nearly all fruit shippers have since become for-profit businesses. Some distribute only crops that others grow, while others—grower-shippers—market their own and others' crops. Since there have been many more growers than shippers, and growers must sell the crops in which they have invested or lose money, growers generally have had little bargaining power relative to shippers. Shippers have thus tended to set the terms of contracts with growers, including the prices they will pay for produce and the quality they expect. Even marketing cooperatives have tended to set high standards and often low prices.[31] Furthering their position of strength relative to growers, some shippers have developed capacity at both ends of the supply chain: selling inputs to farmers and then buying the fruit back for marketing. This remains the business model of Driscoll's, which fairly long ago ceased farming except for research purposes.

Suppliers and buyers are not the only ones that eat into growers' profits. To access the capital to pay for their up-front investments, growers must borrow from banks or others, and they must pay interest on those loans. To obtain the vital condition of production called land, they must pay either rents to landowners or mortgage payments to banks, unless they are so lucky as to have inherited land. And of course to access the labor of others to tend and harvest the fruit, they must pay wages—and sometimes a lot of wages, given the labor intensity of a fruit like strawberries. In short, growers must pay suppliers, creditors, landowners, and workers, while they are subject to what are often the low prices received from buyers, along with quality standards that force growers to discard crops that buyers deem unsellable.[32] Under these circumstances, it is no wonder that growers feel they have little choice but to use chemicals in the (anodyne) name of "crop protection." Without chemicals, their quality and yields may decline to the extent that their sales are inadequate to pay their expenses. Mortgage bankers and insurers may even stipulate that growers use pesticides.[33]

In effect, as agrarian political economists have argued, farmers are squeezed between suppliers and buyers.[34] Growers in fully capitalized farming systems, as California strawberry growers are, must purchase their farm equipment, fertilizers, seeds or (in the case of strawberries) starts, irrigation infrastructure, and agrochemicals from suppliers and then, once crops are ready, be subject to buyers' exacting quality standards and prices. Under these conditions, growers are attracted to yield-enhancing technologies in hopes that additional sales will help their income. Plant breeding, in that way, has come into play as a technology of not only risk reduction but also yield improvement, making plants grow bigger and faster.[35] The folly here is that as innovative farmers adopt yield-enhancing technologies, others join in, and production increases even more. Scholars have referred to this process as the technology treadmill, and strawberry growers are implicated in it as much as anyone. In the end, overproduction is a fool's game, since excess supply in the market generates competition and the poor prices that ensue—unless markets themselves are expanded.[36] Such competitive dynamics are beneficial to consumers, however, when they result in lower supermarket prices.

What I have described thus far is how most strawberry growers see their predicament: if they do not fumigate, they risk significant crop loss, and they may not survive economically. As individuals they of course vary in the degree of risk they are willing to face, prices they are able to obtain, and concern with the health risks of fumigants. But without a scalable alternative, or an economic cushion from elsewhere (for instance an inheritance), this is the bind they face. If they don't adopt the latest yield-enhancing technologies, they lose out, too, as others will, and prices will fall regardless. But they do not make these decisions on their own. They have turned to universities and other supporting institutions for both the development of these technologies and advice about when, where, and how to use them.

Agricultural Scientists and Institutions of Repair

In *Brazil and the Struggle for Rubber: A Study in Environmental History* (1987), environmental historian Warren Dean writes of an agricultural industry that never came to be due to the fungus *Microcylcus ulei*, endemic to the Amazon region. Although tapping wild rubber trees was of critical importance to Brazil's entry into the world economy, with rubber once comprising 40 percent of its exports, efforts at rubber cultivation were unsuccessful. The action of the fungus, discovered in plantations in Trinidad, defoliated the trees and decreased yield tremendously when it did not cause death. Wild trees were apparently spaced sufficiently apart to prevent a buildup of the pathogenic inoculum (the disease-inducing plant material), but not so plantation trees. Yet, as Dean argues, it was not the fungus per se that thwarted the development of the industry. Rather, human ignorance of a means to stop or ameliorate fungus attacks rendered rubber cultivation uneconomical.[37] This ignorance rested on insufficient institutional investment in investigating the problem and, hence, developing solutions. And so rubber cultivation was taken up elsewhere, where the fungus was not endemic, inexorably altering Brazil's economic development. The ill fate of the Amazonian rubber industry can be contrasted with that of the banana industry, which was also plagued with disease in its earlier years. Here the culprit was a strain of *Fusarium*, specifically *Fusarium oxysporum f. sp. Cubense*, responsible for what came to be known as Panama disease. Rather than thwarting the industry's development, as environmental historian John Soluri tells it, the appearance of fungal disease gave rise to new scientific endeavors, ultimately creating an institutional apparatus that would become part of the banana industry.[38]

Managing the biological characteristics of production to ensure the integrity of the end product is essential in any agricultural production scheme.[39] Since management requires knowledge, and since organisms, as well as inorganic elements, cannot speak for themselves, science has served as both a mediator of understanding as well as a means of improvement. Yet science is both a collective endeavor and a costly one, requiring not only instrumentation, laboratory space, and biomaterial but also collections of past knowledge to draw upon.[40] As a result, individual growers, in need of scientific expertise, have rarely been able to manufacture it alone. Instead, the development of agriculture and agro-forestry industries has required pooling resources and enlisting science and the state to do what private capital would or could not do by itself.[41]

And the state has for the most part complied. In the United States this support has largely come from the land grant universities, agricultural experiment stations, and research and extension services, all founded on the proposition that the government should support the practical professions.[42] Cooperative extension services, especially, enshrined in the Smith-Lever Act of 1914, were created to teach farmers the latest in agricultural techniques designed and tested by the agricultural experiment stations. With farmers as the primary clientele of research and extension, the agenda for agricultural science was set with farmers' interests in mind, and scientific findings have been translated in ways that are applicable to farmers.[43] Therefore, many extension scientists engage in the business of what the historian of science Christopher Henke calls "repair." Repair connotes the work of maintaining a system in the face of constant change—and sometimes crisis.[44]

In general, land grant university scientists have been highly responsive to the needs of farmers, and this is certainly true in California. In California fruit production, specialized shippers produced specialized growers whose economic advantages lay in specialized equipment and know-how. But the resulting monocultures attracted insects with no natural enemies. When pests threatened these emergent agricultural industries, the land grant universities stepped in to support the industries, first experimenting with biological controls and later aiding in the development of new pesticides.[45] These institutions, until recently supported almost entirely by the taxpaying public, have therefore provided a significant subsidy to agriculture.

The strawberry industry was one such beneficiary of the University of California's largesse. When industry leaders called upon UC for help in the 1920s, UC responded first with attention to identifying the diseases that were plaguing the industry, then with significant investments in plant breeding, and eventually with the development of fumigation. The success of fumigation freed the university to develop other techniques that would further improve productivity and shipping, including the use of plastic tarping, drip irrigation, cold storage, and much else. Indeed, it was the university's initial success in repairing these problems that brought the industry great success. As a mode of pest control, however, repair in the form of fumigation may have undermined future conditions of production, in no small part because of the role of the pests and other nonhuman actors, including, soil, plants, changing climates, and, for that matter, human bodies as biological entities.

More-Than-Human Assemblages

In his 2002 essay "Can the Mosquito Speak?" political theorist Timothy Mitchell wrote about how a set of seemingly unrelated elements coalesced to produce major outbreaks of malaria in 1940s Egypt, where the disease had not existed before.[46] These elements included the hydro-engineering of the Nile River, which created new habitat for the *Anopheles gambiae* mosquito. It included wartime-induced fertilizer shortages, as ammonium nitrate was diverted to war uses, resulting in famine and malnutrition that made humans more vulnerable to infection. It included sugarcane juice. Workers in the nascent sugar industry would chew on cane, not knowing that sugar worsened the effects of malaria. And, of course, the feeding patterns of the parasitic mosquito that requires human hosts to complete its life cycle figured importantly in these outbreaks.[47] Mitchell's purpose for this essay was twofold. One was to show the limits of sociological explanations when incommensurable, heterogeneous elements and conditions, many nonhuman and operating at different time frames and spatial scales, together helped bring the malaria epidemic into being. The other was to show the limits of human intention, action, and technics to fix these problems of hybrid nature, when interventions themselves depend on working with nonhuman actors.

In explaining the emergence of malaria, Mitchell essentially described what others have come to call a socio-natural assemblage, a constellation of heterogeneous elements and forces that in coming together are consequential.[48] Assemblage thinking enriches the explanation about how the strawberry industry came to rely on fumigants, as well as how this same reliance ultimately undermined the conditions of strawberry production, leading to iatrogenic harm. For one, assemblage thinking goes beyond agrarian political economy and its treatment of nature as a source of somewhat passive constraints or opportunities. In assemblage thinking, nonhumans play an active role of bringing phenomena into being. The implied agency of nonhumans is intended to connote not intentionality, but rather an object's capacity to produce an effect on another object.[49] A second is that it attends to the role of *multiple and disparate* objects, bodies, and forces that together produce phenomena, effectively acknowledging "distributed agency."[50] A third is that it recognizes the influence constituent elements have on one another; indeed, it is their "intra-action" that makes for the dynamism of assemblages.[51] In an assemblage, that is, constituent parts articulate so that perturbations in one area can affect others and even reverberate throughout the whole. In that way an assemblage "is not a mere collection of entities and things, but a complex and dynamic process whereupon the collective's properties exceed their constitutive elements."[52]

Some readers will recognize that assemblage thinking has affinities with actor network theory (ANT), and for that reason some have argued that assemblage thinking and agrarian political economy are incompatible. The fracture between these two approaches is significant. Political economy draws on a critical realist ontology that takes abstract and generalizable dynamics, tendencies, and concepts, such as capital accumulation or racism, as having as real and significant force in the world, whereas actor network theory imagines reality as based solely in the tangible material of the world, with the social constituted by the practices and conventions that translate this tangible material.[53] I follow those who use assemblage more ecumenically, who consider both the material elements and the abstract dynamics that come together in world making.[54] For the strawberry story, this explanatory heterodoxy is important. Not only have the intra-actions of plants, soils, fungi, chemicals, climate, and human bodies shaped the conditions of possibility for strawberry production, but so have tendencies, dynamics, and institutions like profit appropriation, land speculation, regulatory mechanisms, and university science.[55] Critically, sometimes these abstract forces work in tension with the tangible, material world. As we will see, land as property, a social relation, operates differently in the strawberry assemblage than land as soil, a material condition of production, just as labor as a factor of production has different valences than living, breathing, laboring bodies.

Assemblage thinking has been usefully employed by other scholars besides Mitchell who have sought to trace the emergences of diseases and blight.[56] Through this framework, states of disease or blight are *not* caused by a pathogenic entity that invades an otherwise-healthy body or plant. Neither the so-called pathogen nor the host have a particular essence. Instead, states of disease are imminent, and emerge as an effect of intra-action among the multiple agents that come to constitute the disease assemblage.[57] Writing on livestock diseases, geographers Steve Hinchcliffe and company take it a step further to suggest that it is the intensities of assemblage intra-action that create virulence—what they call a topological understanding of disease.[58] An example they provide is the proliferation of *Campylobacter* bacteria, a major source of foodborne illness. A variety of conditions make broiler chickens in mass-market production susceptible to the bacterium. Genetic uniformity, confined housing operations, and the purposeful repression of competing microflora such as *Salmonella* all seem to produce "the necessary physio-chemical conditions for the bacterium to spread both within the body and throughout the concentrated population." Yet things reach a tipping point when birds are sleep deprived and have their feed and water removed for twenty-four hours before slaughter. It is then that they are sufficiently weakened to become ill, making the disease not caught as much as incubated.[59] In discussing how the pathogens came to occupy the strawberry and preoccupy the strawberry industry, I draw on this topological understanding of disease.

Assemblage thinking has also been used to trace how efforts to intervene in disease situations are consequential—and not according to human intentions. Here Timothy Mitchell's comments are again useful, as he notes that human attempts to solve problems depend on collaborating with nonhuman actors whose actions often remain beyond human control. Notably, these nonhuman actors are not only other species, but may include nonliving actors such as chemicals.[60] So, for example, the use of DDT in malaria eradication campaigns had far-reaching consequences. These campaigns were highly effective in some areas because of DDT's ability to persist in the environment. But not only did the mosquito develop resistance to its use, DDT's accumulation in fat tissues spread its by-products throughout the food web. Its apparent ability to disrupt hormonal function, which was its strength, was passed on to other organisms. The pesticide, as Mitchell says, "had purposes of its own, well beyond the intension of research scientists and the eradication teams." So did malaria the disease, as it became resistant to the quinine drugs used to treat it.[61]

Anthropologist Alex Nading's work on dengue fever in Nicaragua uses something like assemblage thinking to make a similar case, but in addition includes how interventions can affect the bodies of workers in the business of intervention. Dengue fever is not only an outcome of complex entanglements that implicate such disparate elements as viruses, international trade, mosquito habits, water infrastructure, indoor/outdoor housing structures, and public health priorities.[62] Since *Aedes aegypti*, the mosquito responsible for dengue, lays its eggs in the tiniest of water fixtures, including flowerpots, tubs, puddles, old tires, and empty cartons, health care workers have been charged with visiting homes and attempting to eliminate mosquito breeding sites by slipping granules of a highly toxic chemical into these small bodies of water. As these health workers are exposed to these toxins through the skin, their own bodies are entangled in the dengue assemblage.[63]

These transitive harms, if you will, between chemical, target organism, and nontarget organism also occur in the strawberry assemblage, where efforts to control pathogens materialize into new pathogens and harm to the humans intimately involved in the assemblage. A "sad irony of pesticide use," geographer Ryan Galt wrote, "is the unintentional displacement of disease from one organism (the crop plant) to another (humans)."[64] But this book is about more than chemical toxic exposure to human bodies as a so-called unintended consequence of fumigation—as bad as it is in that respect. The changes wrought by the solution of fumigation, extolled in the opening epigraph, have extended into many other realms of strawberry production, creating something akin to what some theorists now call chemo-sociality, the relationships and emergent social (and ecological) forms that arise from widespread and unavoidable chemical exposures and dependencies.[65] Unfortunately, these new problems are not easily addressed through existing institutions and their ways of knowing.

Iatrogenic Harm and the Limits of Repair

A sine qua non of assemblage thinking is that assemblages are provisional: "relations form, take hold, and endure, but they also may change or be disrupted."[66] Often what holds them together, to achieve a kind of stability, is a great deal of human effort, taking the form of repair. For example, the Norwegian domesticated salmon assemblage requires constant checking, tinkering, checking, and repairing to hold it together. Despite this care, not only have parasitic sea lice proliferated, but previously unknown viral and bacterial diseases have appeared in the highly managed Norwegian fjords.[67] Norwegian salmon farming, while of quite recent provenance, is probably more elaborated than California strawberry production, which has been developing for over one hundred and fifty years. In any case, both employ an array of practices that constitute "cutting-edge" production schemes that build upon one another, thereby creating path dependencies for future operations. And yet, these over-evolved infrastructures, with their veneer of stability, can be a perhaps surprising source of fragility. With industry practices and infrastructures so rigidly developed, that is, the nonhuman aspects of the assemblage that escape human control become highly disruptive. Like plants that are pampered by not being exposed to pathogens, these highly pampered infrastructures are not resilient to perturbations, to immanent pathologies.

Assemblages are also vulnerable to changes among their constitutive elements. It is well established that pests can evolve to resist their chemical treatments and become more virulent, inducing farmers to increase the amount and frequency of treatment or seek even more powerful chemicals. This is the phenomenon referred to as a pesticide treadmill, but which some call a pest treadmill, denoting that the pests become the stronger.[68] Yet other nonhuman elements can play a role as well. For instance, in Costa Rican vegetable production, crop production in the cloud belt is much more susceptible to blight because of the relatively high temperatures and humidity levels. In particular, *Phytophthora infestans*, responsible for the highly potent fungal disease called late blight, depends on leaf wetness for ten to twelve hours to reproduce. Therefore, the resource-poor farmers who are relegated to land in the cloud belt are most prone to overuse pesticides, which in turn further deteriorates the soil, effectively undermining the future conditions of production as well as their own livelihoods.[69]

Crucially, many actors affect the conditions of possibility for strawberry production, and not all are subject to scientists' technical interventions—indeed, many exceed the capabilities of technologies to control them. In addition, many forces that bear on the conditions of production are simply outside the scope of institutions of repair.[70] As we will see, climate change is contributing to the appearance of novel pathogens and financial markets are raising land values, all affecting the future of strawberry production. Such dynamics are concatenating with one another as well, in ways that further constrain the ability to farm without fumigants. Truly, a range of socionatural threats bear on the strawberry assemblage, and any could break it apart. I will examine some in great detail as the book progresses.

Here, though, I want to introduce an additional, often overlooked, source of assemblage fragility, and that is a lack of scientific attention to the ecological changes the assemblage has produced, making the institutions of repair not well equipped to "repair the repair."[71] One impediment centers on the limited range of solutions that crop science can offer, given its scientific remit and its predilections for technologies that increase productivity or limit crop loss. University science has been much better at addressing underproduction than overproduction, the latter more a marketing and policy problem.[72] But again, overproduction has been a perennial source of problems for farmers, who lose out when gluts cause prices to fall. If success is measured by the ability to keep farmers in business, the record is not very strong for that reason alone. More generally, many problems have not been amenable to technological solutions (for instance, labor shortages until the introduction of robotics), and even those that appear to be (for example drought) may animate solutions not designed with cognizance to how they reverberate through the rest of the assemblage. Scientific specialization, while not the sole source of unexamined consequences, does not help.

A second impediment is that extension science has been caught up in a wave of privatization that threatens to make important innovations less accessible. Plant breeding, a major arena of repair, was never solely in the hands of the university. The development of hybridization allowed the "biological patenting" of plants, since hybrid cultivars do not "breed true" and must be clonally reproduced. This would send growers to the nursery or seed company every year, where their purchases would compensate the producers of these hybrid cultivars. The Plant Patent Act of 1930 gave additional protections for the inventions of private breeders. So despite the role of the land grant universities in improving hybridization technologies, much plant breeding was taken up by private interests.[73] In the case of strawberries, private breeders were involved in cultivar development from the get-go. Even after the University of California stepped in to became a major force in creating the so-called university varieties, their efforts were quickly matched by the Driscoll family and their breeding efforts. But the role of the universities in plant breeding became even more complex when universities began to see funding shortfalls due to public disinvestment and became more interested in industry collaborations that could support university activity. The Bayh-Dole Act of 1980 specifically extended patenting and licensing privileges to university inventors to encourage technology transfer to industry, while allowing the university and the inventor to retain revenue. So although the agricultural colleges, unlike the non–land grant sectors of the university, had long been transferring technology to the private sector, this new context creating salable technologies was encouraged by the university.[74] Since the Bayh-Dole law allowed university scientists to earn personal revenue from their inventions, it unleased an additional dynamic: university researchers, now making money on their inventions, saw opportunities to make even more money outside of the university. One of the crises threatening the industry is the deepening of these proprietary behaviors, as knowledge itself is increasingly squirreled away while growers have to pay for it in the form of license fees and high price inputs.

A third impediment centers on "undone science." The term "agnotology"—the study of ignorance—was coined by scholars investigating the deliberate suppression of scientific findings that would challenge state or corporate interests, such as science that disparages cigarettes or, for that matter, agrochemicals.[75] But other scholars have suggested additional reasons, besides deliberate suppression, why science can remain undone. Sometimes the research questions are not of interest to those commissioning it, or sometimes scientific norms constrain the development of certain forms of knowledge.[76] Sometimes science remains undone owing to the character of the institutions assigned the production and dissemination of the science. The fragmentation of disciplines can itself be a source of ignorance.[77] Soil science, for example, is a subfield of agronomic science, while soil science itself has many subfields. With the vast array of cropping systems in California, extension scientists specialize not only in agronomic subfields, but also in particular crops.[78] Therefore, agricultural scientists and extension scientists tend to be narrowly trained and focused, which undermines their ability to address the connections among various parts of agricultural assemblages.

Yet perhaps the most significant aspect of undone science is that past technologies of repair have created ignorance about the problems they induce. Given their mission to serve farmers, university agricultural scientists tend to ask only those questions that can lead to already imaginable and easy solutions.[79] Easy solutions, in turn, obliterate the need for further study, creating what some call "unknown unknowns."[80] Historian of science Frank Uekötter's work on the fate of biological approaches to soil fertility in postwar Germany is illustrative. As he writes, biological methods such as biodynamics remained in vogue in the 1930s, and much research was conducted on soil microbiology. Yet the answers this science produced were uncertain, and scientists investigating these approaches "faced stiff competition from the fertilizer industry and its army of advisors." So farmers abandoned complexity and embraced the easy fix of agrochemistry.[81] The fate of UC's Division of Biological Control followed a similar path. Even though the use of predator bugs to control pests proved reasonably successful at controlling cotton cushiony scale on oranges, ultimately growers found chemicals simply easier. Nevertheless, the easy solutions may have caused problems down the road that more refined approaches might have prevented.[82]

To be sure, the embrace of easy solutions often confounds understanding of how problems arise along with how the cures actually work.[83] Mitchell writes that engineers in Egypt could see that DDT was effective at eradicating the *Anopheles gambiae*, but they didn't know how.[84] Uekötter argues that the availability of fixes such as synthetic mineral fertilizers, chemicals, and machinery made investigations of much more complex ecological relations seem unnecessary. What answers might have arisen could never compete in their expediency.[85] Ignorance became strength, as Uekötter puts it, among farmers who wanted definitive answers and advisors who wanted to please their clientele.[86] Now, however, such ignorance can be added to the list of threats. With their tendencies to simplify ecologies through eradication, chemical solutions are particularly prone to introduce iatrogenic harm by altering the ecosystems they are supposed to mend.[87]

TELLING AN ASSEMBLAGE

Wilted tells how an industry grew out of many of the advantages that nature offered, including the heterozygosity (genetic variability) of the ancient strawberry, mild climate and sandy soils, the fumigant action of chemicals, and the energy of working bodies. These were intimately connected with many of the political-economic advantages the industry also had at its disposal: loose pesticide regulation, publicly supported agricultural science, undeveloped land markets, and politically constructed labor surpluses. But efforts to control emerging pathogens led to the singular solution of fumigation, which in turn ramified throughout the assemblage, changing breeding priorities, land use patterns, the cost of doing business, marketing needs— and the strawberry ecosystem itself, in ways yet to be countenanced. Additional threats emerged when those earlier political-economic advantages inverted into stricter regulations on fumigant use, tighter land markets, labor shortages, and an increasingly proprietary scientific apparatus. As the assemblage became more pathological, both ecologically and economically, the future of the California strawberry industry became more tenuous. Indeed, the strawberry assemblage became much like the strawberry itself: perishable, fragile, easily rotted, needlessly big because it is "bred" for the wrong things, and not particularly resilient.

I am not the first scholar to note that the intransigence of the socio-natural entanglements of California strawberry production has become a source of frailty. As sociologist Brian Gareau wrote in 2008:

> Some agro-industrial complexes, such as those tied to California's strawberry production, are built around particular historico-geographically constituted production conditions that are difficult to change. California strawberry production relies on certain technological innovations (e.g., strawberry varieties dependent on certain chemicals to combat plant pathogens), the creation of certain ecological conditions (e.g., climatic, soil and hillside conditions that make water-soluble chemicals difficult to apply), and a consistent labor supply (i.e., seasonal Mexican and Mexican-American laborers). Without these specific production conditions, the system would likely fail due to foreign competition.[88]

I expand on Gareau's comments by giving a detailed empirical account of those constraints, revealing the degree of entanglement and entrenchment of the strawberry production system, and suggesting that these conditions have continued to evolve in relation to one another and to human intervention. This has turned many of these erstwhile advantages into threats, regardless of foreign competition. In addition, I emphasize the role of science in attempting to improve the conditions of production but, owing to the nature of productivity-oriented applied science, failing to recognize, much less address, the socioecological changes that scientific interventions have bequeathed.

In telling the story I will use the term "industry" when referring to a set of human institutions and actors connected to the business of growing strawberries for a profit. Today that includes a collection of growers (about three hundred in California, reduced from prior years) who farm anywhere from a half an acre to more than a thousand; dozens of businesses that provide goods and services to the growers, such as nurseries, chemical applicators, pest-control advisors, and farm-labor contractors; and a relatively small number of buyers (grower-shippers, stand-alone shippers, and processors). I will separately refer to the institutions of repair that facilitated the industry's development, and include not only universities and the scientists they employ, but also grower cooperatives and growers' advocacy organizations. I will use the term "assemblage" when referring to the constellation of material things and social forces entangled in strawberry production: the organisms, bodies, plants, chemicals, soils, climatic conditions, border policy, land rents, and more. Even though industry actors and scientists are also part of the assemblage, treating them as distinct will reveal an industry operating according to capitalist, productivist logics yet enmeshed with organisms, infrastructures, and political conditions outside of its control, and closely tied to institutions of repair that have tried to bring those things under control.

I have chosen to organize the text around a few key elements of the assemblage—which, strikingly, I first identified through growers' laments. I begin with the soil pathogen *Verticillium dahliae*, the first of several soil-based diseases to have appeared in California strawberry fields. Soil pathogens are not the only pests that threaten the strawberry industry. Strawberries are affected by a range of other diseases and pests, including anthracnose brown rot, powdery mildew, bacterial angular leafspot, and mites.[89] Yet it was soil pathogens that first induced the nascent strawberry industry to call on the University of California to help, and it was soil pathogens that were addressed by fumigation. Telling the story of the pathogen thus provides the basic melody for the book, on which all following chapters build symphonically, as I show how the key elements interfaced with one another.

The four chapters that follow bring in four erstwhile advantages that inverted into threats: the heterozygous and readily bred strawberry plant; largely unregulated and effective chemicals; plentiful land well situated for strawberry production; and plentiful and easily exploitable labor. These chapters treat these elements as both ecological and institutional, material and abstract, although not always simultaneously. Again, land as soil works differently than land as property. Bodies and labor markets, plants and patented varieties, chemicals and regulatory regimes are different things as well, and the substance of those pairings can be in tension as well as aligned. Science plays a more significant role in chapters on pathogens, plant breeding, and chemical fumigation because those are the areas where university science was better suited to intervene than land and labor.

A subsequent chapter addresses the threats to the industry qua industry, focusing on how its evolution has positioned some actors quite well for an uncertain future while others stand to lose quite badly. I will give special emphasis to Driscoll's, with its eye for competitive advantage and its tendency to seize opportunities that have led to industry dominance. As with other chapters, the beginning of the chapter provides important historical background, but while other chapters read more as environmental histories, this one reads more as an institutional history.

The final chapter turns to the future of strawberry production in California. Here I review the current menu of approaches to farming without chemical fumigants. The underpinnings of these approaches are quite distinct, and I will draw these out, as well as discuss what they each might mean for the future of the industry. Despite its current fashion among ecologically minded social theorists, an approach involving "living with" pathogens is hardly inconsequential and may not create the hopeful emergent ecologies that they imagine. I conclude with further thoughts on the challenges moving forward.

. . .

In what follows I draw on three main sources of data. One is the primary data I collected while interviewing growers, farmworkers, and other industry actors about my original research questions. As a reminder, these questions were about the regulatory battles over soil fumigants and how the industry would move forward with tighter restrictions on chemicals. It was only toward the end of this research that I realized that the many things I had witnessed that exceeded the scope of these questions were connected—and those connections were the inspiration for this book. Those interested in the original research questions may find my published journal articles helpful.

The findings herein only touch on farmworkers' spoken perspectives. Although the original plan was to query farmworkers about their perspectives on methyl bromide and its alternatives, they mainly discussed their experiences with pesticide exposures more generally and not the chemical solutions that drive this narrative, which are fumigants. There is some minor discussion of these interviews in chapter 6, but they are mainly reported on elsewhere.[90] In addition, since the interviews with farmworkers were conducted early in the project, before I learned of growers' laments of labor shortages and long before this book was conceived, unfortunately I was not able to obtain much data on how farmworkers view the putative labor shortages. And yet, the scenes I describe of workers racing to pick berries as if their lives depended on it reveals something about the political and economic vulnerability they experience despite growers' claims that a labor shortage has given farmworkers the upper hand.

As for data from growers and the industry, it was not exactly easy to obtain. The industry is understandably anxious about social science researchers given the latter's tendency to tarnish the former's reputation. Some potential informants were hostile on the phone; some agreed to an interview and didn't show up; some asked me questions out of apparent concern that I would reveal their proprietary secrets and ended up never scheduling; and at least one used my left messages to initiate a federal Freedom of Information Act to obtain my research proposal. The proprietariness and lack of transparency, I want to suggest, are part of the industry's public-relations problem. Of those who agreed to talk to me, some seemed to feel it was their obligation to speak to a University of California researcher, while others gave the impression that they had dirt to reveal. But the vast majority I interviewed hoped my work would be helpful to the industry by at least telling their side of the story. I am grateful to the one hundred or so growers and industry representatives who took the time to meet with me, and I have endeavored to faithfully illustrate their predicaments while also providing robust scholarly interpretation. I can tell only as much as I learned, so those who chose to be elusive should not wonder why their perspectives are missing.

I also draw on several historical accounts of the industry. Of the three books that provide core comprehensive data, two were authored by men with close ties to the industry: Stephen Wilhelm, the lead author of *History of the Strawberry: From Ancient Gardens to Modern Markets* (1974), was a plant pathologist at the University of California and was involved in many of the strawberry industry's early innovations. When he retired he became a raspberry breeder for the Sweet Briar Company, which later became a part of Driscoll Strawberries Inc. Herbert Baum, who wrote *Quest for the Perfect Strawberry* (2005), was a principal at Naturipe, the first strawberry marketing cooperative that was to become a rival of Driscoll's, and served as a leader of the California Strawberry Commission. George Darrow, author of *The Strawberry: History, Breeding, and Physiology* (1966), was a pomologist and small fruits breeder for the US Department of Agriculture. I also draw from the web pages of various historical societies, along with other already-compiled historical documents. Virtually all of these histories are highly anthropocentric. They are stories of the men and a few women who made the industry—and of course they are champions of the industry even though intra-industry rivalries are evident in their writings. I have resynthesized these histories to accommodate a different conceptual framework.

Finally, I draw on interviews with about twenty regulators and research and extension scientists. I also rely on many of their scientific publications, as well as publications by past research and extension scientists. Some of these informants have simply observed the industry, while many are actively involved in the business of repair for the industry. Writing in the vein of science and technology studies (STS), I don't always take what they say at face value and seek to situate their comments and observations within the context of the kind of science they do—applied science on behalf of the industry. Some of my informants, as well as scientific readers more generally, might take issue with my interpretations and speculations, especially those ideas informed by what some call the new postmodern synthesis in biology and its focus on the plasticity of organisms in relationship to their environments.[91] While all the scientists with whom I spoke were very generous and didactic, not altogether surprisingly I found ecologists and integrative biologists more interested in thinking about unexplored species intra-actions than narrowly trained specialists, and I found academic scientists more knowledgeable about potential connections than applied scientists. That goes to my arguments, and that some of my use of science in this book is based on the speculations of scientists not tied to the industry is also part of the point.

TWO

Emergent Pathogens

The strawberry plant is native to a conifer forest. The conifer forest has a pile of organic matter *that* high [gestures], and in that organic matter are these naturally occurring fungi, bacteria that provide a protective zone around the root system. They can survive *Fusarium* and *Verticillium* there, but they can't when you take it out of that environment and try to grow it in the desert. And we have all these people we gotta feed, so we have to grow it, you know, here.

Interview with a pest control advisor, 2014

I've always joked with our farmers that this [methyl bromide phaseout] will separate the men from the boys. For a long time, they just wouldn't listen. And so now what's happening is we're seeing soil diseases that haven't been prevalent in fifteen, twenty years, coming back. It's very similar to immunization programs in the United States with children. . . . When you start compromising that system, these diseases are going to come back and become more prevalent.

Interview with a strawberry shipper, 2015

Disease . . . is rarely simply a matter of spatial incursion of pathogens from a diseased elsewhere making their way over and into a healthy population. Being healthy may not simply mean being free from pathogens, but a matter of immunocompetence—that is, the ability to live with a variety of other organisms that are always in circulation.

Geographers Steve Hinchliffe, John Allen, Stephanie Lavau,
Nick Bingham, and Simon Carter, "Biosecurity and
the Topologies of Infected Life: From Borderlines to
Borderlands," 2013[1]

You don't need to know the biology if you can kill it.

Discussion with a fungal biologist, 2017

WHEN SANDY[2] AND I PULLED into the driveway of a Santa Maria farm, we had a sense that it would be an interesting interview. The grower we were visiting had been friendly on the phone, especially after I had informed him that we were researching how the strawberry industry is faring with tighter regulations on soil fumigants. Unlike others with whom we spoke, he greeted us by our car, and as soon as we parked, with little formality, he began to talk. He was particularly eager to show us the plants in one of his field's buffer zones. A buffer zone is a strip where regulations prohibit fumigation treatments because of proximity to houses, workplaces, or other peopled sites. His farm was not too distant from a school, what California's Department of Pesticide Regulation (DPR) deems a hard-to-evacuate site. Due to DPR regulations, this grower therefore had to ensure that his fumigation treatments were at least a quarter-mile away from the school's boundaries when school was in session.[3] His was a large farm of several hundred acres, abutting other large farms, and looking in one direction there were strawberry plants as far as the eye could see—robust, green plants, loaded with bright-red berries. Yet there in the foreground, in the first two or three beds, many of the plants were brown and shriveled, if not dead, and berries were few and far between. Although the grower hadn't yet tested the soil, he suspected that *Macrophomina* had caused the damage in the buffer zone. *Macrophomina phaseolina* is a "novel" pathogen that has been showing up in fields since growers ceased using methyl bromide. It appears even in fields fumigated with combinations of chloropicrin and Telone, the mixture many growers had begun to use. Together, *Macrophomina phaseolina* and *Fusarium oxysporum f. sp. fragariae* (named for the host species), another "novel" pathogen, along with *Verticillium dahliae*, which has long been affecting strawberry plants, pose a major threat to the strawberry industry. The threat has become existential with the regulatory crackdown on chemical fumigants that since the 1960s have seemingly controlled these diseases and other pests.

Conventionally defined, a pathogen is a living microorganism such as a bacterium, virus, or fungus that causes disease in another organism. The soilborne fungus *V. dahliae* is one such microorganism, almost always discussed in relation to its pathogenic qualities.[4] And pathogenic it can be, especially as recounted by plant pathologists whose scientific discipline is defined by identifying the causes of plant disease. In the strawberry, *V. dahliae* causes disease by colonizing the vascular tissue, preventing the plant from taking up water and nutrients. The result is wilting, then collapse, and often death. During decomposition of the affected plant material, the fungus

FIGURE 2. Suspected *Macrophomina phaseolina* outbreak in a buffer zone, Santa Maria. Photo by author.

releases sclerotium, a hard material that allows it to go dormant. The sclerotia persist in the soil for years, awaiting a fresh host to renew their life cycle.[5]

Strawberries are not *V. dahliae*'s only target, either; hundreds of plants play host to the fungus, with few showing even modest resistance to it, making *Verticillium* wilt an endemic problem for many crops.[6] But here's the thing: *V. dahliae* is evidently not always pathogenic. Writing on the history of *Verticillium* diseases, British plant scientist G. F. Pegg noted that usually it has been simply parasitic, and sometimes it has even been commensal (living with and bringing no harm to the plant in which it resides). Moreover, the fungus has been present in soils for a very long time and may even be present in fields where no disease is evident.[7] Current-day plant pathologists say it may take years for colonies of the fungus to reach the point where plants begin to show symptoms of disease.

What, then, makes *V. dahliae* a pathogen? In their book *Pathological Lives: Disease, Space and Biopolitics* (2016), geographer Steve Hinchcliffe and colleagues argue that pathogenicity doesn't inhere in an organism (or "causative agent") but in a *situation* in which heterogeneous elements and forces "intra-act." In their view, disease outbreaks are less a product of invasions of hostile species crossing space—what they refer to as a topographical understanding—than convergences of events that intensify relationships, bringing immanent qualities to the surface. They call this a topological understanding of disease, and they draw particular attention to the "intra-action" among organisms that stimulate disease emergence.[8] We can see a skeletal form of intra-action if we look more closely at how *V. dahliae* becomes pathogenic. Infection, I am told, is utterly dependent on the activity of the host. It is the host root's exudate—the nutrients oozing out of the root—that stimulates the fungus to germinate. Then, employing its primary defense mechanism, the host plant inflates its xylem (water-conducting tissue), which then facilitates the pathogen's inoculum (biological source of infection) to spread through the plant's taproot and up into its vessels. As the vessels fill with material from the fungus, the host plant has more difficulty taking up water and nutrients, and that is what causes it to wilt and potentially die.[9]

Still, a topological understanding of disease considers much more than the intra-action of host and pathogen, and how the host brings the pathogen into being, as it were; it attends to more distal forces as well. At first glance, this approach appears roughly akin to what ecologists call a disease triangle model, which holds that plant disease can emerge from any number of changes in the host plant, the disease agent (in this case the fungus), or the environment, any of which can act on another node within the triangle.[10] But a topological understanding of disease takes this model quite a few steps further. In this rendering, all nodes in the triangle are in a state of immanence and only have effects in relation to one another. Moreover, the environment becomes a more capacious category, potentially involving a plethora of actors as disparate as soil bacterium, farmers, crop prices, regulatory decrees, climate change and more—in other words, a global assemblage of actors.[11] In the case of strawberries, a topological approach may consider how forms of cultivation typical of agro-industry such as tilling may have spread the organisms, and how mono-cropping may have aggravated their antagonistic tendencies.[12] Or how other ecological stresses such as periodic drought may have made the strawberry plants less able to fight them off. Even then, becoming a pathogen entails not only changes in material practices and conditions, but also new or different knowledge and values.[13] The making of a disease situation, that is, also depends on human experimentation and identification of the substance causing plant damage and a sensibility that such damage cannot be lived with.[14]

Attempting to forward a topological understanding of *V. dahliae*, in this chapter I thus recount the history of *V. dahliae* as both agricultural pest and object of knowledge, including its pathogenic becoming in the context of California's strawberry industry. I then discuss how the devastation it wreaked there led to a particularly violent means of repair. With the advice and support of scientists at the University of California (UC), the industry eventually came to rely on fumigation with a mix of toxic chemicals to eradicate the pathogen. Although effective for a long time, fumigation began to backfire, not only in the regulatory arena—the topic of a different chapter—but possibly in abetting the appearance of the so-called novel pathogens of *Fusarium oxysporum f. sp. fragariae* and *Macrophomina phaseolina*.

That fumigation might have contributed to the appearance of these novel pathogens is not, however, the standard explanation for their appearance. Agricultural scientists involved in the strawberry industry tend to think that they arose from the regulatory loss of the industry's favorite chemical, methyl bromide, along with new fumigation regimes for the remaining fumigants. Drawing on ideas emerging from the so-called postmodern biological synthesis, with its emphasis on the multiple pathways of developmental alteration, I want to suggest that that these novel pathogens may have begun appearing as a result of too much fumigation rather than too little.[15] Yet the lack of concern with this more topological possibility—what one agricultural scientist deemed "a fantasy"—speaks to how the industry and scientists working on behalf of the industry came to know the pathogen's ecology. Once it was seemingly controlled, they learned very little.

STARTING WITH THE SOIL

To understand how *Verticillium* the pathogen came into being, it is useful to start with the soil. In textbook terms, soil is comprised of mineral and organic solids, liquids, and gases that together can support plants. Relative to the mineral particles filled with air or water, a small portion of soil is organic matter. For most of the twentieth century, most soil scientists treated soil just as reductively as the textbooks, as evidenced in the differentiation in soil science subfields. UC soil scientist Hans Jenny was one of the first to see soil more ecologically. Writing in 1980, toward the end of his life, he characterized soil as a living system, rather than a container of organisms.[16] Only recently have conventional understandings of soil begun to incorporate Jenny's insights. The US Department of Agriculture, for example, has shifted its emphasis from quality soil to healthy soil, signifying an interest in soil biota as beneficial to plant growth.[17] This stems from increasing knowledge of the critical roles that a wide diversity of microbes, fungi, and noncellular matter play in keeping soil vibrant, providing all sorts of feedback on how organisms behave—whether they are constitutive of the soil or external to it. Pushing even further, a school of soil scientists less concerned with agricultural productivity have been conceptualizing soil as an assemblage of mineral and organic communities. This ontological shift stems in part from recognition that not all of this lively material is made up of singular organisms with clear boundaries. Fungus, for instance, is the largest organism in the world because it is extensive and unbounded.[18] This school also draws on post-humanist thought that refuses anthropocentric utility as the fundamental measure by which to engage soil. Instead, it recognizes the inherent importance of soil liveliness as more than a receptacle for crops.[19] For feminist science studies scholar Kristina Lyons, for example, "a sustainable approach to soil management is not a radical enough proposal for farmers who do not seek to isolate, 'correct' and employ an entity for the sole production of human food and profit."[20]

Regardless of the merits of this latter sort of approach for crop production, I raise it to highlight the ontological malleability of even pathogenic fungi. Mycologists say that fungi are remarkably polyvalent in both forest and cropping systems, sometimes mutualistic, often parasitic, and occasionally deadly. Fungi live by either decomposing dead organic matter or living inside a living host, and the intimate associations they enjoy with host plants makes them symbionts, not necessarily pathogens. Take for example the oft-vaunted mycorrhizae (a functional classification), without which plants would have never colonized the terrestrial environment. Mycorrhizae penetrate the cell walls of their hosts so as to make nitrogen and phosphorous readily available to the host while the host plant feeds the mycorrhizae.[21] Other fungi allow other plants to take hold and survive. Then there are the fungi that live off of dead or dying plant tissue. Mycologists call these saprobes—also a functional classification. Saprobes aid in decomposition and returning inorganic matter to the soil. In digesting dead plant material, wood, and occasionally even rocks and turning them into nutrients (a process called mineralization), they allow other species to develop roots and thrive. They may also eat away at excess plant life, ensuring adequate dispersal of trees in forests.[22] There is even evidence to suggest that saprobes helped early woods strawberries thrive in diverse habitats, from forests, meadows, and rocky cliffs to sandy dunes.[23] These death-loving fungi (literal necrophiles), in other words, have been critical to making life live, albeit not always in keeping with human intention.

Verticillium dahliae is only occasionally saprobic, living off of dead tissue only in one part of its life cycle. Most of what it gains in colonizing tissue is accomplished during the parasitic phase, when it feeds off the living host's exudate, letting the host's defense mechanisms do the killing.

It is not hard to see how the virtuosity of certain fungi in killing and decomposing other plants has made them a problem for cropping systems where biodiversity is not the aim.[24] Put somewhat differently, *Verticillium* has *appeared* pathogenic to humans only in agricultural contexts. In his accounts of the impact of *Verticillium* diseases in agriculture, Pegg says as much. He notes that the fungus had been in existence long before 1816, when a botanist first erected the genus *Verticillium*. Tellingly, when this botanist did so, it was based on unique features of the fungi's form—not its effect on other plants. It was not until 1879 that scientists describing wilt on potato plants named the causal agent *V. albo-atrum*, and then in 1913 identified a second, morphologically distinct species causing wilt on dahlia and named it *V. dahliae*. Although a taxonomic debate about the distinctiveness of the two species ensued and continued into the 1970s, together what distinguished these species was their ability to cause plant wilt, not true of all in the genus.[25] Nevertheless, Pegg rejects the idea that pathogenic variations of the genus arose *de novo*, about the same time of its human discovery. Drawing from comparisons of the form of present-day strains with their fossil ancestors, as well as knowledge that nonhosts such as weeds act as "reservoirs of infection" without noticeably dying, he deduces that *Verticillium* species "have been parasitic on crop plants during the whole history of their evolution and cultivation." Only with the introduction of susceptible host plants—those that reacted defensively to the fungus—did *Verticillium* became pathogenic in the sense of causing disease—albeit never acting alone.[26]

What, then, are the explanations for the enormous spread and severity of *Verticillium* wilt diseases in many different crops (in the hundreds) after 1879? And how did they reach epidemic proportions in the 1940s and 1950s? And why were there so many apparently inexplicable outbreaks of wilt in previously uncropped soils?[27] A topographical theory of how this fungus became more virulent in the course of its evolution rests on its movement away from its areas of origin, carried by a symptomless host. Even though the historical record includes only a few specific instances of the spread of the species by symptomless hosts, the world-traveling potato was one of them. Its use by pioneers, sailors, explorers, and troops as a convenient food would have made "the distribution of potentially infective material ... widespread," especially during World War I (1914–18), when potato plantings were ubiquitous.[28] According to some scientists, the relatively high genetic diversity and thus plasticity of *Verticillium*, compared to the many fungal plant pathogens associated with the infamous potato blight, Dutch elm disease, and chestnut blight pathogens, would have made it particularly predisposed to virulence. It would have been more likely to hybridize between different species, the upshot of which would be transfer of pathogenic traits, or even the emergence of new plant diseases and the ability to infect hundreds of hosts—from crop plants to trees, wild and cultivated.[29]

Yet this theory of invasion only goes so far in light of scattered evidence that *Verticillium dahliae* was endemic in many temperate regions of the world, including California—or at least had existed there for a very long time.[30] A topological explanation of disease emergence would then have to account for its innocuousness prior to becoming a problem. Before intensive cultivation of specific hosts, the tendency of the fungus to induce wilt would hardly have been noticeable, of course; blighted plants would likely have blended into the landscape. But it is also possible that *Verticillium* may have coexisted even with crop plants because it lived in what agronomists today call suppressive soils. Suppressive soils have a very low level of disease development even when a virulent pathogen and susceptible host are present. The operative theory is that the presence of a wide variety and mass of microbes outcompete potentially pathogenic organisms for nutrients, or even that other organisms have a direct antagonistic effect on the pathogen or induce host resistance.[31] Biodiverse soils, in other words, might have kept *Verticillium* from doing much damage.

A topological understanding would then have to articulate the changes that made *Verticillium dahliae* actually appear as a pathogen. In addition to speculating about topographical spatial incursions, Pegg gives much credence to "trends in Modern Agriculture"—a range of practices that increased the intensity of intra-actions between host and (potential) pathogens. These include increased agricultural acreages—what agrarian political economists call extensification—and increasing use of monocultures of cultivars with a narrow genotype—what agrarian political economists might call intensification. Both would have led to the buildup of "vast levels of inoculum on crop residues . . . providing long-surviving reservoirs of infection for future plantings." Both would have also increased selection pressure in the production of new strains.[32] The issue here is that parasitic fungi such as *Verticillium* tend to be nonpathogenic in nonagricultural settings because they pay what ecologists call a "fitness penalty" for damaging the host and diminishing the resource on which they feed (it's a bad evolutionary strategy to kill your host). The situation differs in agriculture because humans reintroduce the plant every year, taking away the fitness penalty for the fungi to destroy it. And so humans inadvertently encourage the development of strains that cause problems for crop plants.

Other trends mentioned by Pegg include increased use of agricultural machinery such as cultivators, harvesters, and pesticide applicators, all of which could have spread the fungus and damaged roots, making them more prone to infection.[33] He also mentions the cultivation of land where weeds and nonhost plants such as cereals and grasses previously grew, all of which could harbor the (potential) pathogen as a parasite without themselves being affected. Once mowed down, they would have left a reservoir of *V. dahliae*. Finally, Pegg notes the importance of previous cropping with susceptible crops that also left a "reservoir of infection." Along with potatoes, he mentions tomato and salad crops, both of which were once often grown in rotation with strawberries.[34]

Here it is important to acknowledge the speculative nature of much thinking on the origins and dynamic character of *V. dahliae*. The extent to which knowledge of the evolution and polyvalent functionality of the species is hard to come by goes to my point. Such species became objects of study only when they became problems—and problems to the species that were important to humans. Without intensive cultivation of specific hosts, *Verticillium* wilt diseases would have not have drawn much attention, and, hence, *Verticillium dahliae* would have not been isolated and defined as a pathogen. To be sure, it garnered research attention in California only after it had made its presence felt. So the fact that this fungus is often defined as a pathogen says something not only about an instrumental and reductionist regard for the soil, but also how such fungi first come to be known.

THE PATHOGEN AND THE STRAWBERRY

When breeders first tinkered with strawberry varieties in the seventeenth century, the presence of *V. dahliae* in the soil would have been hard to detect, not only because these breeders lacked the necessary technologies of detection and knowledge. Strawberries weren't planted then in the massive monocultures that made the pathogen more virulent and, for that matter, more visible in its effects. Moreover, breeders may have used practices that inadvertently counteracted the pathogen's behavior. One method for growing strawberries, for example, involved sowing seeds in moss "placed over a light soil," which "served not only to hold moisture but . . . to provide aeration and light essential for germination." It is possible that "the moss may have also helped to control . . . a disease caused by injurious soil-borne fungi."[35] It is also possible that the berries were planted after crops of brassicas such as cabbage or mustard seed, both of which have since been found to suppress the pathogen by enhancing microbial populations or emitting natural chemicals antagonistic to the fungus.[36]

It was not until strawberries were well established in California agriculture that scientists identified *V. dahliae* as one of several diseases afflicting California's strawberry plants with "lack of vigor and declining yields." Research on other diseases led them to this discovery, however. Observers had noted "small, yellow, upward-cupped leaves" that "spread through entire fields, often within one year, and rendered the affected plants . . . worthless."[37] In 1915, a grower from Watsonville sent a specimen of the blighted plant to the Department of Plant Pathology at UC Berkeley for analysis. Scientists there named the disease strawberry blight—or yellows, for short. The yellows disease spread in the years following its discovery, causing heavy losses for California's strawberry industry. It was most prevalent on the coast, where the Banner, the prized variety at the time, dominated, and nearly absent in areas where other varieties were grown. Soon enough, scientists determined that this particular disease was caused by a virus and spread by aphids, and bestowed on it the more scientific-sounding name of xanthosis.[38]

By 1912, growers were noting wilt in field crops such as potatoes, and by 1920 it had shown up in strawberries. Already-established industry leader Richard Driscoll and his partner Joseph Reiter had moved their operations to Palo Alto. The Barron Ranch, where they grew, had been previously planted with tomatoes, and in the next few years large amounts of acreage succumbed to what they called "brown blight."[39] They suspected that grading the land in preparation for strawberries had spread an already-present soil infestation.[40]

Yellows and brown blight were not the only possible causes of the stunting, wilting, and root spotting that strawberry growers often witnessed. Nematodes, "threadlike worms" that are invisible to the eye, were attacking both strawberry roots and the above-soil parts of the plant, including buds and leaves, and producing similar symptoms. They were particularly destructive in sandy soils that were otherwise ideal for strawberries. Then there was red stele root rot (*Phytophthora fragariae*), another soilborne disease, first discovered in 1920 in Scotland and found for the first time in California in 1930. In the dry season, diseased plants "may die before blossoming; if the season is wet they may blossom, but die before the fruit ripens." "Affected plants at first show the dull, bluish-green color of wilting plants."[41]

In these circumstances, attention to soil disease became urgent. The director of the newly formed Central California Berry Grower Association (CCBGA) "ceaselessly warned" of the dangers. He gave special emphasis to "a new disease in the Watsonville district which threatened to wipe out the industry."[42] At his urging, in 1924 UC began research on soil disease. One of the first findings was that xanthosis remained with the planting stock (the material used to propagate new plants). In an effort to find clean stock, UC scientists found that the brown blight was soilborne.[43]

Harold Thomas, who was born in the strawberry district of Watsonville, joined the UC Berkeley Plant Pathology department in 1927. He was highly attentive to reports about the various diseases that affected the roots of strawberry plants and, noting their similarities in appearance, was curious about their common origins. He was most keen to isolate the problem of brown blight, having a hunch that it could be *Verticillium* wilt. At the time there were no reports on the spread of *Verticillium* wilt into strawberries except one from Holland, of someone who had cultured it from a dead strawberry plant. This person believed the fungus had caused the disease but had not conducted any inoculation experiments to prove it. Thomas had meanwhile observed "a striking correlation" between the occurrence of brown blight on strawberries and land that had been previously cropped in tomatoes.[44] Knowing that tomatoes were highly susceptible to *Verticillium* wilt, he deduced that that was the disease afflicting strawberries. To prove his hypothesis, in 1931 he conducted an inoculation experiment with steam-sterilized soils, injecting half of the plants with a serum containing the isolated material. Finding that 50 percent of the inoculated plants showed symptoms of the disease while none of the controls did, he renamed brown blight *Verticillium* wilt. Although Thomas felt that "brown blight" was descriptive enough, "wilt" was the better term, since it was commonly employed by pathologists.[45] As it turns out, he was incorrect about the species. He had ascribed the pathogen to the species *V. albo-atrum*; others later figured out it was *V. dahliae*.[46]

FIGURE 3. *Verticillium dahliae* dieback in an experimental field, San Luis Obispo. Photo by author.

Thomas also conducted a number of experiments and observations to learn of differences in the way *Verticillium* affected strawberry varieties, looking for signs of resistance to the pathogen. He noted that in some varieties all the leaves would brown and the plant would gradually collapse and die, whereas in other varieties, the younger leaves in the center would remain green. Once he identified what he thought were susceptible varieties, he noted mortality rates of 75 percent in susceptible varieties, with very few such losses in what he thought were resistant varieties, such as the Banner. Alas, these were opposing results from his observations of xanthosis, which badly affected the Banner.[47]

Unfortunately, identification of a pathogen was not itself the cure, and by the 1930s, *Verticillium* wilt had become a major production problem for the strawberry industry, with plantations established in infested soils often suffering 50 percent or greater mortality.[48] The problem continued well into the late 1950s.[49] Strawberry planting that followed tomato, potato, and cotton crops fared particularly poorly. Strawberries also had difficulty in the presence of *Solanum sarrachoides Sendt*, a common weed of California muck soils and a nearly symptomless host of *V. dahliae*.[50] *Verticillium* wilt had become a true scourge.

The determination that *V. dahliae* was a virulent pathogen naturally precipitated efforts to control it. Plant breeding for resistance was one sort of repair, discussed in the following chapter. But growers' first strategy was simply to relocate to new ground. Driscoll's moved to Palo Alto to avoid planting strawberries on land previously cropped in strawberries, and many other growers followed suit, leasing ground wherever it was available.[51] After Thomas discovered the strong connections between tomato production and *Verticillium* diseases, he warned strawberry growers to avoid land with any history of tomato production. As other experts found that crops such as cotton and potatoes were also hosts of *Verticillium* wilt, they instructed growers to avoid rotating with these crops as well.[52] And given findings that weeds could be refuges for the pathogens, they admonished growers to weed their fields often and carefully.

Keeping the same strawberry plants in the ground for several years in a row contributed to *Verticillium* outbreaks as well, despite that this practice was encouraged as early as 1793 in Johann Weinmann's *Rules of Strawberry Cultivation*.[53] In 1955 Victor Voth, a pomologist at UC, began to suggest replanting on an annual basis, but few growers took his advice seriously until a disastrous year in 1957, after which 20,700 acres of production dwindled to 9,000 acres within five years. The shift to annual planting that ensued was somewhat of a breakthrough because the soil could be free of plant hosts to the pathogen for a short period.[54]

Annual planting affected the demand for nursery plants, as did other productivity advances that ensued thereafter. Readers may wonder why strawberry growers used nurseries at all. As I will discuss in the next chapter, there were multiple reasons. Yet the advent of nursery propagation was also a means to control many of the strawberry diseases. Once it became clear that xanthosis had affected nursery stock and could not be controlled by any cultivation practices, the industry recognized the need for clean stock.[55] Learning that certain pathogens traveled with seeds made maintenance of disease-free seed of paramount importance. This awareness came with a price, however. In 1933, UC breeders, including Harold Thomas, had moved their experimental plots from San Jose to Zayante, deep in the Santa Cruz Mountains. There, away from the main strawberry-producing regions, they witnessed red stele root rot wiping out their plot. This experience led Thomas to write, "It appears future breeding work must be carried on in some place even more isolated ... from commercial strawberry production."[56] Moving the nurseries to the far north of the state, "two hundred miles from the fruiting areas," made good sense.[57]

This experience also inspired a law to mandate disease-free nursery stock. Thomas had first suggested guidelines in 1936 for the certification of strawberry plants. This idea became California law in 1940. The law created the Strawberry Plant Certification Program, to be administered by the California Department of Agriculture. Specific regulations mandated that nursery plantings "be removed from any other strawberry plants ... at a distance of no less than 500 feet"; that "plants must be set at least 8 feet apart in the row and all runners from each plant must be prevented from intermingling with those of any other plant"; and that "rows must be at least three feet apart."[58] Preventing cross-contamination through both temporal and spatial separation of planting thus became the favored approach to managing nursery pathogens.

The industry was not averse to more carefully tuned cultivation practices to manage the pathogen. When growers saw wilt, they tended to assume insufficient moisture. And so they would irrigate more—a practice that aggravated the condition.[59] As early as 1932, Thomas noted the importance of well-drained soils for strawberry cultivation and began to counsel against overwatering. Later UC scientists would recommend more attention to soil nutrition, noting that too much nitrogen would increase vegetation and prolong the life of the plant, making it vulnerable, while too little would also weaken it.[60]

Ultimately, though, fumigation—what Pegg tellingly calls chemotherapy—appeared to solve the problem of the pathogen. Fumigation came about from a collaboration between Stephen Wilhelm, another plant pathologist at UC, and farm advisor Edward Koch, who had been working on possible controls of *Verticillium* wilt. In 1953 Wilhelm and Koch used a handheld gun to apply a gaseous chemical akin to tear gas on about half an acre of land. In 1955 they followed up with a successful trial of a machine application of this same chemical: chloropicrin.[61] They found that fumigation with chloropicrin not only controlled the pathogen, but also somehow made the plant more vigorous. Their 1956 research article in *California Agriculture* declared: "*Verticillium* Wilt Controlled."[62]

During the late 1950s, university scientists had also begun experimenting with polyethylene mulch—what less technically minded observers would call plastic tarps. They found that clear mulch increased soil temperature by at least ten degrees, increasing the potential to plant earlier in the year.[63] However, the

FIGURE 4. Fumigated field covered in plastic mulch, Watsonville. Photo by author.

use of clear plastic mulch also stimulated weed production. They then began experimenting with a combination of chloropicrin and methyl bromide, the latter of which had largely been used to fumigate aboveground (the history and use of both chemicals will be detailed in chapter 4). The first experimental results, reported in 1961, indicated that methyl bromide not only augmented the fungicidal properties of chloropicrin, but also gave excellent weed and nematode control. The plastic tarps were doubly useful to keep the fumigant from dissipating in the air. Growers were happy to adopt this practice, but in the first applications they found that it was only possible to fumigate 50 percent of the field because the ends of the tarps had to be covered. With the introduction of a taping technology, they could fumigate all at once.[64] These technologies paved the way to drip irrigation, starting in 1967. Drip irrigation solved a different kind of problem: strawberries' high sensitivity to salinity, a problem that furrow irrigation had exacerbated.[65] The move to drip irrigation in turn led to experimentation with new bed shapes and fertilization regimes. A four-row bed, serviced with two drip lines, affected fertilizer placement, which then contributed to yield. In short, an entire suite of practices sprang from efforts to control the pathogen, which together worked to "revolutionize the industry"—as well as embed its infrastructures.[66]

These accompanying technologies created marginal improvements in productivity, yet it was fumigation with the two chemicals that proved the singular thing that controlled *Verticillium* wilt. By the end of the 1960s growers had widely adopted the practice. Industry-wide productivity increased sharply and consistently. Yields of three to five tons per acre of years prior jumped to twenty to thirty tons per acre.[67] Most formidably, fumigation solved the replant problem: growers could now stay on the same ground, as "new plantings on properly fumigated soil respond[ed] about like plantings on good soil never planted to strawberries."[68] Fumigation, that is, solved an analogous problem for soil pests that organic and inorganic fertilizers had solved for nutrient management: removing the need to rotate crops for pest management.[69] By the same token, fumigation also allowed growers to rotate strawberries with other major *cash* crops, such as lettuce, creating a win for strawberry and lettuce growers alike (see chapter 5).

Readers would be hard-pressed to find anything in the scientific literature of that day, and going forward two to three decades, that did not herald fumigation as the salvation of the strawberry industry.[70] Writing in 1984, Pegg, for example, noted that plant breeding could play a role in control of most *Verticillium* species, but that the efficiency of fumigation on *Verticillium* species in strawberry is such that it would necessitate a breeding program that would generate yield and resistance "10 times greater than that existing at present to achieve the current yield and with no guarantee of success or durable resistance."[71] So effective was a combination of methyl bromide and chloropicrin in ridding the soil of pathogens that breeders were able to put aside their goals of breeding for pathogen-resistant cultivars and turn to other qualities like beauty, size, taste, shelf life, and yet more productivity (see chapter 3).

Of course fumigation also left the soil bereft of most beneficial organisms, the kind that might be found in suppressive soils. Even that had its advantages. For one, it stimulated mineralization of the soil—turning (now dead) organic matter into minerals and nutrients like phosphorus and nitrogen.[72] Two, it may have been responsible for an additional yield boost, as it curtailed the suppressive action of competing microbes. As simplified soils began to cause other problems, though, farm advisors simply borrowed methods from organic farming and began to promote the use of soil amendments following fumigation, as if that would make it whole again.[73]

Widespread reliance on fumigation introduced a certain complacency within the industry, at the same time that it also created the potential for even more problems with pathogens down the road. Growers like to say that fumigation "sterilizes" the soil, but that is a misnomer. Rather, it disinfests the soil more effectively than nonchemical treatments, but not always entirely; some species have been known to survive methyl bromide fumigation. Moreover, without the huge diversity of species active in nonfumigated, potentially suppressive soil, the biological vacuum left by fumigation makes the soil ripe for colonization of more aggressive species that thrive in disturbed environments.[74] Among fungi, they may be *Trichoderma*, which are opportunistic but not necessarily pathogenic, or *Phytophthora*, a pathogen that often makes its appearance within minutes of a fumigation, I am told.[75] Fumigation also leaves dead plant material that may be attractive to saprobic organisms, perhaps encouraging their recolonization.[76]

While many in or close to the industry would concede that fumigation was imperfect, far fewer have contemplated whether and in what ways fumigation itself might have led to the appearance of the "novel plant collapse" problems that have beset the industry since the mid-2000s. One of these novel pathogens is *Macrophomina phaseolina*, what most feel is the more pernicious of the two newbies. Like *V. dahliae*, it was recognized as a species long before it became a pathogen in California's strawberry fields. Also like *V. dahliae*, there were some debates about its taxonomic classification. The first specimen of what was to become the genus was collected in 1901. The naming of the genus *Macrophomina* came in 1923, following a collection of dried specimens collected in Philippines in 1921. After some jockeying back and forth, plant taxonomists finally settled on *Macrophomina phaseolina*, the only species within the genus.[77] Recent research, however, suggests the possibility of a distinct strain for strawberries, as scientists have found that strawberry isolates do not affect other crops sensitive to *Macrophomina*.[78] Whether this resulted from a mutation, horizontal gene transfer, or some other plasticity-inducing process remains unknown.

As a fungal species, *Macrophomina* shares many of the same qualities of *V. dahliae*, including high genetic diversity, frequenter to hundreds of hosts, and ability to persist in the soil for years. However, *Macrophomina* does not colonize the plant in the same way that *Verticillium* does. Scientists refer to *Macrophomina* as a necrotrophe: it kills the cells it colonizes, but not right away. As a result, it doesn't cause symptoms until late in the season. Also unlike *Verticillium*, which kills by activating the host's defense mechanism, *Macrophomina* works directly on the plant, leading to decay and eventually the disease called charcoal rot. Notably, *Macrophomina* has been associated with saprophytic colonization, entering fields following herbicide treatments of soy, suggesting that it likes scorched-earth environments.[79]

The other culprit is the fungus *Fusarium oxysporum f. sp. fragariae*, a subspecies of *Fusarium oxysporum* specific to strawberries—indeed, identified by its deleterious effects on strawberries.[80] *Fusarium oxysporum* appears to have origins that go far back in evolutionary time, and its native distribution is widespread globally. It is the same species responsible for Panama disease. There is also evidence to suggest that it has nonpathogenic origins and continues to be nonpathogenic in certain contexts.[81] For example, it has been found in grassland soils in California's Central Valley, where it is not damaging because it doesn't draw many nutrients from the plant. It has even been observed to play a protective role for certain plants by suppressing other diseases, including pathogenic strains of the same species.[82] The tendency for humans to select for pathogenic strains of the fungus by reintroducing the same crop host every year seems particularly clear for *F. oxysporum*, which has developed different subspecies for different host plants. The subspecies associated with strawberries acts in very similar ways as *V. dahliae*: colonizing the root and then growing invasively into the plant and entering into the xylem (the water-conducting tissue).[83] Like *Macrophomina*, *F. oxysporum* can act as saprobe at times, too.

Reports of *Macrophomina* infection of strawberries came as early as 1985, from Egypt, followed swiftly by reports from Spain, France, and other regions that had ceased to use methyl bromide.[84] Beginning around 2005, California strawberry growers reported problems with collapsing strawberry plants, even though they were still fumigating with chloropicrin and other chemicals. Symptoms consisted of foliage wilt, plant stunting, and drying and death of older leaves. Plants eventually collapsed and died. When plant crowns were cut open, observers found various shades of brown tissue, strongly associated with *Macrophomina*. In the first years, the *Macrophomina* problem was restricted to the southern part of the state (Ventura and Orange Counties), where growers had begun to curtail methyl bromide use but also where plants tend to undergo more stress from dry and saline soils.[85] By 2010, cases of charcoal rot had moved into the Central Coast and inland counties.[86]

The original reports of *Fusarium*'s infection of strawberries came much earlier: in 1923, following a case of root rot in Mississippi. Yet it did not appear to arrive in the California strawberry fields until 2006, after *Macrophomina* had first appeared. Compounding the detection problem, the symptoms were identical to those of *Macrophomina*, even though the method of infection is much closer to *Verticillium*'s.[87] Through culturing tissue, scientists were available to identify *F. oxysporum sp. fragariae* as the culprit. Patterns of dispersion seemed to mirror that of *Macrophomina*. It was first found in Ventura County, yet by 2010 it had migrated to Monterey County strawberry fields. Notably, other pathogens, such as *Verticillium*, *Phytophthora*, and *Colletotrichum* species, were not isolated from the affected plants, suggesting that fumigation was either continuing to be effective at staving those off or that they were simply returning at a slower rate.[88]

Most observers within the industry attribute these outbreaks simply to the end of methyl bromide use. Those following the research are aware that changes in fumigation regimes were also involved: more growers began to fumigate within beds rather than throughout entire fields, both to take advantage of smaller buffer zone incentives and to save costs.[89] Extension scientists hypothesize that the pathogens remain in the furrows between the beds, allowing the fumigated soil to be easily recolonized when growers tear down the beds and mix the soil. After witnessing the outbreaks, a team led by Steve Koike, a farm advisor with UC Cooperative Extension, conducted a set of field experiments in which they either did not fumigate or fumigated in beds with a mix of Telone and chloropicrin. This is a mix known as "Inline" that many growers have adopted, especially those farming in hillier areas. Koike and his team witnessed little plant mortality until March, but as the season progressed and temperatures increased, plant collapse began to occur in all of the cultivars tested. Through laboratory tests, they found that all cultivars were individually positive for either *Macrophomina* or *Fusarium*, although some were not as profoundly affected. Importantly, they did not find statistically significant differences in die-off between bed fumigated and untreated plots. In their published reports they suggested that the appearance of these two pathogens was related to changes in disease management practices, leaving it open to interpretation whether these changes were a result of the new chemical mix or the practice of bed fumigation.[90]

Surely, plant pathologists are correct that the change in fumigation regimes has in a proximate sense caused "novel plant collapse." Yet others think there's more to it. A farm advisor who did not want his identity disclosed showed me a field that had seen *Fusarium* after yearly fumigations of chloropicrin and Telone and had even been recently broadcast fumigated with methyl bromide. The field was in bad shape. "When you see the same diseases year after year in fields that have been treated," he said, "you need to look at other options"; "fumigants are not the end-all be all."

TOPOLOGICAL SUPPOSITIONS

The appearance of *Fusarium* and *Macrophomina* in California's strawberry fields, where they had supposedly never been before, returns me to questions about how these fungi came into being as pathogens. If it was solely a change in fumigation regimes responsible for these recent outbreaks, one would have to ask why these two diseases were not apparent before the advent of fumigation. In keeping with a topographical understanding, some have theorized that they were introduced, presumably after growers regularly began fumigating. However, the evidence for this theory is stronger for *Fusarium* strains than for *Macrophomina*.

Pathologists working on the *Fusarium oxysporum* species tend to think that the strain pathogenic to strawberries was introduced, based on observations that this species is prone to creating new strains through processes such as horizontal gene transfer. Horizontal gene transfer can occur when two different but closely related strains fuse to the extent that the nuclei come together to create a nucleus of both. Of the three pathogenic strains of this subspecies found in California, one of the isolates looks much like one from Japan, where a pathogenic strain was first found.[91] They thus believe that the strawberry strain came from Japan, likely arriving with host plants. In contrast, there is nothing in *Macrophomina*'s genetics or distribution to suggest a recent introduction.

Others posit that both of these fungi were present all along, part of the soil fabric, but escaped detection until recently. Their reasoning follows a topological understanding of disease. Although the approach more generally considers the conditions of possibility that allow a disease to appear, the key supposition is that the *intensity* of interaction acts as a tipping point to make the disease salient. Surely such intensities are more likely with monoculture more generally, but there are additional changes post–methyl bromide that might have been the tipping point for these two diseases. One is that the plants have become increasingly stressed, while the fungi do particularly well in the abiotic conditions that stress the plants. *Macrophomina*, for example, is particularly adept at surviving under adverse environmental conditions, such as low soil nutrient levels and temperatures above 30 degrees Celsius.[92] Researchers in California have found that diseases from all of these pathogens are often more severe if the infected plant is subject to weather extremes, water shortages, and poor soil conditions, including salinity—precisely the conditions that California was seeing in the 2010s, when several years of drought had set in. Interestingly, they have also found that these infections are more severe when the plant carries a heavy fruit load.[93] The irony that the much-desired productivity that fumigation enhances also causes stress seems to have thus far been lost on many a scientist and grower. But the point hasn't been lost that global climate change may be intra-acting with the rest of the assemblage to bring new pathogens into being.

Other possible tipping points are simply outside the scope of current agricultural science but not lacking plausibility. Given that some fungi also exchange genes in nonreproductive encounters, it is possible that *Fusarium* and *Macrophomina* have inherited genes to make them more virulent, or the host has received genes to make it less resistant.[94] It is possible that virulence is worsening through epigenetic changes—methylation processes that make the strawberry plant more vulnerable or the fungi more determined. It is also possible that the plant's always changing and highly variegated microbiomes (referring to the microorganisms that live in or around bodies and are increasingly believed to play a role in health) have made strawberries more susceptible. And therefore it is certainly possible that the scorched-earth approach of fumigation itself has promoted those genetic, epigenetic, or microbiomatic changes. These are exactly the sort of suppositions that the "postmodern" biological synthesis and post-genomic science propel.[95] Once, that is, we understand evolution to be much more indeterminate and organisms much more plastic than previously thought, any transformations to the soil ecology, pathogen, host, or even chemical reactivity could alter the entire assemblage, creating many paths to the virulent presence of ostensibly novel pathogens.

When I discussed these possible explanations with a renowned fungal biologist in 2017, he added several more.[96] And yet, when I posed questions along these lines to the applied scientists with whom I spoke, some were intrigued, but several gave responses ranging from "that's an interesting thought but of no practical importance," to "that's unlikely," to the most piquant suggestion that my speculations were fantasy.[97]

That these suppositions not only have been unexamined but have even been dismissed by agricultural scientists does not meet the bar of deliberate suppression of knowledge. To the contrary, much of this science remains "undone" because the questions were not even evident for a long time.[98] If the pathogens themselves were immanent in both their material form and hence as objects of knowledge, questions of how they arose and interacted ecologically were necessarily "unknown unknowns." Once the fungi became pathogens and objects of knowledge, however, the science remained undone because of the nature of agricultural science and the business of repair. They were studied largely by plant pathologists, whose function was not to understand their coevolution, but to manage them on behalf of an industry. And once chemical fumigation became the way to manage them, understandings of their complexity became irrelevant. Writing on German soil science in the early part of the twentieth century, historian of science Frank Uekötter argues that very thing. He shows how the availability of easy fixes such as agricultural chemicals made unnecessary and even wasteful more time-consuming investigations of complex ecological relations that would inevitably yield less definitive answers—and would simultaneously frustrate the clientele of agricultural science.[99]

So, just as the fungi were not pathogens until they noticeably killed things of value (in this case cultivated plants), knowledge supporting alternative explanations for pathogenesis was not produced or circulated because it wasn't useful to an industry that became wedded to fumigation. The availability of effective fumigants, in other words, precluded the search for ecological understandings.[100] As quoted in the opening epigraph, "You don't need to know the biology if you can kill it." Unfortunately that seems no longer the case, especially as the primary means to kill the pathogen may have indeed changed it.[101]

Meanwhile, few would disagree that becoming a pathogen has relied on intra-acting with a susceptible host. Arguably, this host plant lacked resilience and wasn't ready to withstand the onslaught. This plant was wasn't born this way, though; it was bred.

THREE

Curiously Bred Plants and Proprietary Institutions

No other plant bears fruit earlier in the spring nor as soon after planting, nets more profits per acre in so short a time, nor thrives in as many different climatic zones of the world.

Agricultural scientists Stephen Wilhelm and James E. Sagen,
History of the Strawberry: From Ancient Gardens to
Modern Markets, *1974*[1]

In my experience, making crosses is almost like a shotgun approach. That's why you have to look at thousands of different seedlings. We're pretty conservative when we're just looking at ten thousand.

Interview with a commercial strawberry breeder, 2017

If I need to put more armor on, I can't be carrying more guns.

*Interview with a strawberry geneticist referring to
breeding priorities, 2016*

I WAS DELIGHTED TO BE INVITED to view a breeding field, but I must have sounded very naive when I met the breeding team at one of a handful of strawberry nurseries operating in California today. I recall getting out of my car on a warm June morning in California's San Joaquin valley, fairly far from the fruit-growing regions on the coast. It was a soft, dusty field—the kind that farmers' trucks navigate well but not so much little hybrid sedans—sequestered between homes on a somewhat residential street. I approached the team, which had just finished supervising the planting and was packing up to go— I was late due to the horrendous traffic that now routinely occupies roads between the urbanizing valley and the fully urban San Francisco Bay Area— but they were still willing to answer my questions. After a round of introductions, the first question out of my mouth was: "Starting with the seed, can you describe the process of strawberry production?" My question was not exactly dismissed. But it was met with a stream of terms that were new to me ("meristem," "bare root"), a show-and-tell of what they had just planted (not seedlings but fibrous material—that's the bare root), and explanations of the breeding process that were more confusing than revealing. In short order I saw that I was ill prepared for the interview, even as I reminded myself that you have to begin somewhere. By the end of the next day I had visited strawberry nurseries in several locations—including in other low elevations in the far north of the Central Valley (the Sacramento valley), and in the high elevations to the northeast of the volcanic Mount Shasta. I had collected hours of audio recordings, which were later to serve as a reminder of my poorly conceived questions as much as they contained patient answers. It was not until I had transcribed, collated, and reviewed the notes, and triangulated them with websites and UC Davis publications, that I began to grasp the nature of the breeding and plant propagation processes and see their profound relevance to the fate of the strawberry industry. I also came to see why my opening question appeared so off-base—and yet hinted at a bigger truth.

In his classic text *First the Seed: The Political Economy of Plant Biotechnology* (2005), agricultural sociologist Jack Kloppenburg wrote that the genetically engineered seed, "as embodied information, becomes the nexus of control over the determination and shape of the entire crop production process."[2] As packets of genetic information, that is, seeds not only contain templates for the characteristics of bearing plants, such as yield, stature, and time to ripeness, and the produce that derives from them, such as fruit color, size, texture, and durability; they also shape the ability for plants to grow in certain environments and with particular inputs and technologies. It is indeed this latter aspect that has generated some of the critiques of genetically engineered seeds, which thus far have been designed in large part to be used with certain chemical inputs such as the herbicide Roundup. Yet even in more conventional breeding settings (without genetic engineering), seeds embed the priorities of breeders and those to whom they feel accountable for particular production, shipping, and consumption traits, even as these priorities are limited by the plant's evolutionary potentialities.

But strawberries are a different breed, so to speak, starting with the fact that what appear to be seeds bespeckling their outsides are actually distinct fruits, all containing their own seeds, while what we understand to be fruit is actually the stem of the plant. Saving and planting any one of these seeds would be highly unwise if indeed your goal was to produce a plant with qualities like the one you hold in your hand. If you plucked all the seeds off that strawberry and planted them in the ground, the "fruit" that would emerge, if it emerged at all, would be as varied as the number of seeds on that berry. Each would essentially be a distinct variety.

Plant geneticists say that the tremendous variation of strawberries grown from seed stems from the strawberry's exceptional heterozygosity. Many of the thirty thousand to sixty thousand genes present at any one position contain different alleles, or variant forms. The modern strawberry is also an octoploid, meaning that each cell bears eight sets of chromosomes, whereas wild or ancient varieties may have anywhere from two (diploid) to ten (decaploid) sets of chromosomes. The many alleles are primarily responsible for nearly infinite variation, while octoploidy supports adaptability to many environments.[3] By the same token, varieties with lower ploidy levels are of interest because they tend to be highly specialized to specific environments and therefore may contain secrets for specific adaptations.[4] In short, the strawberry genome is evidently far too indeterminate to predictably generate anything like a desirable plant starting with the seed. In addition, since berries can be open pollinated (enabling them to obtain plant material from a nearby plant), even the parents can be difficult to identify.

The strawberry's genetic variability has therefore both enabled breeders and challenged them. On the one hand, it has bestowed upon the plant extraordinary adaptability, providing nearly infinite options for plant breeders to create varieties suited to a multiplicity of conditions and desires. In his comprehensive 1966 book on the strawberry, the late pomologist and plant breeder for the US Department of Agriculture George Darrow said the following about the extraordinary plasticity of the strawberry:

> Because the cultivated strawberry is a hybrid of two highly variable octoploid species, it is possible to raise strawberries profitably under extremely different conditions: from irrigated desert . . . to the semi-tropics; under the continuous light of the Arctic to the twelve-hour day under the Equator; . . . under glass or plastic covers, with concrete blocks, and with plastic ground covers as mulches; as a six month crop, and as a crop occupying the same soil for hundreds of years.[5]

On the other hand, the characteristics of the strawberry have made plant breeding a highly stochastic and costly affair, even with established hybrids. In California, developing an improved cultivar takes no less than three to four years. Breeders first make hybrid seedlings in greenhouses, generally crossing existing cultivars. Once these seedlings germinate, they plant them

FIGURE 5. Meristem propagation in a strawberry nursery, Red Bluff. Photo by author.

in the various nursery regions where plant propagation generally takes place, always trying to replicate the conditions under which they will be grown. Most breeders plant ten to twenty thousand seedlings a year at this stage. After each seedling grows several hundred runners, they dig and freeze the runners. In the following year, they take pairs of seedlings from the runners to the prime fruit-growing regions on the coast and plant them there. After the runners start fruiting, breeders begin making selections, attempting to narrow down the two hundred or so that could be commercially viable. Those selections are harvested and frozen again, and possibly tried again in a fourth year. If and when breeders settle on a variety they like, they patent it, name it, and store the plant material with either the University of California (UC) Foundation Plant Services or private germplasm banks. Thereafter, the plant material is cloned: produced through asexual means to ensure the integrity of the varietal in fruit production. The breeding process is therefore, as put to me by breeders, "like finding a needle in a haystack," as well as "expensive, time consuming, and fraught with peril, . . . requiring a long-term vision." For these reasons and others, very few breeders in California besides those at UC and Driscoll's have developed successful varieties.[6]

FIGURE 6. High-elevation breeding field, Shastina. Photo by author.

Therein lies the deeper relevance of *First the Seed*. For Kloppenburg's primary concern was not the potentialities of the seed per se. Instead, he was interested in how plant-breeding institutions went about making a material readily found in the commons of nature into a commodity, to be bought and sold in markets, leading eventually to the ascendance of the corporate-controlled biotechnology industry. Key to this movement, he argued, was hybridization, the crossing of two genetically different individuals to develop a new breed with improved traits. Hybridization made crops more sturdy and productive—creating so-called hybrid vigor—but seeds from these crops wouldn't "breed true" if they were saved and planted the following year. Farmers who wanted those quali-ties—and few could afford to not want a more productive plant—would then need to purchase seed from suppliers every year.

And yet breeding hybrids could be very costly. So, as with hybrid corn, it was public institutions that first stepped in to play a major role in improving the strawberry, beginning in the 1930s. The UC breeding program became most important, producing a set of cultivars with an array of strengths and weaknesses. It was only after the original investments were made and knowl-edge established that, as with hybrid corn, entrepreneurial interests, starting with Driscoll's, saw potential in hiving off the university's breeding apparatus and moving it into the private sector. No doubt the exceptional heterozygosity of the strawberry was attractive to these new breeding entities, given an industry hungry for many possible traits. And given the necessity that a successful cultivar be asexually propagated, the plant material—and not the seed— became the thing of value, leading to outright theft of plant material from the university on more than one occasion.[7] Unlike the story of hybrid corn, the university remained in the breeding business and came to compete with private entities. The most recent and egregious instances of these plant grabs, described in what follows, came to a head in a monumental series of lawsuits involving UC, the fallout of which remains unclear as of this writing.

As it turns out, the nearly infinite possibilities for plant breeding worked at cross-purposes to developing resistance to the diseases and pests that coevolved with the plant. The strawberry offered so many possibilities for trait improvement, and California's geography offered so much potential for extending strawberry seasons, that breeders had a difficult time setting priorities, at odds with basic principles of selection. Particularly as chemical fumigation came to control diseases, they turned their attention to other qualities and were particularly neglectful of the disease resistance possibly endemic to ancient or wild varieties. Today, with new pathogens appearing concurrently with tighter regulations on soil fumigants, this neglect poses a real threat. At the same time, in a competitive environment unleashed by the privatization of breeding, it also stands as a potentially huge competitive advantage to the first who can retrieve the genes for multiple pathogen resistance.

This chapter takes up one of several challenges to the California strawberry industry arising from the resurgence of pathogens. I focus on the increased proprietariness of breeding at a time when much hope within the industry lies with disease-oriented plant breeding. I begin by laying out the key moments of improvement to the strawberry plant, introducing the actors and institutions that brought both the modern strawberry plant and its breeding mechanisms into being. The latter part of the chapter focuses on breeding priorities and the specific threats of proprietary behaviors.

FRAGRANT ORIGINS

Standing by a modern, large-acre strawberry field, you cannot help but note the intense smell of strawberry essence. During harvest season, when many rotten or under-grade berries lie squished on the ground, trampled by the foot traffic of pickers, the smell can be almost sickly.

The *Herbarius of Apuleius Barbarus*, written in the sixth century, described strawberries as "bearers of fragrant fruit"; Virgil called the strawberry the *humi nascentia fraga*, or fragrant fruit born of the soil; and Pliny the Elder's *Historia Naturalis* refers to the strawberry as *Terrestribus fragis*, literally "ground fragrance."[8] The origin of the English name for the strawberry is less settled, with some botanists suggesting that it originated from the practice of placing straw under plants to protect ripening fruit from the soil, while others say it emerged from the "obvious characteristics of the plant: runner daughter plants, which were strewed (formerly, 'strawed') over the ground."[9] Regardless of whether fragrance or straw best connotes the plant's essence, that essence was pronounced enough that botanists found it fitting to ascribe one genus to a plant that was highly varied and native to several different temperate regions of the world: *Fragaria*.

Fragaria vesca, the woods strawberry, was native to "deep, shaded valleys" and "alpine heights" throughout the northern hemisphere.[10] People and other animals (especially bears!) no doubt foraged the woods strawberry for thousands of years, picking out the choicest morsels for their consumption, and through their defecation in the woods contributed to a natural selection of sweeter, bigger, and perhaps prettier berries.[11] At least since the thirteenth century, "gardeners, botanists, and medics" became more purposive, noting differences in fruit size and habit of growth among wild strawberries, selecting the best and longest bearers," and doing their best to perpetuate them as varieties.[12] In their history of the strawberry, Stephen Wilhelm and James E. Sagen suggest that interest in strawberry breeding owes much to the pliability and responsiveness the fruit displays in variations in its growing conditions.[13]

LEARNING FROM EARLY VARIETIES

Although many imagine that the modern strawberry originated from the tiny, sweet *F. vesca*, plucked from dark forests, the modern strawberry actually owes its provenance to two other native strawberries. The Virginia (*F. virginiana*) was a "large-fruited, abundant, and pleasant-tasting" berry, which had been obtained by the British in their New World exploits and brought back home to cultivate. The Chilean strawberry (*F. chiloensis*) was a large-size berry that produced little fruit yet grew "naturally in the fields" and was successfully cultivated in gardens as well.[14]

Through experimentation, deliberate and accidental, amateur breeders began to bring together the strengths of multiple varieties. For instance, Brittany farmers developed a method of overcoming the barrenness of the Chilean strawberry by planting other varieties in proximity to the Chilean. This practice predated knowledge that the problem with the Chilean is that it lacks pollen. The success of this approach was key to the emergence of a veritable strawberry industry in Brittany, which supplied the markets of Paris and London with fresh strawberries for about one hundred years beginning in 1750. Still, it was an accidental cross between the Virginia and the Chilean that jump-started strawberry breeding in Europe and the United States and ushered in the modern era of commercial strawberry culture. This cross created what was commonly known as the "pine" or "pineapple" strawberry or, in Latin, the *F. ananassa*. Named for its fragrance and flavor, it was also hardy and vigorous enough to endure extreme weather—and it created a large fruit gratifying to gardeners.[15]

Through his studies of other indigenous varieties, the French botanist and strawberry breeder Antoine Nicolas Duchesne (1747–1827) uncovered the mystery of *F. ananassa*'s success: it did not require sexually mixed plantings to reproduce. Hence, the pineapple berry became the first known hybrid. Apparently, the Chilean strawberry hybridized readily when pollinated with other strawberries as well. With this discovery, the Chilean variety began to be used more frequently and became commercially successful around Brest, France, which notably was blessed with a marine environment and mild year-round temperatures, quite similar to coastal California.[16] Much later, well into the mid-twentieth century, geneticists at Friedrich-Wilhelms Universität, Bussey Institution of Harvard University, and the University of Manchester would hone in on the basic principles of hybridization: "Cultivated large-fruited strawberries and the species *Fragaria chiloensis*, *virginiana*, and *ovalis* from which they come, intercross freely, and their hybrids, with certain exceptions, produce fertile seedlings. But the cultivated varieties and the octoploid species from which they come do not readily cross with diploid, tetraploid, and hexaploid species," producing largely sterile seedlings.[17] Even sterile seedlings were important breeding material, however, and breeders went on to develop other methods for transferring desired qualities to large-fruited cultivated strawberries.[18] *The Strawberry in North America: History, Origin, Botany, and Breeding*, written by a professor of horticulture and published in 1917, identifies *chiloensis* and *virginiana* as the "two chief, if not the only" progenitors of the cultivated strawberry.[19]

The turn of the nineteenth century marked the beginning of commercial strawberry production in the United States, although farmers thus far had used only native *F. virginiana* plants. Critical was the 1840 release of Hovey's Seedling variety, a pine hybrid, and "the single major impulse given to both commercial strawberry culture and hybridization in the US."[20] C. M Hovey was a Massachusetts horticulturalist and botanical researcher who collected every variety as it came onto the market. Hovey's Seedling was considered vastly superior to existing varieties, and became "the standard of market excellence" in its time. The 1858 release of the Wilson, a "hardy, large fruited strawberry," marked another improvement.[21] As breeders found that working with hybrids could lead to these improvements, they all but abandoned working with native varieties. Thereafter, "the history of strawberry varieties is comprised of an endless series of substitutions of new and better varieties for old ones, which then disappear, or survive only as the parents of more sophisticated plants which take their place."[22] Alas, this also meant that genetic qualities that might have been useful for as-yet-unforeseen problems were also left behind from hybridization and lack of attention to germplasm maintenance.[23]

CALIFORNIA BREEDING

In 1859, Swain's Ranch in Santa Cruz produced some of the largest berries on record, combining Virginia and Chilean varieties.[24] This, along with the relative success of other "pine" hybrids such as the Hovey in California, hinted that California might be fertile ground for strawberry breeding and cultivation. In Massachusetts, where they originated, they would bear "only a spring crop and in exceptional years a light fall crop." In California "they bore throughout the months of June to January."[25]

California's own native species contributed to the breeding stock: a subspecies of *Fragaria chiloensis* known as the Californian beach strawberry. Observers found *Fragaria chiloensis ssp. lucida* growing abundantly in sand dunes and ocean-facing hills along the foggy California coast from Cape Mendocino to Monterey—that is, thriving in regions that were similar, if not identical, to conditions in which modern production takes place. It was described as "hardy . . . pleasantly aromatic, flavorful, sweet and beautifully, shiny red. They ripen very late—at the end of July—when all other strawberries are finished."[26] Not to be confused with the *F. californica*, which was a woods strawberry of much less interest to the growers and nurseries of Europe, *F. lucida* became popular among breeders and was hybridized with European and English varieties in the 1870s. Before it virtually disappeared, as interest in non-hybrid varieties waned, it furnished "valuable new characters" to the strawberry gene pool, including genes responsible for lengthening the bearing season, allowing "fresh strawberries 'out of season' to become a special feature of English markets."[27] For example, the *F. lucida perfecta* variety, released in 1861, was a cross between the British Queen and Californian. It was vigorous and fruitful, with "large, spherical, orange red, of high quality" fruit that matured from mid- to late season. It also had "Californian characteristics" of white flesh and prominent reddish-brown seeds.[28]

During the late nineteenth century, before the University of California took up the cause of strawberry breeding, California's nascent strawberry industry necessarily relied on varieties developed by private breeders. However, it appears that at least in the beginning, their proprietary behavior was primarily restricted to producing attractive cultivars, as well as providing vanity names to new cultivars. The Cinderella variety, featured in the *Pacific Rural Press* in 1878, was developed by a Mr. Felton of New Jersey. The press described its fruit as "large, conical," and "regularly formed"; its color as "bright glossy scarlet, rendering it surpassingly beautiful and attractive in appearance"; its flesh "very firm, with a mild, rich aromatic flavor" so as to "combine all the essential excellencies of a market and table fruit." It was of "very vigorous and robust habit," to boot.[29] The spread of the Cinderella variety catalyzed the Watsonville–San Francisco strawberry shipping boom, and the newly born Pajaro Valley industry expanded rapidly (see chapter 7).[30] The William Parry, developed by none other than William Parry, was a similar variety, but it never exceeded 10 percent of Pajaro Valley's acreage.[31] By the beginning of the twentieth century, two hundred varieties of diverse origin were available for California growers. Most of these varieties "were chance seedlings," and "introduced to California shortly after their appearance."[32]

Soon enough, private breeders in California began to more purposefully draw on native strains in attempts to use varieties more suitable to the region. Breeders at the Pajaro Valley Nursery, for example, developed the Melinda, which by 1890 had become the dominant variety in its place of origin, maintaining that dominance until 1915.[33] Albert Etter, a particularly entrepreneurial private breeder, was the first to become involved in systemic hybridization with native species. Working circa 1920, he used two *chiloenses* in his breeding, one from the California coast and another from South America.[34] Presciently believing that "plant vigor and hardiness derived from wild sources would be valuable new characteristics for the garden strawberry," he received wild strawberry clones from a UC botanist. Etter eventually developed at least fifty strawberry varieties, with the drought-tolerant Ettersburg 80 considered the best. Etter's berries "served as forerunners to a number of varieties," including several developed by UC that became of immense agricultural value.[35]

It was the Banner variety, however, that brought about California's preeminence in strawberry cultivation, put the still-reigning Driscoll-Reiter strawberry empire on the map, and began to transform the institutional and legal landscape of breeding. As the story goes, Charlie Loftus, a nursery owner from Shasta County in the far north of the state, had purchased the nearby Sweet Briar Ranch. There he found a patch of strawberries left by the previous owner and took notice of their size, shape, flavor, fragrance, and color. Loftus preserved the berry's genetics and named the berry the Sweet Briar. Louise Herman, sister of Pajaro Valley strawberry pioneer Joseph "Ed" Reiter, was a friend of the Hoppe family of Sweet Briar, and not infrequently vacationed at the ranch. One summer morning at breakfast, she was served Sweet Briar strawberries. She instantly recognized the new berry as superior in quality and flavor to anything then grown in the Pajaro Valley. On her return she told Reiter and his friend Richard "Dick" Driscoll of her discovery. This breakfast launched what Joe Reiter was later to call the California strawberry gold rush. Not much time passed before Loftus entered into an agreement with Driscoll and Reiter to protect and propagate the new berry, renaming it the Banner.[36]

Driscoll and Reiter planted their first commercial field of the Banner variety in Watsonville in 1904, and found that "the new berry was indeed superior to any strawberry ever grown before in the Pajaro valley or ... California."[37] The Banner was highly fragrant, flavorful, large, bright, and conical, although its softness limited its shipping capacity to San Francisco markets. Significantly, Banner plants "were large and grew vigorously," and were resistant enough to *Verticillium* to "produce satisfactory crops on lightly infested soils"—at least for some years. Their open crowns also deterred cyclamen mite, a pest that affected the strawberry industry badly in 1935.[38] However, the Banner did have one weakness that betrayed its origins in the colder, northern part of the state. It "runnered prolifically when grown in northern California" but was "incapable of vegetative reproduction in Watsonville." Not to be deterred, Driscoll and Reiter recognized that the usefulness of the new plant could be maintained only if initial plant production was carried on in the north, where winter cold conditioned the plants to send runners. That, along with disease issues, were the primary rationales for locating strawberry nurseries, where plants would be propagated, far from strawberry farms, where fruit would be produced.[39]

Until 1912, Driscoll and Reiter were the only commercial producers of the Banner variety, but other growers copied the unpatented variety so that it steadily eclipsed others in California's Central Coast growing region.[40] Apparently, this experience encouraged the Driscoll-Reiter partnership to become more proprietary in its efforts, among other things working to pass the Plant Patent Law of 1930 that redefined breeders as inventors. With patents, breeders could prevent others from using the plant material for a set amount of time and profit from their substantial investments in plant improvement.[41]

HYBRID INSTITUTIONS

Up until 1920, there was little public involvement in plant breeding, despite UC's status as a land grant institution with the imperative to serve agriculture and agriculturalists. UC's role in strawberry cultivation had first been limited to maintaining collections of strawberry varieties and species, "true to type" (having stable varietal qualities), and distributing them to other interested breeders and experiment-minded growers. It was in this context, for example, that UC botanists had distributed the beach strawberry to Etter. UC scientists also studied the suitability of already-existing varieties to California growing conditions. For example, from 1895 to 1897 the Carolina Superba was grown in UC tests at the Southern California Agricultural Experiment Station in Pomona, where observers noted high production, but poor fruit quality.[42]

The xanthosis scourge discussed in the last chapter proved severe enough to precipitate UC's direct involvement in plant breeding, a program that was "urgently needed by a declining strawberry industry."[43] Dr. Harold Thomas of UC's Department of Plant Pathology was charged with the program, and he hired the experienced plant breeder Earl Goldsmith to head up the Santa Clara experiment station. There Goldsmith began crossing and raising seedlings for selection. Thomas and Goldsmith's first experiments undertaken in 1929 and 1930 were followed by "systematically building up the desirable characters toward an ideal type."[44] In 1937, the California Packing Corporation generously made its ranch at Wheatland available to Thomas and Goldsmith, at no cost to the university.[45] This resulted in the 1945 release of five new "university" varieties: the Shasta, Lassen, Sierra, Tahoe, and Donner. Readers familiar with California will recognize these names as drawn from California's mountain regions, just like the Banner.

Due to the heterozygous nature of the strawberry, breeding continued to be serendipitous, although university breeders clearly were most interested in meeting the needs of their clientele: growers. Growers were delighted with the Shasta for its "large, firm, attractive berries and plants which are somewhat tolerant to virus diseases" and its ability to fruit from April through November, despite that it lacked flavor and was "only fair for freezing."[46] (At the time the market for frozen and processed berries was much more significant than that for fresh berries; see chapter 7.) Shasta was therefore well suited to the coast, where temperate temperatures prevailed throughout the summer. Lassen was popular among growers as well for its "short rest period requirement" and "high production of large berries" even though it was "too soft" for shipping, "not adapted to freezing," and had "only fair flavor." It fared better in Southern California because of its low chilling requirement, salinity tolerance, high productivity, and "wide adaptation under a variety of planting systems." As for the others, Donner had qualities of "high flavor," and the Sierra variety was "unique" in its resistance to *Verticillium* and therefore "could be grown where other varieties failed from the disease."[47]

Despite the original rationale for the plant-breeding program, to combat disease, growers opted for productivity as opposed to disease resistance—or taste, for that matter. Hence, it was the Lassen and Shasta varieties that were responsible for the tremendous expansion in acreage, yield, production, and farm value. In 1945, 20 percent of California acreage was in UC varieties, the vast majority of which were Shasta and Lassen; ten years later, 95 percent of the California acreage was planted with these new varieties, from which growers "profited immensely," and the value of the California crop rose from about $2 million to more than $30 million annually.[48] As these things go, disease, poor fruit quality, and excess supply led to somewhat of a reversal of fortune in 1957 and a few years that followed. Productivity continued to climb, but acres in production dropped considerably.[49]

During the 1930s, meanwhile, Dick Driscoll's son E. F. "Ned" Driscoll had been "cooperating" with UC's efforts by experimenting with some of these developing varieties.[50] Perhaps chastened by the experience of having the Sweet Briar / Banner widely adopted, in 1944, just before the university varieties were released, he created the nonprofit Strawberry Institute of California, stationed in Morgan Hill. The stated objectives of the new institute were to develop new and better strawberry varieties, to conduct research on all aspects of strawberries, and to provide scientific assistance to service grower members.[51] But it is clear that Driscoll also wanted to keep these developments in the family. He immediately hired away Goldsmith from the UC experiment station to make him the institute's chief breeder. A year later Thomas resigned from UC to join Goldsmith and become chief pathologist and director of the institute.[52]

This was not the last time a one-way door between UC and the private breeding sector was opened and shut. Yet "what remains unknown, perhaps to anyone but The Strawberry Institute, is the quantity of planting and nuclear stock, previously developed at the university, which almost certainly became Strawberry Institute foundation stock for further nursery and commercial growing." Writing on this incident, former industry bigwig Herbert Baum insinuates that Thomas would have had to have taken some stock to make the Strawberry Institute successful. But since no one had done an inventory before the pair's departure, it remains unknown whether the institute took the *only* copies of the stock, which would have dispossessed the university of much of the genetic material it had collected for future breeding.[53] Regardless, by 1959, the Strawberry Institute had established its own nursery, directed by Thomas, oriented toward "propagation of the highest grade plant stocks for growers."[54] Varieties developed at the nursery were patented for use by institute members exclusively, namely Driscoll-associated growers.[55]

Thomas and Goldsmith went on to develop many proprietary varieties for the institute that many would claim had better size, taste, and appeal to the fresh market than university varieties. Among them was the Z5A, or Goldsmith, named for the breeder himself and released in 1958. Growers chose Z5A for its productiveness, "especially in the summer and early fall along the California Coast," and "good shopping quality," given consumers' predilection for "large, firm, glossy, and attractive berries" notwithstanding their "fair flavor" and weak suitability for freezing.[56] The Z5A became dominant and established the newly formed Driscoll Strawberry Associates (DSA) as the premier shipper of strawberries.[57] The breeding team was bolstered in 1955 by Harold Johnson, a UC-trained plant pathologist who had worked as manager of the McCrea Seed and Chemical Company to support the Sheehy Berry Farm, one of the earliest growers for DSA—later renamed Driscoll's.[58] Under Thomas's leadership, the institute also "pioneered new cultural methods" and "supported research on control of soil borne diseases by fumigation with methyl bromide chloropicrin mixtures in the crucial years when that idea was being developed."[59]

Meanwhile, the strawberry industry and UC forged a more formal relationship when in 1956 the California Strawberry Advisory Board (CSAB) began funding research on the Davis campus, a program that continued when it became the California Strawberry Commission (CSC). In the course of this relationship, the CSAB, which had been holding the patents on varieties developed by UC researchers, handed them over to the university. The board wanted to rid itself of the paperwork of filing patents and collecting and disbursing royalties. Yet they granted these rights to the university on the condition that California strawberry growers would get first rights to all the varieties produced by the university, and they would not be able to distribute any other plant material outside of California for the first five years that it was in commercial production. That way California would always have an advantage over Florida, Spain, or anywhere else growing large volumes of strawberries.[60]

To revitalize its vacated breeding program, UC hired Victor Voth and Richard Baker, although Baker left seven years later to be replaced by Royce Bringhurst. Bringhurst was a pomology professor and became head of UC's strawberry research, charged with remedying the many ills of the industry.[61] Under his leadership, UC developed five new varieties from crosses of existing ones. These included the Fresno, introduced in 1961 to replace the Lassen in Southern California. Fresno's advantages included its "low chilling requirement" and that it was "more attractive, larger, firmer," which made the calyx or "caps" easier to remove.[62] Torrey, introduced that same year, had similar qualities to Fresno but an even lower chilling requirement. Tioga, released in 1964, was "10 percent larger than Lassen and larger than Fresno, Shasta and Torrey" and also more productive than Lassen. Solana, introduced in 1957, had "high flavor, short rest period requirement, and tolerance to salinity and virus diseases," and thus was more suitable to the drier climates of Oxnard and Fresno.[63] The team also developed the hygienic-sounding "Wiltguard," selected to be both resistant to *Verticillium* wilt and flavorful.[64] The near dearth of information on this variety, save for one 1961 broadside from Bringhurst and Voth announcing its release with the Fresno and Torrey, suggests that it was not widely adopted, if adopted at all.[65] The disinterest may well owe to the fact that at this same moment Stephen Wilhelm and Edward Koch were pioneering new cultural methods, including controlling soilborne diseases with methyl bromide and chloropicrin. To be sure, the revolution in fumigation, bedding, irrigation, and so forth that transpired in the 1960s allowed these new university varieties to have their "best performance" and, as we shall see, to have the university all but abandon breeding for pathogen resistance.[66]

With hindsight on Thomas's and Goldsmith's departure from UC, industry leaders, through the CSC, sought out a new generation of breeders to replace Bringhurst and Voth as they neared retirement age. They selected Douglas Shaw in the Department of Pomology at UC Davis, and even financed his 1986 hiring prior to Bringhurst's retirement to ensure adequate transfer of pomology knowledge and experience. When Voth retired in 1991, UC was reluctant to replace him, but the CSC successfully pressured the university to do so with Dr. Kirk Larson, also of that department.[67] By 2004 Shaw and Larson had developed thirteen new university varietals, including the Pajaro, Chandler, Seascape, Diamante, Camarosa, Ventana, and Albion. The Diamante had a long bearing season, but issues with taste and coloring, and was largely discredited. The Ventana also had quality problems. The Albion, in contrast, was a hit due to its excellent taste (a quality to which I can attest) and became a favorite among growers selling in regional markets and directly to customers in farmers' markets.

Ostensibly Shaw and Larson's breeding program emphasized not only improved production attributes, but also resistance (or tolerance) to pests, pathogens, and other environmental factors or stresses—a smorgasbord of desirable traits.[68] After 2004, about when the methyl bromide phaseout was starting to become a reality, they supposedly even doubled down on efforts to breed for resistance, and it appears from their own publications that *Verticillium* was their first target.[69] Of the several varieties they developed after 2004, including the Palomar, Benicia, Mojave, Merced, San Andreas, and Petaluma, the Petaluma, released in 2014, was showcased as having good resistance not only to *Verticillium*, but also to *Fusarium* wilt and crown rot caused by *Macrophomina*.[70] As we shall see, though, it is not altogether clear that pathogen resistance was the priority in breeding, nor even possible given the number of other qualities they sought.

In any case, the inconsistency of the newer varieties indicated trouble afoot. Only the San Andreas was being widely adopted, at the time mainly for its flavor. Problems with the Diamante and Ventana, as well as growers' return to the Camarosa after others had been released, caused some to cast doubt on the quality of the UC breeding program.[71] By 2012 Shaw and Larson were voicing their own frustrations, publicly questioning the university's commitment to plant breeding and calling out the CSC's 1997 cut in funding to UC's program.[72] They and others charged that UC was refocusing on more basic research such as developing gene pools and using genomics to identify certain characteristics like disease resistance. This was an ironic charge given the history recounted by Kloppenburg—that beginning in the 1970s universities were asked to step away from the more profitable aspects of agriculture and turn to a more traditional division of labor between the universities and privately held companies.[73]

But something else was brewing, and it is worth a pause to discuss what was at stake financially in UC's breeding efforts. Much has been made of the neoliberalization of public universities, in the United States and elsewhere. Universities, bereft of revenue thanks to the withdrawal of much state support, have had to create profit centers and otherwise make themselves especially useful to those who might fund their research and education missions. Among the many changes this has brought is decreased emphasis on the arts, humanities, and humanistic social sciences and more on STEM fields— science, technology, engineering, and mathematics.[74] Many universities have thus forged new alliances with the private sector in which university research expertise is purchased with large grants and contracts, in expectations that funders will receive some sort of return. Although the university's arrangement with the California strawberry industry long preceded the era of neoliberalization, it cannot be ignored that licensing of its strawberry varietals had become a major revenue generator for the university. Indeed, licensed varieties earned the University of California $50 million in royalties from 2004 to 2013, while costing only $16 million, placing strawberries among the UC system's top money-earning inventions. Roughly a third of those royalty payments, $18 million, went directly to Shaw and Larson.[75]

Nevertheless, Shaw and Larson also saw more lucrative possibilities elsewhere, finding that amount nominal compared to what private breeders were receiving for their inventions. So in 2014 they announced their intention to leave and form a private company, California Berry Cultivars LLC (CBC), a partnership with strawberry grower and former secretary of the California Department of Food and Agriculture A. G. Kawamura of Orange County Produce.[76] Other partners included California Giant, Lassen Canyon Nursery, Eurosemillas (the master licenser in Europe of the UC varieties), and two individual growers who took only small stakes in the company.[77] Notably, Shaw and Larson did not themselves invest in the company, presumably foreseeing ample income from royalty payments. Following the announcement of their departure, the university stopped taking the CSC's

$350,000 in annual research payments, casting further doubt on the university's commitment to the breeding program.[78]

Of enormous significance is that in creating their new proprietary program Shaw and Larson took a collection of germplasm of about fifteen hundred varieties of strawberry plants, owned by UC, from which they would breed new varieties. Shaw even told his investors that he had total ownership of the varieties.[79] The problem is that plant patent law specifies that non–patent holders cannot use material of any patented plant in breeding.[80] Therefore breeders without legal access to patented plants are left to use material from plants that are wild or with expired patents. The legal gray area at issue is that Shaw was indeed the patent holder while at UC. However, he had access to the germplasm only as a UC employee and had signed an agreement with UC that he would not appropriate the physical material. An additional legal gray area is that Shaw allegedly acquired the plant material in Europe (presumably with assistance of the partner Eurosemillas), produced progeny, and brought it back to California.

In the wake of Shaw and Larson's 2014 announcement, the CSC filed a lawsuit against UC. The lawsuit specifically accused UC Davis of preparing to abandon its plant-breeding program, in violation of the agreement between the two institutions. In addition, it claimed that UC Davis was letting the two breeders privatize their publicly funded research and walk out the door with vital trade secrets. To prevent further incidents, CSC also requested their own copy of this "genetic library."[81] To address these charges, the university promised to replace Shaw and Larson and make duplicates of the fifteen hundred plants. Through their own countersuit, however, UC Davis refused the CSC's demands that they turn over a copy of the germplasm. To the contrary, the university sought rulings declaring that it was the sole owner of the germplasm and that it alone controlled patents on nine (at the time) strawberry varieties developed over the years by the two departing scientists.[82] The 2015 settlement of this particular lawsuit decreed that UC, not the CSC, was the sole owner of the germplasm. UC did go on to hire a new plant breeder, Steven Knapp, and created a stakeholder committee to oversee the breeding program henceforth. Although the program is moving in the direction of incorporating advanced genetic tools, including conducting DNA fingerprinting on the entire UC collection of germplasm, it continues to be involved with cultivar development as well.[83]

Surprisingly, Shaw and Larson were not named as defendants in either of these suits, even though their actions triggered the kerfuffle. But then, the university asked Shaw and Larson to surrender their copies of the germplasm and subsequently refused to license a copy of the germplasm for their new company, CBC, because that would have bestowed on them an unfair competitive advantage over others in the strawberry industry. UC also asked that Shaw and Larson submit a patent for unreleased cultivars to keep them as property of the university.[84] Shaw and Larson retorted that they had been denied the rights to their life's work and that UC was attempting to suppress competition. Based on those claims, CBC went on to file a suit against UC, going so far as to say that UC was contributing to the dominance of Driscoll's over the rest of the industry. In their countersuit, UC claimed that Shaw was illegally breeding with the (unpatented) cultivars that rightfully belonged to UC, and that he was trying to destroy the public breeding program to enrich himself and his friends.[85]

In May 2017, jurors unanimously decided that Shaw and Larson willfully infringed UC patents, breached loyalty and fiduciary duties, and used plant material owned by the UC Davis Public Strawberry Breeding Program to develop berries for CBC. Crucial to the ruling was testimony from an expert in plant genetics at Yale University, who conducted DNA analysis on CBC plants grown in California from material imported from Spain. He found that virtually all of CBC's seedlings contained genetic material from five university-patented varieties that had not been released at the time they were bred and nineteen that had never been released.[86] While still employed by the university, in other words, Shaw and Larson had indeed used UC breeding stock to make crosses in Spain for CBC—hence the Eurosemillas connection—and sent them back to California to produce pedigree plants without the university's knowledge or permission. However, the judge in the case was less convinced that only Shaw and Larson were to blame. After the jury rendered its verdict, the judge stated that UC Davis was as guilty of "bad conduct" as the two scientists, and so the parties had to agree to settle. In the final settlement, on September 15, 2017, they agreed that Shaw and Larson could continue using some of the plants they developed at UC Davis, but had to return others. They also had to forfeit $2.5 million in royalty payments for work they did at Davis.[87]

To the extent that the suit prevented public breeding stock to enter entirely into private hands, it could be read as a victory for public breeding—and the land grant mission of the university. It is not as though UC's program is entirely public, however, and the judge in the case clearly had picked up on some of UC's own proprietary behavior. In a 2014 broadcast, a National Public Radio reporter captured the true hybridity of UC's breeding program, in addition to the fruits it breeds. It surely differs from private programs by making the fruits of its labor (plant varieties) available to strawberry growers for a reduced royalty price. Yet UC Davis has made clear its disinterest in simply advancing the field of genetic breeding and sharing its findings and prototypes with others. On the contrary, this reporter noted, UC Davis's focus was on developing commercial varieties that it could patent and profit from.[88] The germplasm, that is, was the basis of any competitive advantage UC would have in plant breeding and thus collecting royalties, even as the mission of the institution remained to serve a broader public.

Today there are several private breeding companies with their own stock competing with UC, including Driscoll's, which began it all. Given that some buyers mandate use of their proprietary varieties, including Driscoll's, this has insignificant implications for the structure of the industry, as I will discuss in a later chapter. My question here is to what extent these institutional structures have and will continue to shape the possibilities of breeding for pathogen resistance, especially given how early use of the plant narrowed these possibilities considerably.

DID THE PATHOGEN SPEAK?

Time and time again in early breeding histories we see the lauding of varieties that could take advantage of California's cool summers and mild winters to bear fruit for several months. Commenting on the California strawberry industry as early as the nineteenth century in *The Natural Wealth of California*, one author noted that "there is not a month in the year in which strawberries are not to be had in San Francisco. They are plentiful during five months beginning with April."[89] Others later noted the marketing advantages of the strawberry's early arrival in the spring, ripening before all other fruits.[90]

Many breeding efforts were thus geared to making those seasons even longer, by creating plants that not only bore berries over ever more extended periods, but also could be complementarily grown in different regions of the state. This could enable the shippers, if not the growers, to supply strawberries nearly year-round in the market. The breeders at Driscoll's recognized this early, forcing UC to catch up.[91] In smoothing the season of the berries, moreover, breeders could at least attempt to address the inevitable decline of prices when the gluts of late spring arrived. Therefore, rather than reproducing ever-bearing varieties—those developed in the US northeast that do not

produce fruit until the weather warms and then have a short season—
California strawberry breeders worked on both short-day types and day-
neutral varieties. Short-day varieties, those that fruit before the days lengthen,
could allow early-season entry into the market; they would generally be
planted in the south (the development of these varieties is what opened up
the southern regions). Day-neutral varieties, those that can withstand colder
temperatures, could continue to grow and produce late into the fall; they
would be used more in the north.[92] Today, the mix and match of these traits
means that annual strawberry production begins in winter in Southern
California or Mexico (where several large players have operations), works its
way to the south coast of Oxnard and Santa Maria for shorter seasons, and
then ends up in Watsonville and Salinas for the long season that ends when
and if the late fall rains come.

It is not as if plant and soil diseases were entirely ignored in the development of the modern strawberry. As we saw, the Banner was at first valued for
its *Verticillium* resistance, and UC breeding was initially very much focused
on resistance to the diseases that began to plague the strawberry industry in
the 1930s. Some historian-participants of the industry claim that disease
resistance was a primary goal of the first five university varieties developed in
the 1940s.[93] However, other qualities in addition to long seasons became
more important. A price crash in the slump of the early 1930s "made it emi-
nently clear that California's position in strawberry cultivation demanded
not only virus-resistant, high-yielding varieties but also berries of the highest
quality and flavor."[94] Reflecting on the situation of the early 1960s, Darrow
wrote in his history, "For the strawberry to continue as a major crop with a
high per acre return, some basic requirements are: still higher yields, moderate production costs, and more appealing berries for consumer demand."[95]
Meanwhile, shippers were interested in firm berries less prone to rot, and
berries holding higher flavor for several days in shipment. In short, it wasn't
long into UC's program that productivity, beauty, durability, and occasionally flavor began to trump disease resistance. Nutrition, to my knowledge,
never became a breeding priority.

When discussing varietal choice in my interviews with growers, I often
asked what traits they looked for in a cultivar. Commonly the response was
a long list of traits. I began to wonder whether a multiplicity of breeding
goals—productivity for growers, durability for shippers, beauty and taste for
consumers, and the long season for all—could have been achieved with
breeding for disease resistance as well. Apparently, the answer is no, and as

early as 1966 Darrow noted a possible inverse relationship between *Verticillium* wilt resistance and desirable performance traits.[96] In my private discussions with strawberry breeders, they also acknowledged that breeding for certain traits would make it more difficult to breed for others, including resistance to multiple pathogens. Within an organism, there is only so much carbon, they say, and when it is directed toward one function it generally detracts from another. As put by one of my informants, "If I need to put more armor on, I can't be carrying more guns."

To the extent it was ever found, this important lesson was lost in the 1960s revolution in cultivation practices, most notably the intensive use of soil fumigation with methyl bromide and chloropicrin that made breeding for *Verticillium* resistance a lesser priority. Researchers may have tried to understand, screen for, and even select for pathogen resistance, but it no longer needed to be a primary focus of breeding.[97] And there is some evidence to suggest that many breeders disregarded resistance altogether. Using genomic technologies, current UC researchers have been tracing pedigrees of all UC trials and found that that the *Fusarium* resistance gene—one for which there is a definitive marker—is spattered randomly in the attempted varietals, suggesting that *Fusarium* pathogen resistance, at least, was never a priority. Recent field tests demonstrate that the recent Shaw-Larson varieties don't exhibit much resistance at all except, perhaps, San Andreas. Other breeders have commented on the reason for these poor results: even when Shaw and Larson were putatively breeding for resistance, they had bred in fumigated soil and hadn't tried inoculating the plants with the pathogens.[98] And yet, even current UC breeders are continuing to breed for productivity and shippability, knowing that growers and shippers are their primary clientele, even though they recognize the importance of pathogen resistance.

Through my discussions with growers, scientists, and others, I began to wonder whether fumigation somehow weakened the breeding stock. Growers like to say that clean, fumigated fields make for a "robust" plant, echoing the ideas of many of the scientists past and present who have worked on behalf of the industry. Yet some of the scientists with whom I spoke suggested otherwise. They said that fumigation made the strawberry plant more susceptible to pathogens—that what was effectively "pampering" the plant made it less resilient to external stressors. They additionally noted that plant breeding conducted in fumigated fields, as virtually all was, may have inadvertently selected for plants that only survive well with fumigation, contributing to the entrenchment of the fumigation assemblage.

A related, even overlapping question I had is whether modern crop breeding itself weakened the plant.[99] Writing in 1966, Darrow noted that "intensive selection for desirable traits seemed to have made most of the selected clones susceptible [to disease], even when both parents were resistant."[100] Writing fifty years later, UC Davis researchers who study nematodes suggested that "modern crop breeding may have inadvertently selected against traits that fed beneficial microbes and encouraged their establishment," also making it more difficult for the plant to withstand pathogens.[101] Another informant stated quite plainly that high-yielding hybrids have little chance of surviving in diseased environments. Corroborating this notion, and following reports that plants that carry a heavy fruit load are more susceptible to diseases such as *Macrophomina*, another informant suggested that productivity itself stresses the plant. Nevertheless, none of these suggestions have been demonstrated. As with the origins of pathogens, it seems that the scientific focus on productivity and marketability has left such questions to speculation—ironically, given that these foci might have contributed to the problem. More undone science.

The challenges moving forward are formidable. While breeders say that *Fusarium* is a relatively simple disease, *Macrophomina* is complex one without much natural resistance. Significant research gaps in its evolutionary background and genetic character have further challenged resistance breeding.[102] Meanwhile, scientists have scarcely considered that some cultivars may be changing epigenetically or microbiomatically in relation to the diseases that haunt them. Since it would be difficult to observe or predict these changes, breeders see it as of little practical value, despite what clues that line of research might hold for managing the strawberry ecosystem. At the same time, many in the industry now believe that resistance breeding is the best, and possibly the only, hope for the strawberry industry in a context of enhanced regulation and threatened phaseout of remaining soil fumigants. How, then, is California's breeding mechanism set to deal with this brave new world?

PROPRIETARY THREATS

There is little question that plant breeding has contributed resoundingly to the strawberry industry's success, and that much of that success has derived from the capabilities of the strawberry. The strawberry's heterozygosity—having a great many genes that can be combined in an infinite variety of ways—provided many opportunities for breeders to fiddle and create different qualities, adaptable to a wide variety of conditions.[103] California native varietals, especially, revealed characteristics of long seasonality, which, again, were highly attractive to breeders. The opportunities the strawberry germplasm provided were in many ways as important to the growth of the industry as the ability of industry to overcome the challenges of breeding. Today, however, modern strawberry cultivars have a relatively narrow germplasm base, derived as they are from the octoploid "Pine," to the neglect of potentially valuable traits found elsewhere, including those in varieties with lower ploidy. Given the qualities for which the Pine hybrid was valued, most modern varieties are therefore concentrated with alleles or copies that confer industrial traits.[104]

It is not an entirely new insight that reliance on the hybridized Pine while ignoring wild varieties created a sort of path dependency in the evolution of the strawberry's genetics. As early as 1966, the pomologist Darrow wrote that some wild Pacific Coast *chiloensis* selections "display resistance to many diseases, including Red-Stele, *Verticillium* wilt and mildew" and wondered whether these desirable characters could be bred into the modern strawberry. For that matter, he also wondered whether beneficial qualities could be found in other wild strawberries.[105] For a long time, however, such possibilities were all but ignored. The problem now is that true forms of such ancient or wild varieties have been virtually lost by both unintended and intended inbreeding.[106] Therefore so has the potential resilience these varieties might have imparted.

At the same time that hope for breeding for a post-fumigation strawberry industry lies with wild or unimproved varieties or even old heirloom cultivars, obtaining and using this germplasm therefore presents major challenges.[107] Besides the problem of inbreeding, some has never been collected. It is true that many old hybrid varieties are retained in germplasm banks. In the US, these include UC's collection at Foundation Plant Services, as well as the USDA collection held at the Federal Plant Repository in Corvallis, Oregon. Yet the problem with these collections is that they continue to expand—often exponentially, considering the crosses produced each year, making them very difficult to use. Current UC breeders have fingerprinted UC's entire collection and found that 57 percent has never been used in a breeding pedigree or an experiment. Although they think that most of it is of little use and could be discarded—that utilizing it would be "like throwing darts"—they also recognize that preserving the entire collection is the fail-safe approach.

Meanwhile, finding wild varieties are random and occasionally dangerous events: one breeder told a story of having cut and pocketed a strawberry plant growing on a rocky seaside cliff. He lost it out of his back pocket when a large wave came along unexpectedly. Other breeders found *Macrophomina* resistance residing in the old beach strawberry. But they found it in a national park, where collection permits are required. In addition, the park wanted to retain intellectual property rights to the material used.

Finally, as we have seen, some of the germplasm has been lost or stolen by private institutions. Even that which might have been obtained within the bounds of the law, and for which public institutions retained copies, has been further developed by private breeders who are highly unlikely to grant public access to the plant material.

Thus, the threats to the strawberry industry of concern in this chapter not only inhere in the difficulties of obtaining ancient varieties, or even in the formidable challenges of breeding resistance to complex, coevolved pathogens. They also inhere in the avarice that has pervaded strawberry breeding in California. In improving the strawberry, breeders saw opportunities for patent revenues and royalties; growers saw opportunities to profit on either yield or aesthetics; and shippers saw opportunities to gain competitive advantages by being first to release berries of certain qualities. The particular qualities of the strawberry seem to have only intensified this proprietary behavior. The potential of creating long-bearing varieties well suited for California's climate led to early competition among breeders. Its heterozygosity and octoploidy made breeding full of promise but also risky and costly, so that breeders wanted to recoup their investments—and more. The indeterminate character of strawberry seed made plant material of existing cultivars the thing of value, inciting UC researchers and others to peel off and take the germplasm with them—more than once! Notably, this proprietary behavior came far before the putative shift from public to private science that many claim to be a hallmark of neoliberalism.

When UC researchers first left to join the Strawberry Institute, the transfer of knowledge and germplasm catapulted the ascendency of Driscoll's into industry domination.[108] Today Driscoll's is clearly positioning to do more. Scientists there are proud of the company's leadership in organics and sustainability more generally. In breeding, Driscoll's has long emphasized pathogen resistance, and its chief breeder, Phillip Stewart, seemed to have been ahead of the crowd in recognizing the importance of ancient germplasm.[109] In addition, he has been selecting in nonfumigated soil for many

FIGURE 7. UC field trials at breeding field day. Photo by author.

years to expose potential varieties to pathogen pressure, a practice that Shaw and Larson neglected. While the foresight of Driscoll's is certainly to be lauded, the fact is that these developments will only be used by its own growers. This could create an insurmountable competitive advantage for the company once fumigants are finally disallowed. And because proprietary varietals are more expensive, these developments will contribute to an already-existing shakeout of low-resource growers, even those working with Driscoll's (see chapter 7), potentially leading to an economic monopoly by one farsighted company.

Since the lawsuits began, UC has revitalized its breeding programs, and pathogen resistance has become a major priority for researchers in all major breeding programs as well. A major grant received in 2017 from the USDA to develop and deploy cultivars resistant to soilborne and aboveground pathogens could even make UC great again.[110] In carrying out the grant, UC intends to cooperate with other breeding programs, including private institutions like Driscoll's even as UC itself will be patenting its varietals and collecting royalty payments. Whether UC will be doing so to generate funds in a neoliberal funding climate or simply to keep private breeders from competing against them is hard to say—and it's not clear that those are substantially different rationales. If the goal, however, is to allow quasi-public breeding to persist, it is unclear that this is a substantial enough counter to the threat of monopolization posed by Driscoll's, in a context in which chemical fumigation may indeed disappear and the entire production system has already been bred into the strawberry.

FOUR

Chemical Solutions and
Regulatory Pushback

Methyl bromide had this negative connotation, you know, it's
an ozone depleter. But what people don't know is how well it
worked. It was great! It worked really well! And if you want to
reduce your carbon footprint, you need to get as high of yields as
you can. If you want to reduce the amount of water that it takes
to grow a crop, you want to get the highest yields you can.

Interview with pest control company employee, 2014

Fumigants do the job they were created to do—kill everything.

Interview with crop advisor, 2015

There are very few industries outside of the agro-industrial complex where toxicity is a use value.

*Geographer Adam M. Romero, "'From Oil Well to
Farm': Industrial Waste, Shell Oil, and the
Petrochemical Turn (1927–1947)," 2016*[1]

ONE FINE DAY IN JULY, I drove up to the Driscoll's research facility at
Cassin Ranch, situated on an alluvial bench just above the Pajaro Valley.
Cassin Ranch is vaguely reminiscent of an upscale country inn; it consists of
a collection of wood-framed ranch-style buildings, replete with surrounding
porches and lovely Mediterranean-style gardens. There are no ranch managers at Cassin Ranch, since Driscoll's does not commercially grow berries. But
as part of their sustainability programming, they do conduct a great deal of
research on farming without fumigants, which they share with their contracted growers. Jenny Broome, whom I was there to meet, is one of the forward-looking scientists that Driscoll's has employed and leads their global
plant health research department. I had met Jenny years ago, when she was
the associate director of the Sustainable Agriculture Research and Education
Program at the University of California (UC). Previous to that, she had

worked at California's Department of Pesticide Regulation as a research scientist. She knows her stuff. We sat down in her office for what turned out to be a wide-ranging and intellectually generous discussion. At one point, she showed me a remote sensing map of some fields. One side of the paper was dominated by a deep blue hue. She explained that the dark blueness represents a high value on what is called an NDVI (normalized difference vegetation index), which indicates the amount of vegetation.[2] The deeper the blue, the more luxuriant the plant cover. That side of the map is the fumigated side, she explained. The other side of the map was dominated by lighter blues, fading into white. This represented the nonfumigated side, where they had found substantial disease. Jenny is no fan of fumigation, and certainly doesn't find it a long-term sustainable solution. "But yeah," she said, "it's a good snapshot to show there is a reason why people fumigate."

Strawberry fields, which are always fumigated before planting, are indeed green, and the plants look robust. Many growers thus say that fumigation begets vigor—a word that Jenny used, too. Since there is no denying that fumigants are used to kill, some have likened fumigation to sterilizing sheets after a tubercular patient has died and vacated the hospital—a sort of good killing that allows life to flourish again. Along similar lines, some suggest that fumigation helps the agricultural environment. By keeping fields "clean," growers say, they can enhance and preserve other nutrients and properties in the soil that are beneficial to crops; they can reduce the amount of fertilizer they use, thus reducing nitrates in the water supply; and they can viably implement integrated pest management plans whereby less pesticides are used and beneficial insects are encouraged. (They say this although strawberries continue to top the Environmental Working Group's "Dirty Dozen" of crops with high pesticide residues.)[3] They can even reduce the footprint of farming by ensuring bountiful yields. Fumigation, in other words, is often defended on behalf of life.

In this chapter, I trace the history of three soil fumigants that together helped produce the strawberry industry's success, with a special focus on its much-favored methyl bromide. Although their origins differed, these three all came of age during the mid-twentieth century, when many technologies born of the automobile and war industries were repurposed for agricultural uses.[4] They also came of age when an American public largely accepted that chemistry could be deployed to improve lives and solve a multiplicity of modern problems—the age, that is, of "better living through chemistry."[5] Yet the same qualities that made fumigants effective for managing pathogens also made them dangerous to the people who made them, applied them, and lived, worked, or learned in close proximity to their use. With new scientific knowledge, it became increasingly untenable to believe that the chemicals so effective at killing soil microorganisms would somehow not affect humans. In that way, the pushback that led to the ban on methyl bromide, thwarted methyl iodide from seeing much of the light of day, and engendered tighter use restrictions on two remaining fumigants—much in the name of human health—may have been unwelcome on the industry's part, but it was not unreasonable. Nevertheless, these tighter restrictions, which this chapter also describes, pose another major threat to the industry, as the reemergence of soil pathogens appears directly attributable to the loss of one chemical and the change of fumigation regimes precipitated by restrictions on the others. A total ban, which the industry has some reason to fear, would of course threaten the entire assemblage, with its foundational dependence on fumigation. By the same token, and as I have suggested in previous chapters, it is possible that in having precipitated changes in plants and pathogens that the industry is poorly equipped to understand, fumigation itself has become a threat to the assemblage. This chapter also briefly returns to that question.

FUMIGANT ORIGINS

Of the three chemical compounds that became mission critical to the strawberry industry—methyl bromide, chloropicrin, and 1,3-dichloropropene (1,3-D)—none were originally designed for agricultural uses.[6] But by the early twentieth century, food production and distribution were in dire need of treatments that were effective at killing the enemies of plants, yet were seemingly safe enough to handle and not detrimental to future plant growth after treatment. Fumigants appeared as newfound solutions to rather old problems. Although they were first put to work to address aboveground pests, agriculturalists were also anxious about subterranean insects. Wireworms, in particular, were a big nuisance, and entomologists were contemplating the potential benefits of treating *soils* with chemicals. In 1913, personnel from the USDA Bureau of Entomology conducted the first known fumigation-like treatment by injecting soil with sodium cyanide.[7] Deeming that chemical too dangerous, thereafter entomologists began experimenting with various others to control the wireworm.[8] One of the first they found effective was chloropicrin, first tried in California in 1933.[9]

FIGURE 8. "The American Soldiers in Presence of Gas—42nd Division," Essay, France, September 20, 1918. OHA 80 Reeve Photograph Collection. Otis Historical Archives, National Museum of Health and Medicine.

Chloropicrin—from Lethal Weapon to Warning Agent

In 1848, a Scottish experimental chemist first prepared chloropicrin by adding a water-based solution of picric acid, then used as a dye and an explosive, to bleaching powder.[10] Yet chloropicrin remained a "laboratory curiosity" until World War I, when it was widely deployed as a weapon by both sides. As a noxious lachrymatory (tear gas), it caused vomiting when inhaled. Soldiers would remove their gas masks so they could vomit, thereby exposing themselves to the full spectrum of lethal gases.[11] Its insecticidal potential became apparent when people noticed a scarcity of insect pests in some of the war-torn regions.[12]

In the United States, an economic entomologist tested chloropicrin's insecticidal qualities in 1917. Though effective, he found it volatile and extremely toxic, specifically "283 times more toxic" than carbon disulfide. He also found it prone to deter seed germination in agricultural settings, although he reasoned that cultivation after fumigation might mitigate the problem.[13] Large stocks of tear gas left over from World War I precipitated additional research into the possible peacetime uses for the chemical.[14] Researchers from the Chemical Warfare Service and the USDA's Bureau of

Entomology included chloropicrin in a test of four different chemical weapons on lice, hoping to find "a gas which can be placed in a chamber and be experienced safely for a short period of time by men wearing gas masks and which in this time will kill all cooties and their nits." These same agencies were concerned not only with lice; they researched the efficacy of war chemicals against dozens of insect species. Most of the gases fell short, but chloropicrin, "the compound that chemical warfare had lifted from obscurity," proved highly effective. Because of its "warning properties" (referring to its smell and propensity to cause tearing), entomologists deemed it not exceedingly dangerous even though it "harmed civilian exterminators as readily as enemy soldiers."[15] Agriculturalists also experimented with its use as a nematicide and saw great results. In pineapples it not only killed the little worms but also stimulated plant growth.[16] Later experiments showed that it was effective in killing soil bacteria and fungi.[17] According to a bibliographer of the chemical, the advantages of chloropicrin as a fumigant included "its high toxicity to many species of insects and rats; fungicidal and bactericidal properties; complete freedom from fire and explosion hazards; low solubility in water; ability to penetrate bulk commodities; non reactivity . . . ; and a pronounced odor and lachrymatory effect which usually effectively warn[ed] of its presence." Its disadvantages included its slowness in action, tendency to be harmful to living plants and seeds, difficulty in removing its odor, and its "nauseating effect upon the operator."[18] Moreover, it was very expensive to manufacture because of the cost of picric acid.[19]

Chloropicrin first came into commercial use as an insecticide during the 1920s and 1930s.[20] In 1937, Innis, Speiden, and Co., later to become Larvacide Products Inc., began manufacturing and marketing the chemical as a soil insecticide and nematicide, although its primary customers were owners of greenhouses and nurseries.[21] It still remained too expensive to use on a large scale and therefore only made sense for valuable crops.[22] But soon enough chemists figured out they could make chloropicrin more cheaply from nitromethane, using petroleum as their carbon source.[23] In the meantime, its malodorous quality proved highly useful when it was combined with other chemicals that were effective and relatively cheap soil disinfectants.

1,3-Dichloropropene—from Petrochemical Waste to Nematicide

The invention of the automobile with its many multiplier effects, among them the production of tires, roads, and suburbs, created enormous demand for oil in the 1930s. This, together with new refining technologies and abundant and cheap oil in California and Texas, paved the way for the petroleum industry's longtime economic dominance. Soon enough, petroleum proved itself useful for a vast array of industry and consumer goods, beyond automobility.[24] With the important exception of petroleum-fueled agricultural machinery, however, agricultural uses were a bit of an afterthought for most petroleum-based products. Once they were repurposed, as geographer Adam Romero argues, these new uses resolved an enormous problem for the petrochemical industry: getting rid of its abundant and toxic by-products. The toxicity of the waste is precisely what made it "potentially useful raw material for industrial agriculture."[25]

D-D, a combination of 1,3-dichloropropene (1,3-D) and 1,2-dichloropropane, was one such by-product of the petrochemical industry and was put to immediate use in the soil. The former contributed the toxicity, and the latter facilitated soil dispersion and penetration (much like the combination of chloropicrin and methyl bromide were later to do).[26] The discovery of this potent combination began when the lead scientist at the University of Hawaii's Pineapple Research Institute conducted a series of soil fumigation experiments with a recently acquired batch of organic chemicals. Like strawberries, pineapples had seen dramatic declines in yield from soil pests to the extent that some were predicting the end of the Hawaiian pineapple industry. His goal was to restore "the virginity of the soil," observing that pineapples (like strawberries) planted in previously uncultivated soil did not have the same pest issues as those in older plantations. In 1936 he assayed soil injections of a number of common pre–World War II poisons, including sodium cyanide, formaldehyde, paradichlorobenzene, and carbon disulfide. But he found these compounds either ineffective or causing severe growth problems. He then tried chloropicrin and was impressed by the initial results, although like others thought it too expensive for commercial uses. One lucky day, in late 1939, he received "a shipment of fifty-five-gallon drums filled to the brim with organic chemicals," all waste products from Shell Oil's Emeryville-based petrochemical research and development subsidiary. Through experimentation, he found that a fifty-fifty mixture of D-D was highly effective on a pineapple field heavily infested with nematodes. Punching holes every fifteen inches, pouring in the chlorinated liquid, covering the holes, and allowing the chemicals to do their work yielded dramatic results. "[D-D] spread through the soil like a lump of sugar. Fumes shot out in a circle, killing every worm they reached."[27] Encouraged by these results, the institute extended its experiments across the Hawaiian Islands.

Pineapples in the D-D-treated plots were large, robust, and vibrant, "as if there were no nematodes in the soil at all," while "the plants on untreated plots were stunted and diseased."[28] Not only was it effective: derived from waste products, D-D was incredibly cheap.[29]

Hearing of these results, and supplied with chemicals from Shell and support from Shell's technologists, USDA scientists, and the War Production Board, a UC Davis entomologist began actively exploring D-D's utility in California to address the huge infestations of pathogens and other soil pests that had overtaken much arable farmland. In 1944, a good portion of this work was done in the Salinas Valley, where many lettuce fields were so "sick" that production was near impossible. When yields were tested the following year, fields treated with D-D and fertilized with anhydrous ammonia (another Shell by-product) were eight times higher than controls. "The yields were not just back to normal; they were better than they had ever been."[30] D-D also turned out to be highly effective on the wireworms that had continued to plague everything.[31]

As an end product of Shell's manufacturing, D-D had not drawn sufficient interest from the company in separating the two compounds. Chemists from Dow, however, had learned of the impressive field results and soon figured out that it was the 1,3-D component that was doing the heavy lifting. So they went on to develop a process that allowed them to produce "a relatively pure dichloropropene." They named the chemical Telone, and marketed it exclusively as a soil fumigant. It turned out to be more effective at eliminating nematodes than pathogenic fungi like *Verticillium dahliae*, or seeds of weeds, so for fumigation purposes it was later combined with chloropicrin in various formulations, including the popular product Inline.[32] Today growers recognize it primarily as a nematicide that is not all that helpful against pathogens. Indeed, fields that have been turning up with new pathogens have a history of being treated with 1–3,D instead of industry favorite methyl bromide.

Methyl Bromide—from the Ocean to the Ozone

Industry lore about methyl bromide is that it comes from the sea, as if its natural origins belie its toxicity. Apparently, a wide variety of marine creatures, including sponges, corals, sea slugs, tunicates, sea fans, and seaweed, do in fact participate in producing the hundreds of identified organobromine compounds.[33] Still, as STS scholar Stefan Helmreich discusses in *Alien Ocean* (2009), many microbes from the sea, while certainly "natural" and potentially important for a range of functions, can be highly toxic in other environments.[34] Moreover, recent science suggests that most naturally occurring atmospheric bromomethane has terrestrial rather than marine origins, and that microbial degradation in soil may account for a large proportion of the atmospheric burden. Some scientists have even suggested that the ocean has more significance as a sink than as a source for the compound.[35] Evidence that substantial fluxes of methyl bromide enter the atmosphere from coastal salt marshes suggests, however, that both origin stories are worth their salt.[36] In what may well be an acutely politically motivated account, coauthored by one of the scientists who obtained a patent on methyl iodide, one study pointed out that the brassica plants in particular are adept at taking up ubiquitous bromine in the soil, producing methyl bromide through methylation, and releasing the chemical into the atmosphere.[37] Since brassicas are the favored bio-fumigants of organic farmers, the subtle implication is that organic farmers may be contributing to the ozone depletion for which methyl bromide has been damned.

Claims that methyl bromide is fundamentally safe because it is found in nature echo a common contention among manufacturers and users of industrial chemicals. There is no difference between synthetic and naturally derived chemicals, the argument goes; they are all just molecular assemblages of atoms and bonds, and their benefits and hazards are the product of this molecular structure, not their production in a field or a factory. *A fortiori*, a synthetically produced compound that also exists *in* nature—like methyl bromide—can't be a hazard *to* nature. The defense is reasonable enough, as far as it goes. After all, "methyl bromide" literally means a molecular structure—specifically, a molecule of methane with one hydrogen atom replaced by a bromine atom— and molecules with this structure have the same properties regardless of where they come from.[38] But chemicals used as treatments, even if found in nature, are made in labs at high concentrations, and then are applied to create reactions. Surely methyl bromide wouldn't have become so important had it not been effective at wreaking havoc on its target materials and organisms.[39] Methyl bromide's particular effectiveness, as a chemist explained to me, centers on breaking chemical bonds and creating reactive groups, including the addition of methyl groups to certain enzymes, affecting their function. Unfortunately, some of these same capacities cause methyl bromide to break down in the atmosphere, producing the free bromine atoms that cause ozone depletion.

Methyl bromide was originally synthesized in Europe in the 1880s. Although its first reported use in 1884 was for medicinal purposes, it quickly became more common in the preparation of synthetic dyes.[40] Its first non-dye use was in the 1920s as a fire extinguisher on airplanes. It was not until the 1930s that scientists began testing its insecticidal properties. Looking for a gas that would kill insects without retarding future germination of grains and without imparting a flavor, French scientists experimented with fumigating grains and live plants.[41] Having corresponded with these French entomologists, in 1935 a UC agronomist who often worked with the state's Department of Agriculture first introduced methyl bromide in California. He was looking for a nonflammable, nonexplosive, and non-injurious fumigant to be used on the potato tuber worm, since other fumigants he had been testing were highly explosive.[42] In 1936, another scientist at the UC Riverside Citrus Experiment Station was likewise searching for a fumigant that was safe to handle but toxic to the codling moth larvae found in the burlap sacks that walnut packing houses returned to citrus growers.[43] Laboratory studies conducted by a Dow Chemical scientist found methyl bromide to have toxic qualities equal to chloropicrin.[44] Owing to the success of these experiments, methyl bromide began to see commercial use, largely as an aboveground fumigant for food storage and live plants. It was used to fumigate dried fruit and grain, destroy the Japanese beetle on nursery stock, control the potato tuber worm, treat succulent produce such as celery, and much else.[45]

Following an opposing trajectory to chloropicrin, in the early 1940s methyl bromide was repurposed for war-related uses. During World War II it was widely used on troops and others to rid clothing and bedding of typhus-carrying lice. By the end of the war, tens of thousands of methyl bromide fumigation units had been shipped overseas, particularly to the Pacific theater.[46] Its success at disease prevention was quickly overshadowed by the introduction of DDT.[47] But it continued to be of interest for aboveground post-plant insecticide uses, including control of cyclamen mite on strawberry starts and strawberry fruit, although it was quickly becoming redirected to subterranean uses.[48]

Methyl bromide's fate as a soil fumigant is somewhat tied to another fumigant derived from a bromine compound. While at Iowa State University, a soon-to-be Dow employee learned that ethylene dibromide was an active insecticide when administered to soil. Although Dow had manufactured this compound since the 1920s as a lead scavenger—a substance used to prevent the buildup of lead deposits in internal combustion engines—UC Davis scientists began experimenting with it as a soil fumigant in 1944. In 1946 Dow commercialized it as a soil fumigant for nematodes, which "helped put over the concept" of controlling soil pests with an underground treatment.[49]

The 1940s had already seen some experimentation with methyl bromide as a soil fumigant, although it had primarily been used to control beetles.[50] Even after UC scientists had isolated *Verticillium* as the cause of wilt and had begun their experiments to manage it for strawberries, their goal for methyl bromide was to kill the weed seeds that had proliferated with the use of plastic mulches to warm the soil. Plus, chloropicrin had already proven effective at controlling *Verticillium* wilt. Nevertheless, it turned out that methyl bromide was also effective at killing fungi, especially relative to ethylene dibromide. It also cost less than chloropicrin, even though it was still expensive enough not to be economical for low-value crops. Observers also lauded the chemical over other soil fumigants for its fast penetrating ability, the speed with which it leaves the soil, and its low phytotoxicity, making it possible to treat soil in close proximity to growing plants.[51] So even though UC scientists Stephen Wilhelm and Edward Koch, in their efforts to repair the strawberry industry's problems with soil pathogens, first had success with chloropicrin alone as a pre-plant soil fumigant in 1953, they shortly found it could be used most efficiently when combined with methyl bromide. The two chemicals together turned out to be the true silver bullet for the strawberry industry even though methyl bromide got most of the glory—that is, until chloropicrin use became threatened, too.

SYNERGIES

Early on, laboratory studies of methyl bromide noted that the chemical lacked a distinctive odor when used at doses that are toxic to higher forms of animal life. Its action could therefore be "more or less insidious," particularly when exposure was protracted, so that those in the vicinity of application, including applicators, could be at considerable risk.[52] Chloropicrin, with its strong, pungent odor but higher cost, seemed the perfect solution. In one of the first commercially available formulas, Dowfume, MC-2, chloropicrin was included primarily to serve as a warning agent for the highly toxic but odorless methyl bromide—a role it continued to play as long as the two chemicals were mixed together.[53] Over the years, chemical and fumigation companies created dozens of other formulations of the two, in addition to combinations of Telone and chloropicrin.

While these formulations became widely used, two other chemicals were waiting in the wings as soil fumigants: methyl iodide and propargyl bromide. In 1938, methyl iodide was one of several chemicals tested as a fumigant and showed much promise as a fumigant of stored grain, despite being expensive at the time.[54] Throughout the years additional experiments confirmed its efficacy.[55] And Dow had conducted field tests on propargyl bromide and found great results. But in the process of blending propargyl bromide with chloropicrin, an explosion occurred. The project was shelved and all but forgotten. Plus, neither compound received much attention because methyl bromide and chloropicrin made for a winning combination.[56]

The key to success was how the qualities of the chemicals combined. As frequently noted by strawberry field advisors, the combination went beyond an additive effect; there was a definite synergism, a *je ne sais quoi* quality. Methyl bromide did well at controlling weeds, which, among other things, tend to act as reservoirs of *Verticillium* inoculum. With its very high vapor pressure, the fuming action of methyl bromide was superior as well. As put by a modern-day field advisor, "It goes through soil clods, it goes through the trash. You know what, if you don't prepare your field quite right, it goes through everything." So, according to fumigation experts, methyl bromide was primarily responsible for distributing the two chemicals throughout the soil, while chloropicrin carried the load in terms of the suppressive action on fungi in the soil. The latter chemical mysteriously even gave the plants a boost, making them "strong and vigorous," as growers were later to say. Together the combination of methyl bromide and chloropicrin had results that many deemed miraculous, with productivity climbing nearly fourfold in California between 1950 and 1978, even before some other yield-enhancing innovations were introduced.[57] Previous to fumigation, strawberry yields were two to four tons per acre; by 1969 they averaged sixteen tons.[58]

The role fumigation played in expanding the strawberry industry went far beyond improving yields just by minimizing plant loss. To be sure, this was an important risk to minimize, given the enormous up-front investments in strawberry production, even before any berries are picked. Fumigation became a kind of insurance against crop loss, to the extent the banks would not even lend to growers who didn't fumigate.[59] Yet the security of fumigation encouraged growers to plant even more acreage, making production more *extensive*. At the same time, as fumigation reverberated throughout the production system, it also made production more *intensive*. As we saw in the previous chapter, fumigation eliminated the exigency to breed for resistance, allowing breeders to focus on other qualities, including yield and day neutrality. And as I'll discuss further in the chapter that follows, fumigation allowed growers to forgo fallows and rotations with other crops, maximizing value on individual pieces of land. So notwithstanding the role the chemical industry has played in encouraging the use of chemical fumigants, a proposition that others have argued, extension agents and, hence, growers, became their biggest champions.[60]

The expansion and intensification of the strawberry industry, enabled by fumigation, could not have taken place, however, without a regulatory system very kind to chemical use in agriculture and elsewhere—one unwaveringly supportive of Dow's famous slogan about better living through chemistry. At the time that fumigation became de rigueur for strawberry growers, pesticide laws were minimal. The United States had passed its first pesticide law in 1910, the Federal Insecticide Law, yet that law's raison d'être was to quell the concerns of farmers concerned with mislabeled and ineffective products. Congress overhauled the law in 1947 in enacting the Federal Insecticide, Fungicide, and Rodenticide Act (FIFRA), with an expanded mission of protecting the environment and public health. Still, the regulatory purview of the new law was limited to requirements that all pesticides be labeled and registered with the Department of Agriculture.[61] With very little in the law to curb or prevent the use of toxic pesticides, only cost, inefficacy, acute hazard, or bad press kept them at bay. Much of that changed, of course, when things took a left turn. Most famously, Rachel Carson's 1962 publication *Silent Spring* alerted the public about the long-term and hidden effects of agricultural chemicals, especially DDT. A newly inspired environmental movement pressed for reform, giving rise to the regulatory agencies that today oversee the use of agrochemicals, however imperfectly.

The US Environmental Protection Agency (EPA) was established in 1970, following the uproar around DDT. Regulatory oversight over pesticides, including the implementation of FIFRA, was transferred to the new agency. A law passed in 1972 reformulated FIFRA and authorized the EPA to do more than just register pesticides, but also regulate their testing, labeling, sales, use, and disposal. Specifically, the new law stipulated that pesticide manufacturers and formulators would have to register the active ingredient of each pesticide with the EPA prior to shipment or sales. In considering an application, EPA would require "evidence that the product will perform its intended function without adverse effects; will not cause unreasonable harm to non-target organisms, including humans, crops, livestock and wildlife, or to the environment; and will not result in harmful residues on food or feed."[62] The EPA would thus evaluate all new pesticides for health and safety and set precise requirements for how chemicals deemed acceptable would be used—the so-called "label" requirements.

At this same watershed moment, California passed its own landmark legislation that required pesticides to be thoroughly evaluated for both environmental and health effects prior to being registered for use. It designated a separate agency, the California Department of Pesticide Regulation (DPR), to oversee this process and established one of the most far-reaching pesticide surveillance systems in the world. According to DPR regulations, every time a grower plans to use a pesticide, the grower must file a permit with a county agency specifying exactly which chemical they intend to use on what crops, how much they intend to use, and exactly where (by geographic coordinates). Counties compile these permits and make them publicly available. Though not all counties made it easy, the Pesticide Use Reporting System (PURS) became a source of tremendous data on pesticide use, invaluable for researchers like me.

While most growers and their allies have claimed these laws are overreaching, scholars and activists have found them more performative than meaningful. Practically, the EPA has been notoriously slow at reviewing chemicals, especially addressing the 1972 mandate to review the thirty-five thousand registrations that came before its advent.[63] Both agencies have been plagued by revolving-door politics such that many of the political appointees heading them up hail from industry and were installed to undermine agency mandates.[64] Substantively, the pushback on environmental regulation that led to new assessment practices such as quantitative risk assessment and cost-benefit analysis have forced large amounts of uncertain and ambiguous data into singular, bright-line measures of acceptable use that more often than not have allowed highly questionable chemicals to be registered.[65] Rather than justifying their removal from commercial applications, the resulting benchmarks have been used to manage chemicals—through, for example, mitigation measures that mandate how and how much they can be applied, not whether they can be applied.[66] Outmoded understandings of toxicity, such as "dose makes the poison," and lack of attention to interactive and cumulative exposures have also led to what critics believe to be insufficient regulation, as has the problem of imperceptibility of chemical emissions.[67] A lack of attention to inert ingredients in often proprietary chemical formulations, those that presumably are not active, has garnered critique as well.[68] As anthropologist

Suzana Sawyer has written, that certain substances in chemical formulations are thought to be inert and benign says less about their essence and more about the lack of relevant techniques to evince their potentialities.[69] All of this is to say nothing about how pesticide regulation is under-enforced, another critique that scholars have consistently made.[70]

Other scholars have gone a step further to critique chemical regulation on ontological grounds.[71] Historian of science Evan Hepler-Smith, for example, critiques the convention of naming chemicals according to their molecular structure. Notwithstanding the importance of seeing elements as building blocks for practicing chemists, in the regulatory arena this convention has yielded information that is precisely ordered but unintelligible to all but trained scientists. Accordingly, not only has it made chemical regulation appear objective and systematic; it has served to shift the arena of regulation from the fields that had proven tractable to political and ethical claims—clean air, clean water—to the less concrete and more technical terrain of chemicals in general, and thereby to depoliticize chemical regulation. Molecular bureaucracy, as Hepler-Smith calls it, which reviews chemicals compound by compound rather than functionally, allowed the chemical industry to turn to close substitutes when certain compounds met regulatory resistance. For agriculturalists, however, the logic of substitution is impeccable: a close substitute fends off the need to reconfigure the infrastructural regime in which a certain chemical is applied.[72] In other words, it allows the assemblage to remain intact.

Critically, molecular governance has also allowed continued regulatory neglect of how chemical substances act in concert with the materials and organisms with which they come into contact. As philosophers of chemistry Bernadette Bensaude-Vincent and Jonathan Simon have implied, chemists tend to locate the capacities of chemicals in abstract and absolute structures, rather than in situated and contingent relationships.[73] Yet in the real world chemicals are never pure molecular substances with stable capacities. They are always mixed with other materials, and the manner in which they are manufactured will affect the products that contain them. They may be dilute or concentrated. They may penetrate certain kinds of bodies and not others. This may depend on the chemicals they are mixed with, including so-called inert ingredients in the case of manufactured chemical products. All of these factors affect the intensity of chemical exposures—whether and how much of a chemical comes into contact with other chemicals, including chemicals in the bodies of humans and other organisms. What makes manufactured chemicals effective as cleansers, solvents, dyes, pesticides, and more is their relationships with the chemicals that constitute bodies, materials, and environments. Like the pathogens themselves, context is everything.[74]

And yet, in agrochemical regulation, the chemicals, rather than the intra-actions, remained the primary objects of concern. As a case in point, I found few toxicology studies that discussed how chemicals actually act on their target and nontarget organisms. Instead, most confined themselves to laboratory and epidemiological studies showing correlations between exposures and harm—the type of study required for regulatory review. Thus an additional "synergy," if you will, for the strawberry industry was a regulatory system that for the most part facilitated the use of these chemicals even when they received greater scrutiny with the founding of the EPA and California DPR. This produced a paradox of more regulation with less protection.[75] As such, all remained well for the strawberry industry—that is, until scientists discovered that the problem with methyl bromide was that in addition to propelling active ingredients through just about everything, it could also propel them into the ozone layer.

TOXIC TROUBLES

The formative battle over soil fumigants did not begin with either the EPA or the DPR, but in the international arena, in the 1970s, when a series of scientific reports documented how chlorofluorocarbons were creating a hole in the ozone layer. Scientists worried that depletion of the ozone would lead to an increase in ultraviolet radiation at the Earth's surface, which would result in an increase in skin cancer and damage to crops and marine phytoplankton. The result was the Montreal Protocol on Substances That Deplete the Ozone Layer, signed as a treaty in 1987 and put into force in 1989. Although methyl bromide was not originally on the chopping block, reports that followed the original signing singled out substances containing chlorine or bromine as contributors to ozone depletion. And although not addressed by the Montreal Protocol, it is important to recognize that methyl bromide was a public health problem beyond its potential to cause skin cancer through ozone depletion. By 1992, the US Agency for Toxic Substances and Disease Registry had definitively identified methyl bromide as a respiratory irritant and neurotoxin.[76] Opponents of methyl bromide raised these more localized health concerns in later hearings, where parties discussed possible exemptions to the Montreal Protocol.[77]

Nevertheless, it was the Montreal Protocol that in 1991 mandated the phaseout of methyl bromide. As a signatory, the United States agreed to stop producing and importing methyl bromide by 2005. The purpose of the generous delay was to allow methyl bromide users to find feasible alternatives. Instead, as the phaseout deadline drew near, industries that depended on the chemical started to push back. The strawberry industry in particular called on the US government to obtain exemptions from the ban.[78] Based on already-existing exceptions to the ban for "essential" uses—those necessary for the health, safety, or the functioning of society—the United States thus began to argue for an exemption to the ban for "critical" uses and, indeed, threatened to withdraw from the treaty if the exemptions were not granted. Critical uses, as defined in the language promoted by the United States, were those for which there are no technically and economically feasible alternatives and that would result in significant market disruption. Crucially, the US case for these critical use exemptions (CUEs) was made on the basis that the ban "would make a significant portion of the California strawberry industry economically unviable."[79]

After the signatories allowed the CUEs, California's strawberry industry became the most significant user of CUEs, allowing for the continued use of methyl bromide well beyond the international deadline.[80] Eventually, though, the phaseout came about. Approved amounts for fruit production declined precipitously each year and the CUEs were finally eliminated altogether by the end of 2016—at least that's the word from the fields. (When the Trump administration's EPA reconfigured the EPA website in 2017, I found shockingly little information on the status of the CUEs.)

For the time being, the strawberry nurseries have been able to avoid the phaseout under a separate "quarantine pre-shipment" exemption. The justification for this exemption was to allow them to meet the certification requirements for California's Clean Plant program, in part to meet the phytosanitary standards of countries importing California nursery stock, as many do.[81] Certification requirements are highly stringent regarding disease infestation, allowing only a few individual plants in any lot or block of nursery stock or on the premises to show any infestation or infection. If more than a few plants show signs of infestation, entire lots can be denied certification. The operating rationale is that selling plants already infected with pathogens is a recipe for widespread virulence; therefore no markets exist for plants that do not meet certification requirements. Methyl bromide thus continues to be used throughout the plant propagation process, which takes place over several years, in both "screen houses" (netted enclosures that keep bugs and debris out) and outdoor fields. This has involved a significant amount of land, regardless of where that stock is eventually planted for fruit growing.[82]

Nevertheless, facing the specter of an outright ban on methyl bromide, nursery growers joined fruit growers in their ardent search for a suitable replacement. Methyl iodide had been waiting in the wings, and the phaseout of methyl bromide provided an opportunity to reintroduce the chemical. A team of scientists at UC Riverside thus sought and received a patent for its use as a soil fumigant. Methyl iodide was touted as a suitable alternative because it shares important qualities with methyl bromide in terms of soil sterilization, but does not make it into the upper atmosphere. Moreover, it could be applied almost exactly like methyl bromide, thus causing no perturbations to the rest of the production system—the logic of substitution at work. These same qualities, however, made it less desirable for those who would come in contact with the chemical, for, as many argued, it was even more acutely toxic and environmentally degrading than methyl bromide, associated with a range of conditions.[83]

Arysta LifeScience, to which the UC researchers transferred the license, first sought to register the chemical with the EPA for commercial use in 2002. Emerging controversy over its high toxicity caused the EPA to at first deny registration. For instance, in September 2007 more than fifty scientists, several of them Nobel laureates in chemistry, delivered a letter to EPA opposing registration. However, eventually the EPA reversed course, granting registration without time limitations in 2008. Registration of the chemical in California was not so easy, however, due to tougher environmental laws, DPR's missteps, and the concerted activist campaign that arose to thwart it. Visible public backlash dissuaded many growers from adopting the chemical, although the availability of methyl bromide via CUEs also played a role in non-adoption. Eventually, a lawsuit regarding DPR's handling of the registration process put the nail in the coffin, and Arysta withdrew the chemical from the market in 2012 because of commercial nonviability.[84]

Tighter state restrictions on chloropicrin use followed right on the heels of the methyl iodide fiasco. The context of these restrictions was a routine reregistration study by the EPA. Based on new studies that showed eye, nose, and respiratory irritation from inhalation exposure, the agency designated the chemical a toxic air contaminant in 2010 and tightened its requirements for application. DPR reviewed additional evidence for claims that chloropicrin is a more potent carcinogen than methyl iodide, perhaps equal to Telone.[85] The

FIGURE 9. Fumigation workers wearing protective gear. Photo by Sam Hodgson.

DPR concurred with the EPA on the toxic air contaminant designation but equivocated on the chemical's carcinogenic qualities. Nevertheless, in 2013, under the leadership of its new director, Brian Leahy, a former organic farmer and Democratic appointee, the DPR proposed mitigation measures that went significantly beyond the revised EPA restrictions.

In the end, the actual rules put into place for chloropicrin did not differ too much from the EPA's, but still were tighter than what existed before. The centerpiece of these new mitigation measures was expanded buffer zones between areas of treatment and other land uses such as schools and housing. At the same time, DPR offered incentives in the form of reduced buffer zone requirements for growers who use totally impermeable film (TIF) rather than virtually impermeable film or no plastic at all (a rare practice). Film presumably keeps the chemical from drifting, and became a requirement in some counties. DPR also reduced buffer zone requirements if growers fumigated in beds rather than across entire fields ("flat fumigation").[86]

DPR also began to regulate Telone more stringently. Restrictions had been in place for Telone since 1995 for its carcinogenic properties.[87] These were primarily township caps, meaning that only a certain amount was allocated to each thirty-six square miles—on a first-come, first-served basis. In 2001 DPR added some flexibility to these caps to address the phaseout of methyl bromide, although, somewhat perversely, growers were able to obtain

CUE allocations where there were township caps.[88] After further review to determine whether the caps were necessary or sufficient to protect public health, in 2017 DPR imposed further restrictions on the chemical. These included less flexible township caps (although sometimes higher than previous ones) and discontinuing the practice of letting growers roll over unused allocations from previous years. DPR also banned use of the pesticide entirely during the month of December, when weather conditions tend to make air concentrations of the chemical increase.[89]

Finally, other substitute chemicals, used less frequently in fumigation, saw more scrutiny. In 2010 the DPR released new permit conditions for metam sodium, metam potassium, and dazomet, primarily involving buffer zones and worker protections. These are chemicals that growers in Ventura County have used, with mixed success, for pathogen suppression.

LIVING WITH RESTRICTIONS

With these restrictions, the strawberry industry saw its most central disease management mechanism curtailed. Unusually, two chemicals were taken off the market. Less unusually, the existing substitutes met enhanced mitigation measures. One question is what that meant for growers; a second is what that meant for those the restrictions are designed to protect.

As it happens, the new restrictions on fumigants did not necessarily reduce the amounts of fumigants applied. As methyl bromide supplies dwindled and remaining inventories became more expensive, many growers increased the proportions of chloropicrin they used in mixtures with methyl bromide until methyl bromide was no longer available; others went cold turkey and eliminated methyl bromide altogether.[90] But to make up for the loss of gaseous power, they began to use chloropicrin at higher rates. Some shifted to mixes of chloropicrin and 1,3-D.[91] Even in nurseries, growers cut down on methyl bromide and substituted it with more chloropicrin.[92]

Some growers nevertheless found the buffer-zone credits that came with the new mitigation measures to be enough of an incentive to move to bed fumigation. They were also motivated by the potential to reduce costs. The cost differences between broadcast versus bed fumigation are significant. Broadcast fumigation, or what growers often call "flat fuming," involves working with a leveled field. The fumigation rig drives through the field and injects the fumigant into the soil, then covers the treated area in plastic—the totally impermeable film (TIF). Because flat fumigation requires higher amounts of chemical to be applied, it is also potentially more dangerous. Those who flat fumigate can only fumigate forty-acre blocks at a time on any given ranch, slowing down the process considerably. With bed fumigation (also called drip fumigation), growers first make the planting beds, then lay down irrigation hoses and cover the beds in plastic. Fumigants are then injected into the beds through irrigation tubing, and the same plastic remains on the bed for weed control. Unlike in flat fumigation, the rows in between beds are never fumigated. Therefore, bed fumigation not only uses less chemical volume per acre, but also reduces the complexity of land preparation processes.

These changes in both fumigant content and method of application were in fact consequential for growers. Although some were satisfied with the switch, many growers said these changes made fumigation less reliable. Field advisors attributed this to chloropicrin's lesser ability to fume through ground. If fields are not adequately prepared—if, for example, they are too wet or too dry—the chemical does not evenly spread. They also noted that the drip system used in bed fumigation was less efficacious in reaching all parts of the bed, so that some sections were not receiving concentrations lethal to the pathogens. Tests showed that the pathogens were also taking refuge in the rows between beds. And so growers were seeing more disease and plant death, including from the "novel" pathogens discussed in chapter 2. They were also seeing difficult weeds such as nut grass.

Sadly, the restrictions did not exactly accomplish their public health ends, either. Sociologist Jill Harrison has written at length about the problem of pesticide drift, referring to the windborne movement of chemicals beyond the fields to which they are applied. She argues that industry has treated these incidents as one-off accidents even as many community members and farmworkers have claimed they occur with regularity.[93] In her research she found that those exposed to drift complain of shortness of breath, headaches, nausea, eye stinging, and rashes, aside from their less visible and unfortunately less provable long-term effects. According to one informant in my study, in the ten years prior to 2013, more than seven hundred people in California had been acutely sickened in twenty-two drift incidents. One 2012 occurrence was directly attributed to pre-plant fumigation with a mix of chloropicrin and Telone.[94] And those were the *reported* illnesses. Harrison's research shows that the vast majority of incidents are not reported mainly because they affect workers who fear deportation.[95]

Moreover, the substitute chemicals have been arguably worse than methyl bromide in terms of the health effects. It is true that methyl bromide is particularly prone to drift. Its ability to "whoosh through the soil," as one farm advisor put it, also made it likely to whoosh through the air—so much so that it made it into the upper atmosphere. The chemicals that have remained allowable, in contrast, are heavier and therefore stay closer to the ground, subject to drifting to nearby areas. That means that the chemicals that have become substitutes for methyl bromide are more dangerous to those in surrounding communities and work sites. Additionally, the qualities that made chloropicrin an effective warning agent when used with methyl bromide— the smell and intense tearing—are compounded when it is used in higher volumes to make up for the loss of methyl bromide.

Ironically, the mitigation measures themselves have contributed to chemical exposures. Consider the plastic film used to cover fumigations, the primary purpose of which is to ensure efficacy and only secondarily to minimize danger. In the early days of fumigation, growers left the plastic on for only twenty-four to forty-eight hours.[96] But that meant the chemical would dissipate quickly. So growers learned that extending the length of time that plastic covered the fumigation, along with using impermeable film, would keep the chemical working longer and the concentrations higher.[97] Yet, as I learned in my research, tarping also requires that fumigation workers help secure the plastic and shovel dirt on the outer edges. And it requires that workers remove the tarps to make beds in the case of flat fumigation or puncture holes in the tarp for planting in case of bed fumigation. Workers who testified at various public hearings about chemical restrictions spoke of the off-gassing they experienced at precisely those moments of removal. As put by a worker speaking at a hearing about methyl iodide, "You have to punch the holes to put in the plants; there are chemical residues, and it comes out at the moment that you open the hole. And at that moment is when you get dizzy." I also found out that the tarps used to secure the chemical are subject to malfunction. Again, public hearings revealed many stories of tarps flapping in the wind and work crews being sent in to fix them. I was told by one grower that he had instructed workers to use duct tape to secure tarps that had torn.

Or consider in-field monitoring for chloropicrin drift—another mitigation measure. If growers opt for monitoring rather than notifying neighbors, they are required to station human beings at the edges of the buffer zone to report any sensory irritation experienced. These monitors function as

FIGURE 10. Fumigation tarps shredded by the wind, Pajaro Valley. Photo by author.

canaries in the coal mine. Moreover, although fumigation workers nearly always use protective gear, the shovelers, tarp repairers, planters, and monitors do not.

Overall, the problem with the mitigation measures is that they exist to enable chemical use, not reduce it. Although their increasing strictness has undoubtedly created challenges for growers, their fundamental purpose has been to manage chemicals in space and time. Restrictions on when fumigation can take place, the timing of field reentry following fumigation, and enhanced buffer zones around "hard to evacuate" sites such as schools and hospitals exist precisely so growers can continue to use the chemicals. Most growers understand this, too, and therefore rarely complain about mitigation measures. As a case in point, one grower pointed out that had the mitigation measures for chloropicrin been in place for methyl bromide, the industry wouldn't have "lost" the chemical.

In short, the restrictions on fumigant use instantiate many a critic's points about the weakness of US chemical regulation: among other things, that it has too easily allowed facile substitutions, that it has been more about managing chemicals than eliminating them, that it has all but disregarded populations with little recourse to contest violations, and that its toxicity assessments rely too heavily on threshold models for individual chemicals that neglect cumulative and interactive exposures. To these I would add the arguments made by those who have suggested a reorientation of chemical regulation from a paradigm rooted in molecular structures to a paradigm that understands chemicals as functional and relational. Such a reorientation would be very much in keeping with the nature of chemistry itself: that it is about combining elements in ways that purposefully induce reactions in order to transform materials and organisms. The hope is that these transformations are beneficial—and some undoubtedly are. Unfortunately, some routinely are not—or, at least, not beneficial to all.

Had such a reorientation been widely adopted, it would have brought to the fore not only the regulatory urgency of evaluating the work of substances, including those considered "inert"; it might have also demanded more attention to how chemicals interact ecologically with the biophysical world they are designed to transform. When attempting to learn more about how methyl bromide acts on target organisms, I came across a publication that generally touted methyl bromide's efficacy, yet mentioned that there have been cases of plant injury or growth retardation for plants grown in soil treated with the chemical.[98] While the author for the most part dismissed these findings, claiming that they were inconclusive, I was left wondering where things might stand if such questions had been pursued. For the strawberry industry, it might have furthered understanding of the coevolution of pathogen, plant, and chemical. Instead the industry and those involved in the work of repairing it have for the most part continued to disregard whether and how chemicals themselves may be making the pathogens stronger and the plants weaker.

Regarding this last point, it is certainly arguable that investigating such possibilities is pointless because there is no going back. As the historian of science Hannah Landecker has written in relation to the prophylactic use of antibiotics in livestock, the antilife capacities of these treatments far exceed the target organisms, so much so that they become part of the environment. In effect, such treatments change the future, along the way making prior knowledge of the target organism and those with which it cohabits obsolete.[99] Exposure to fumigants (and ubiquitous other chemicals), that is, has likely permanently altered what organisms are and what they are capable of, an instantiation of the "planetary scale alternations" of the so-called Anthropocene.[100] Nevertheless, in the case at hand, it is no small irony that fumigation itself may be responsible for declining chemical efficacy and, hence, the undermining of strawberry production—and that the industry's institutions of repair are either ill-equipped or averse to investigating that.

The strawberry industry does have a more immediate problem on its hands. Environmental health science has progressed mightily since these chemicals were born, back when acute danger was the sole concern. Cumulative and interactive effects have received more attention of late, and threshold models of toxicity have come into question for suggesting a bright line of safety or non-safety.[101] As new knowledge about intergenerational (epigenetic) effects continues to emerge and circulate, there may be additional efforts to end fumigation. Indeed, a large cohort project to study long-term effects of agrochemicals is taking place in Salinas, right in the heart of strawberry country. The CHAMACOS project (the name references "children" in Spanish) has already found relationships between mothers exposed to methyl bromide during pregnancy and adverse birth outcomes.[102] Other studies have found decreased fecundity among women who work with chloropicrin.[103] While farmworker lives have nearly been disregarded in pesticide practice and regulation, the calculus my well change with the labor shortages discussed in chapter 6.

The possibility of a complete fumigant ban is not idle speculation. In 2013, under its new leadership, DPR issued its Nonfumigant Strawberry Production Working Group Action Plan.[104] In the report the DPR argued for the need to curtail and eventually phase out fumigants altogether to protect the health of farmworkers, bystanders, and nearby communities. Mainly a call to action on developing alternatives, the report signaled that the regulatory climate will not become any easier for growers who want to continue using fumigants.

Rather than countenancing a future without fumigants, some in the strawberry industry (and elsewhere) have taken a new tack. They no longer dismiss health and safety concerns, realizing that denying and defending is a losing game. Gone are the days when growers claimed they would drink the chemicals! Instead they are acknowledging the concerns through claims and practices to perform care and compliance. Some growers, for example, discuss how thankful they are that the fumigation business has become wholly professionalized so they can turn over the hazardous work of fumigating to an organization that does due diligence. New, voluntary organizations in which growers, pest control advisers, safety professionals, and others get together to share best practices for avoiding accidents and everyday injury are another manifestation of this tactic. The coalition Spray Safe operating in Ventura County is an example. The organization took its lead from Spray Safe Kern County following an incident in which someone filmed a spray rig dousing a group of schoolchildren at a bus stop. The governor's office called the responsible party and gave them a choice to self-regulate or the state would get more involved. So, according to one spokesperson, "We've got to get as good at doing this as humanly possible." I suppose these approaches are not entirely cynical. Many growers have had untoward experiences with chemicals and recognize the importance of safety. Yet, as Jill Harrison has written, voluntary measures only go so far; enthusiasm doesn't generally translate into practice as growers face the risk of crop loss.[105]

It also must be said that these same groups, and growers more generally, think the public has misunderstood the dangers of agricultural chemicals and needs to be educated so they no longer perceive that "these people are out trying to poison us," as this same spokesperson said. They emphasize that today's chemicals are much safer than previous ones and the impossibility of getting a pesticide registered "that is not low mammalian toxicity, non-carcinogenic, non-mutagenic, has a very short half-life so it doesn't persist in the environment.... I mean, things that people have under their kitchen sink would be too toxic for us to use." Fumigants are not included in this new class of chemical, however. As we have seen, they were developed a long time ago, out of highly toxic materials, and they were designed to kill. And so when defending fumigants they take yet another tack, claiming that fumigation makes plants healthier and therefore allows reduction of other fungicidal and insect sprays.

In my view, these strategies are attempts to inoculate the industry (pun intended) against further regulation—an understandable approach given how fumigation has acted as insurance, and certainly given how the rest of the strawberry assemblage has become entangled with fumigation. Nevertheless, these strategies are also highly rearguard, reminiscent of the battles the industry took to extend the CUEs. The industry's efforts on behalf of the CUEs, and the short-term success in extending them, led to a kind of complacency and lack of preparation for the future. Many assumed methyl bromide would never be taken away, and there are still those in the nursery business who think that the quarantine exemption may endure. As I will discuss in chapter 8, even those willing to face the eventuality of no fumigants are hoping for other "drop-in replacements"—substitutes that do not shake up the entire production system—and thus continue to kill. Others are more forward-looking, or at least are hedging their bets by investing in nonchemical alternatives to disease management, including breeding and the technologies that I will discuss also in chapter 8. Such is the case with

Driscoll's and other lesser-known companies and individual growers who hope to gain a competitive edge through innovations that obviate the need for this sort of killing and stray a little further from straight chemical substitution. Yet they will likely not veer too far. For the logic of substitution becomes nearly essential when land values are capitalized on the basis of fumigation and all that it has allowed—when land values, that is, are a crucial part of the assemblage.

FIVE

Soiled Advantages and Highly Valued Land

Agriculture is somewhat special. The land here not only supplies a stock of nutrients to be converted by plant growth and animal husbandry into food and sundry raw materials, but it also functions as an instrument or means of production. The production process is partially embodied within the soil *itself*. . . .

If land is freely traded . . . it becomes a form of fictitious capital, and the land market functions simply as a particular branch—albeit with some special characteristics—of the circulation of interest-bearing capital. Under such conditions the land is treated as a pure financial asset which is bought and sold according to the rent it yields. . . .

The situation changes materially if interest-bearing capital circulates through land markets perpetually in search of enhanced future ground-rents and fixes land prices accordingly. In this case, the circulation of interest-bearing capital promotes activities on the land that conform to highest and best uses, not simply in the present, but also in anticipation of future surplus value production.

Geographer David Harvey, Limits to Capital, *1982*[1]

Strawberry growers have been living in a fantasy land. . . . They are getting artificially high yields from fumigation, while the margins aren't there for other crops. The industry is built on this model, and if you change it, the industry's survival is an open question.

Interview with extension scientist, 2017

ONE SOMEWHAT STRANGE INTERVIEW took place in a suburban office building, a bit of a distance from the strawberry fields of Oxnard. We were to meet Mr. S (a fictionalized name) of S Farms. At the time S Farms had several hundred acres of strawberries in production, at least a third of which were organic, so we were anxious to hear his perspectives on how fumigant regulation was affecting the industry. After being greeted by a receptionist, Sandy and I were led from the reception room into a conference room. As we waited, we both took note of the unusually expensive and somewhat garish furnishings for a farm office. After about fifteen minutes, Mr. S entered. We were relieved to find him affable and forthcoming in his answers. It was an informative and collegial interview, the kind social scientists are glad to obtain with a research population that tends to be skeptical of and even hostile to social science researchers.

After we had finished the formal part of the interview, I stepped out to use the restroom. When I emerged, I heard Sandy speaking with several men, including Mr. S, down the hall. I followed their voices and was surprised to see an office with a sort of men's-club decor, replete with fancy wine bottles and photos of show horses. I couldn't help but note that the men with whom Sandy was speaking were significantly less friendly than the affable Mr. S; one looked quite dour. As we exited the building, Sandy looked a little distressed, and I must have looked bewildered. Amid nervous laughter, we joked that the principals of S Farms were probably running the strawberry operation to launder money.

Over the next few days we began to piece together some of the information we gathered in this interview with stories we had heard from other growers in the area. Several interviewees mentioned that some growers were outbidding others for leases on land with high-quality water—a point that Mr. S himself had made—to the tune of an unheard-of (at the time) $5,500 per acre. This was happening even though fruit prices were down, soil pathogen load was up, and labor shortages were acute in the Oxnard area, seriously threatening the economic viability of many strawberry growers. And, yes, we heard that the ones responsible for these bidding wars were believed to be using their strawberry operations as a front for other businesses.

I introduce this chapter with this story only in part for color—I have no reason to believe that money laundering is a common practice in the strawberry industry. Nor do I merely want to point out the truism that land is an important factor of production in the strawberry assemblage. Instead, I want to impress upon the reader the fact of rising land values in the context of the industry's other challenges. As a result, land functions in opposing ways for this industry: the value of land as fictitious capital—valuable for what it could earn in rents and financial speculation (exchange value)—continues to escalate, even as the value of land as material substance (use value) becomes increasingly compromised. Together with other elements of the assemblage (plant breeding, fumigation restrictions) the increasing divergence of these two different forms of value makes alternative ways of growing in soil virtually impossible. And yet, the option to move to less land-intensive or soilless alternatives is anathema to many in the California strawberry industry, whose greatest asset remains the real estate that contains conditions highly conducive to strawberry production.

While previous chapters have discussed how soil-based organisms became a threat to strawberry plants, this chapter lays out how land as both material and property became a threat to the strawberry industry. It begins by reviewing how both qualities of land were once so advantageous, since pieces of property, even if leased, were plentiful and contained rights to several excellent conditions for strawberry production. Tightening land markets changed all that, however, as did the nonhuman threats that increasingly bore down on the industry.

CALIFORNIA'S ADVANTAGES

California was not always the preeminent strawberry-growing region in the United States. Before World War II, thirty-one states were strawberry producers and California was a "minor participant," with only 4.2 percent of the US crop on 3.4 percent of its acreage.[2] Yet it wasn't for a lack of "natural advantages." Rather, breeding had not yet prioritized shippability, and so strawberry production needed to be close to urban markets. With postwar developments in breeding and refrigerated transportation technologies, California strawberries could be shipped afar, and it became possible to enroll California's manifold advantages in strawberry production.[3] By 2017, California controlled 88 percent of the market, with Florida a very distant second.[4]

In *The Fruits of Natural Advantage: Making the Industrial Countryside in California* (1998), historian Steven Stoll writes that "so-called advantages have nothing to do with the functioning of plants and animals in ecosystems" but only exist in the imagination. "They are the riches that people read into soils and climates and water."[5] I beg to differ. In my view soils, climate, and well-situated aquifers very much contributed to the success of the California strawberry industry irrespective of human imagination and aspiration. Indeed, when growers veered too far from the locations of these advantages, they encountered difficulties. I scare-quote "natural advantages" mainly to signal that they were brought into being as advantages with the aid of human labor, ingenuity, and technology.[6]

Many qualities of California's environment have turned out to be ideal for strawberries, but chief among them has been the temperate weather of its coastal regions. Situated on the west coast of a continent with most of its geography between 30 and 40 degrees latitude, much of California possesses a Mediterranean climate. Summers are warm and dry, while winters are cool and hopefully wet, with about 90 percent of annual rain falling between November and April. The southern part of the state tends to be warmer and drier, with temperatures decreasing and rainfall increasing more or less with increased latitude. About one hundred miles north of the San Francisco Bay Area begins another climatic region altogether, the Pacific Northwest, with temperatures and rainfall patterns more akin to the British Isles. Also benefiting strawberries are the wind circulation patterns, which in the summer blow across the Pacific Ocean to bring cool, moist air in the form of morning fog into the low-lying coastal areas, and in the wintertime keep frosts at bay. Coastal denizens refer to the summer fog as the natural air conditioning of the Pacific Ocean, and it ensures that on summer days the temperature rarely rises above a very comfortable 72 degrees Fahrenheit. The benefits of what some call the "eternal spring" of the South and Central Coast of California are clear: moderate temperatures elicit blossom production and improve fruit quality without excessive vegetative growth. And considering that strawberries are fragile, highly perishable, and susceptible to molds and moisture-generated pests and diseases, the rainless months of harvest season minimize these problems.[7] In addition, the relatively mild climate makes possible a very long harvest season, inviting breeders to develop long-bearing varietals.

A second advantage of coastal California for strawberries are its pockets of sandy loam soils, especially in the alluvial plains where rivers meet the sea. An observer of Brittany's early strawberry industry noted that the Chilean strawberry grew best in sandy soils, only survived in well-drained heath soils, and died in chalky "calcareous soils."[8] Although the soils in strawberry-growing regions are generally not classified as prime soils, they work well with the characteristics of berries and their accompanying technologies. The good drainage afforded by sandy soils is essential to prevent buildup of moisture and salt.[9] It turns out that sandy soils also facilitate fumigation, allowing the chemical to blast through the ground without being deterred as it would by dense clay soils. The concern that sandy soils are not so rich in nutrients has been a minor one, with so many fertilizers and soil amendments at hand.

While temperate weather and sandy loam soil gave the California straw-berry industry its early "productive edge," its preeminence would not have been possible without high-quality water, as well.[10] Fruit is mostly water, which must be taken up from the ground.[11] Some say it takes twenty gallons to produce one pound of berries.[12] California's water politics are legendary, but the major dam projects that have been the source of much controversy do not serve strawberry fruit-growing regions. Only minor rivers course through most of these regions—those very rivers that deposit the alluvial soils. Although the once-high water tables tended to be swampy during the rainy season, especially in the more northern fruit-growing regions, reclamation projects in the 1850s and 1860s converted marshy areas around the sloughs into productive agricultural land.[13] Since then farmers have generally found that groundwater stored in aquifers, drawn out using wells, has been more than adequate to irrigate their plants.[14]

It is not clear that the first strawberry growers in the state sought out these characteristics. It may have been dumb luck that the first strawberry farms in California were located in the sandy loam soils of coastal Alameda County and the Pajaro Valley spanning Santa Cruz and Monterey Counties.[15] While the windward areas of Alameda County have since been fully urbanized, the Pajaro Valley, drained by the Pajaro River, has remained an important center of commercial production. On the other side of the coastal ranges and farther inland, things turned out not to be so sweet for the modern strawberry fruit. Along the entire Central Valley, from Bakersfield in the south to Redding in the north, temperatures are hot in the summer, moderated only slightly in the Delta regions of the Bay Area, where cooler breezes occasionally slip through. Nevertheless, before World War II, when strawberries were still a super-perishable and short-season crop, the hot climate was not much of a liability. Following the industry's nascence in Pajaro, strawberry production spread throughout the state, and at one point 41 percent of production took place in those hot inland valleys.[16]

Eventually, though, fruit production mostly ceased in the inland valleys and turned to the coast, with growers generally opting for land containing well-drained soils within three miles of the ocean.[17] The largest concentrations of strawberry production ended up in the Oxnard/Camarillo area of Ventura County, the Santa Maria/Guadalupe region spanning Santa Barbara and San Luis Obispo Counties (a relatively newer area), Salinas and the surrounding hills in Monterey County, and the Pajaro Valley, whose best-known town is Watsonville. Even as Southern California became defined by its mass suburbs

FIGURE II. Strawberry field close up, Monterey Hills. Photo by author.

of single-family homes and shopping malls, strawberries continued to grow in a few small pockets of undeveloped land in San Diego and Orange Counties, regions that along with Mexico became the main source of strawberries in the winter. Today, only a few highly diversified farmers and recently settled Hmong refugees grow strawberry fruit in the Central Valley.[18] Nevertheless, these inland areas have become an absolutely critical, if nearly invisible, asset of California's strawberry industry. For it is in these inland areas that strawberry *plants* are propagated, a part of the production not incidental to their success on the coast. Strawberries not only draw on a different set of natural advantages in those areas; the relative proximity of those areas to the coast creates a locational advantage that cannot be matched by other inland regions farther afield.[19]

Soil, water, and climate are materially heterogeneous and governed by very different kinds of institutions, to the extent that they are governed at all. Yet access to all can be obtained at once through purchasing or renting a piece of land where they come together favorably. A plot of land, that is, has certain soils and water sources, and is located in a certain climatic zone, yet rights to use that land are conferred through institutions of property. And so it is through land where the tangible material meets the abstraction of property, and therefore where the more-than-human assemblage meets political economy. In the case of strawberries, growers found that having uncertain tenure through leasing property worked well enough to access those material advantages just so long as they remained advantageous, while landowners could treat this same property as fictitious capital (valuable for its financial returns) to their advantage. How that win-win of sorts came to be is not a straightforward story.

A LAND OF LESSORS

California's land history is famously contentious. Following statehood in 1850, much of its land was in swamp, desert, and forest, which exempted it from the US Homestead Act and other land giveaways that were putatively designed to create a nation of farmers. Through highly lenient interpretations of federal land law, and massive giveaways to the railroads, by the 1860s much of California's land was grabbed in huge swaths and then "reclaimed" to create a vast agricultural empire in its Central Valley. Yet nearly all of California's strawberry fields lie on land to which those stories do not apply.[20] Rather, this prime coastal land was virtually all held by wealthy Mexicans of Spanish ancestry—the Californios. As remnants of the original Spanish land grants and missions, once backed by the force of the Spanish crown, the land was granted to the Californios by the Mexican government following the end of Spain's rule in 1822.[21] Generally used to run cattle for the hide and tallow trade, these land grants were the classic ranchos.

The survival record of the ranchos was "unimpressive," however, and many had already broken up by 1843, before the treaty of Guadalupe Hidalgo separated Alta California from Mexico and made it US territory in 1848.[22] What remained in Californio hands was transferred to (Anglo) Americans through a combination of outright purchase, intermarriage, exchange for debt, and excessive legal fees in settling the title disputes that arose out of the Land Act of 1951.[23] For instance, in what is probably a highly sanitized version offered by the Monterey County Historical Society, in the early 1850s a James Bryant

Hill purchased the 6,700-acre Rancho Nacional with the intent of setting up a huge farming project in what is now Salinas, while Jacob Leese purchased the nearby 10,000-acre Rancho Sausal for a mere $6,000 to grow wheat.[24] The final straw of the rancho era was the drought of 1861–64 that resulted in widespread cattle starvation and sale of the remaining ranchos. Such was the case with the Rancho Rio de Santa Clara in the Oxnard plain area. Rafael Gonzales was a former Mexican soldier who had been granted an undivided interest in the 44,883 acre rancho south of the Santa Clara River, adjacent to the Pacific Ocean. Following the drought, in 1864, a land agent, Thomas Bard, acting for oilman Thomas Scott, purchased the property along with five other Ventura ranchos in order to exploit their oil potential. Ten years later he had sold large portions of this land to newly arrived northern Californians, many of German and Irish descent.[25] One of these buyers was Dominick McGrath, who came to Oxnard with his wife in the 1870s and bought 700 acres for as little as 75 cents an acre on the Oxnard plain. His four sons expanded the holdings to 5,000 acres, where they grew hay, grains, and later lima beans.[26] Similar stories were to be had farther up the coast. A John T. Porter acquired 820 acres of the San Cayetano Rancho, which had been owned by General Vallejo, and went into sugar beet farming in what is now the town of Pajaro. He was to become one of the region's first strawberry growers.[27]

The switch from livestock to sugar beets, grains, and beans was fairly typical in California. The gold rush and its multiplier effects had caused urban markets to burgeon, and by the early 1870s landholdings throughout the state had converted to wheat and barley production, including in the Salinas and Pajaro Valleys, as well as the Oxnard plain.[28] Santa Maria, in contrast, was a more of a backwater, especially given its fierce winds and lack of surface water. In 1873 one of the town's founders went so far as to *give* land to those who would settle there.[29] Thereafter, land in that area was given over to cattle, dryland farming, and oil drilling.[30] Wheat turned out to be a gamble throughout California, however, since so many world regions gained footholds in wheat production at the same time, often at the behest of colonial powers. By the end of the nineteenth century, wheat had boomed and busted at least twice, and many of California's large "bonanza farms" had been broken up and sold, or leased out to small "specialty crop" growers.[31]

In the regions that were to become strawberry strongholds, grains were mainly succeeded by sugar beets early in the twentieth century, as major sugar companies set up refineries in Salinas, Watsonville, Santa Maria, and Oxnard.[32] Although some growers were able to buy land, virtually all of the sugar beet production was handled by tenant farmers who paid either cash rents or shares to the sugar companies that often owned the land.[33] Still, land was cheap. In 1915, landowners sold unimproved strawberry land for $100 to $200 per acre or rented it out for $20 to $30 per acre per year.[34] Sugar beet production remained important in Oxnard until the 1950s, when the industry moved elsewhere. Following the decline of sugar beets, growers in Oxnard transitioned to lemons and row crops, mainly vegetables, presumably in a mix of both owned and leased land.[35] Other specialty crops were grown by both tenants and smallholders. In the Pajaro Valley, the cradle of the strawberry industry, many of these new tenants and smallholders planted apple farms; some would intercrop berries while fruit trees were growing, only to move into strawberries alone when demand from the San Francisco Bay Area convinced growers that strawberries might be profitable.[36] By the end of World War II, when Salinas was well on its way to becoming the "salad bowl of the nation," the Salinas Valley was populated by medium-size farms, more than half of which were owned by farmers growing a range of vegetables.[37] By 1962, grower-shippers had taken a larger role; while some owned their land and leased additional acreage, some leased their entire acreage.[38]

It would be impossible to tell the history of land tenure in the strawberry regions without recounting the experience of the Japanese, some of whose descendants are important players in the strawberry industry to this day. The Japanese began to arrive in California in the late 1800s, and many were recruited to be farmworkers to replace the Chinese population that had been decimated by exclusion legislation. With the support of mutual aid societies that generated needed credit, many Japanese farmworkers were able to become farmers themselves, either as tenants or sharecroppers, or through outright purchases. Having arrived with knowledge and skills important for horticulture, many of the Japanese farmers began to compete with white farmers. This competition generated a backlash, resulting in the Alien Land Laws of 1913 and 1920, which prohibited foreign-born Japanese from leasing or owning land. Although some were able to get around these laws by purchasing land in the names of their US-born children or paying citizens to buy land for them, by the 1930s few owned farms and most had instead become sharecroppers on white-owned farms, including strawberry farms. Of course World War II changed this, again, when the Japanese were cast as national security threats and forcibly relocated to internment camps. Following internment, many returned to Pajaro or Oxnard rather than Salinas, where prewar hostilities had been more intense. Some resumed their work as share-

FIGURE 12. Japanese sharecroppers, 1942, San Jose. Photo by Dorothea Lange.

croppers to build up savings, while others asked whites to purchase land in their name. When the state Supreme Court overturned the Alien Land Laws in 1952, they again used mutual aid to purchase land in the less-expensive areas of the region. By the late 1950s farmers of Japanese origin comprised about three-quarters of the strawberry-growing population, accessing land through a combination of owning, leasing, and sharecropping.[39]

Gradually many of the white landholders in the strawberry regions pulled back from farming, either by retiring or having heirs who declined to farm.[40] With the high crop values of vegetable farming (see below), they realized that they could make as much money, with less stress, by leasing out their land—or farming farmers, as some like to quip. No doubt the high interest rates of the 1970s made purchasing new land prohibitive as well. By the late 1970s Salinas vegetable production was almost entirely organized by major multinational corporations that were willing to pay good lease prices for prime land.[41] Oxnard was moving in a similar direction, if the McGraths provide any indication. Eventually there were hundreds of McGrath descendants, and the land had been divided among them. While many sold their allotments to real estate developers or lost them to eminent domain, others opted to rent to farmers. That included Phil McGrath of McGrath Family Farms. Of the three hundred acres he was left with, he ended up leasing 85 percent to Reiter Brothers, associates of Driscoll's, while maintaining a thirty-acre organic farm where he would grow a variety of vegetables as well as some strawberries.[42]

It is safe to say that today most strawberry land is leased. Specific statistics are not available, since the US agricultural census does not collect land tenure data by crop, but according to census data the vast majority of farmland is leased in the five counties where strawberries are one of the top two crops in value.[43] Of the seventy-four growers I interviewed for this project, only nine reported growing entirely on owned land; some both leased and owned, while the majority operated solely on leased land. When asked, many told me that the landowner was either a former farming family or a wealthy investor from the city.

Although the precariousness of tenancy is generally derided as not conducive to stewardship (see below), leasing has in fact worked quite well for strawberry growers. When early strawberry growers witnessed blight in their fields, they would move to another "piece of ground," as they say. Japanese-origin strawberry farmers in the Pajaro Valley, for example, generally farmed one parcel for four to six years, then moved to another farm when the soil was depleted.[44] Indeed, as we saw in chapter 2, before the era of fumigation, university scientists actively advised growers to handle soilborne disease by moving on. The approach worked well as long as land was available and cheap, as grain and ranching land tended to be in the early days. By the 1950s, though, land available for strawberry production was becoming scarce. Here again, the advent of fumigation saved the day, allowing much more production on much less land—and in doing so shored up other advantages of leasing.[45]

MAXIMIZING LAND

The importance of the petrochemical revolution for agricultural productivity has long been recognized. Generally, however, fertility innovations, such as the development of synthetic nitrogen, have received the lion's share of attention. Yet, as geographer Adam Romero has argued, the introduction of synthetic fertilizers may have eliminated the need to rotate crops for nutrient management, but it did not immediately overcome the need to rotate crops for pest management. It was not until the 1940s, with the discovery of cheap and effective petroleum-based soil fumigants (described in the previous chapter), that crop rotations could be done away with, "severing the link between the intensive production of a single crop without rotation and the build-up of commercially destructive pests in the soil complex."[46] Fumigation not only allowed growers to stay on the same field year after year, solving the "replant problem," it also allowed growers to remain in areas with good, sandy soil that would otherwise be unusable due to disease history.[47] At the same time, it likely worsened the presence of disease in these soils by making knowledge of the ecology of soil disease obsolete.[48] In short, it allowed growers to contend with increasing shortages of land that contained California's natural advantages, even as those advantages began to be compromised.

It was not only fumigation that intensified the use of land, extracting more economic value on a per-acre basis. Breeding accomplished that, too. The development of short-day varietals, those that could flourish during the contracted winter days, allowed the warmer Southern California regions to expand and, coupled with fumigation, even made it possible to rotate winter plantings with summer plantings.[49] Farther north, on the Central Coast (Santa Cruz and Monterey Counties), the development of day-neutral varietals could extend the harvest season by months. Growers could begin the harvest as early as March, and if the weather stayed dry, continue it as late as December.[50] (Growers once even left the plants in the ground a second year, but second-year fruit is of notoriously lower quality, and the practice has all but ceased.) Given the substantial fixed costs of land preparation in strawberry production, including the cost of fumigation, the stretching of seasons greatly increased per-acre profitability.[51]

Extending the season created some stickiness in land use, however, because it required devoting the land to one crop for more than fifteen months. Generally, farmers on the Central Coast fumigate in late summer, plant in the fall, and begin harvesting in late winter or early spring. Therefore, extending the harvest into December did not give them enough time to fumigate and plant for the next season. The long-season problem was solved with yet another innovation: rotating with vegetable growers. It is not unheard of for vegetable growers in the Salinas and Pajaro Valley regions in Monterey and Santa Cruz Counties to plant up to six rotations per year of fast-growing leafy greens. With synthetic fertilizers, greenhouse and transplant services, and quasi-mechanical harvesting, for a crop like baby greens it is relatively easy to harvest a field in a day, prepare it and fertilize it for a week or two, and plant it again, for harvest in as little as six weeks. Even for a longer-season

FIGURE 13. Intensive strawberry production, Monterey County. Copyright © 2016 The Regents of the University of California. Used by permission. Photo by Jack Kelly Clark.

crop, say head lettuce, growers may still get in two rotations within eight months. Such arrangements therefore became a win-win for strawberry and vegetable growers (who rarely are one and the same). Strawberry growers could have their fifteen-month season, while vegetable growers, who are generally not allowed to fumigate, could obtain the benefit of fumigation without having to pay for it, report it, or take the public flack.[52] These arrangements were effectively enabling an "off-label" use of fumigants for vegetable growers.[53] But not any vegetable grower. Only growers who could afford the high rents of strawberry land partook, leaving out, say, broccoli growers, whose crop actually helps ameliorate soil disease rather than add to it.

Although strawberry growers needed the land for longer periods, often vegetable growers held the master leases on these arrangements. They certainly had a longer history in the Salinas area and may well have had the support of the major vegetable grower-shippers in obtaining the land. The rub is that many of these growers asked that strawberry growers flat fumigate as conditions of their lease, and until it was phased out, many of these vegetable growers insisted that strawberry growers use methyl bromide. Since leafy greens add to the pathogen load of the land, it was not unusual for vegetable growers to lease out the most infested blocks of land to strawberry growers to receive the benefits of fumigation.[54] Even when strawberry growers held the master lease, many vegetable growers would not rent it without a promise of fumigation. Nevertheless, this arrangement was a better solution for straw-berry growers than leaving the land unused for nine months of every other year while still paying rent.

With fumigation eliminating the need for rotation, growers could plant an annual crop on the land to which they had access. And with fumigation creating a kind of insurance against crop failure, growers would seek out more land, increasing production even further. And so, as we saw, strawberry fruit production burgeoned after 1960, facilitated as well by the marketing of the strawberry, to be discussed in chapter 7. A relatively invisible repercussion of this expansion was the enormous increase in demand for starts. Thus nurs-ery land had to expand as well—and it had to expand a *lot*, given how starts are produced.

STARTS BEFORE THE FRUIT

As I explained in chapter 3, strawberry plants are not grown from seed; they are asexually propagated from clones of the varietals growers wish to culti-vate. Even before developing a marketable start became the baroque, pains-taking, and highly regulated process that it is today, many in the industry recognized the advantages of separating plant production from fruit produc-tion. We saw in chapter 2 that a scourge of red stele disease in a Zayante nursery (Santa Cruz County) alerted the industry to the potential of cross-contamination between nursery and fruit field. And although it may have been an accident (or pure mythology) when Driscoll and Reiter discovered the sweetness of Sweet Briar berry when traveling in the northern reaches of the state, they also noticed the great-looking runners on those plants, gaining an inkling that they might want to grow plants in that area. Still, as told to me by a principal of one of the founding nurseries in Northern California, even though farmers were growing berries in the far north, they didn't start growing nursery stock until 1945. Instead, runner production took place in situ in the fruit-growing regions. Growers would plant sparsely, remove run-ners, and re-set them in the ground to thicken up the planted area. At the time, the land extensiveness of this approach was unimportant; the problem was that this method fostered the spread of pests and disease.[55]

These days plant production is a highly exacting process, designed to keep the plants clean from all diseases, maintain the integrity of the cultivar, and produce a prodigious amount of plant material for fruit growers, certified as disease free.[56] It can take up to four years, adding to the lengthy breeding process for new cultivars described in chapter 3. All certified plants begin as meristems—pieces of plant tissue where cells are actively dividing but not yet differentiating into leaves, stems, or other organs. The meristem process ostensibly both eliminates pathogens and reinvigorates plants. First, however, the meristems must themselves be generated. Nurseries obtain what is known as nuclear stock from breeders, UC Foundation Plant Services, or other private germplasm banks. They plant this nuclear stock, the original "mother," in hanging pots in a greenhouse. The "daughter" runners that grow down the sides without touching the ground produce the meristems, the tips of which are clipped to become the original "mother" plants for plant propagation.[57]

In the first generation of propagation, nurseries tissue-culture these meristems in test tubes. Several steps later, in which they maintain the plant material in highly controlled conditions, they pot a small set of daughters in sterile media and grow them in a screen house (see fig. 5 in chapter 3). Among other things, screen houses protect against insect invasion, especially aphids, which carry disease. Depending on the cultivar, these new meristem mothers produce one hundred to fifteen hundred plantlet daughters, many more than a non-meristem mother would generally produce.[58] At every step of this process, plant material sees inspection and testing for plant viruses and bacterial and fungal pathogens, and receives DNA fingerprinting to ensure cultivar integrity. The California Department of Food and Agriculture (CDFA) can do this testing, or it can be done privately through the California Seed and Plant Lab. If the plant is virus free, CDFA certifies it as part of the clean plant program, a virtual necessity for further sales. The plant material is then ready for transfer to the propagating operations, although the nurseries I visited did the meristem process in-house.[59]

The second generation involves transferring plantlets to what are called foundation blocks, where they become land grabbers of a different sort. Foundation blocks are in fields rather than greenhouses, but they must be located at least a mile from other sites of strawberry production.[60] As with succeeding generations, nurseries plant them in widely spaced rows with the expectation that the runners will spread, grow roots, and become baby plants—all clones of the mother. As they are growing, workers nip blooms in the bud to encourage these plants to run.[61] Each of these (new) mothers generally creates two hundred to three hundred daughters, transforming the field into a solid green (or fall color) mass by the time they are ready for harvest, after which they are dug as bare root, separated into individual plants, and put in cold storage. For the third generation, nurseries plant them in what are called increase blocks, located in conditions that create rapid acceleration of plant material. These, too, are harvested and separated into individual "plug plants." Nurseries may propagate a fourth generation of daughter plants in addition. All fields used in the propagation process are fumigated, almost always with methyl bromide in a traditional mix with chloropicrin, allowable by the quarantine exemptions for nurseries. And due to separation requirements in the foundation planting, as well as the imperative to produce many plants, they take up a lot of land.

But not just any land. Over the years the industry has developed a complex geography of propagation—a geography that is curiously distinct from the geographies of fruit production. Most of the foundation blocks are located in Northern California, at low elevation, in areas around the towns of Redding and Red Bluff. Their placement there may well be an accident of history. After World War II, when nursery production separated from fruit production and fruit production moved to the coast, the handful of strawberry operations in the area were left high and dry, as it were. They had cheap and available land, far from the coastal fruit operations. As it turned out, the hot summer climates of these areas served an important purpose. In the heat the mothers would create many daughter runners, with few blossoms, and thereby expand the plant material in huge orders of magnitude. Later these nurseries expanded their operations into the more southern San Joaquin Valley, in places like Manteca and Turlock. These became the sites of the increase blocks, where, again, plant material would increase exponentially.

Meanwhile, sometime in the 1950s, UC breeders Royce Bringhurst and Victor Voth discovered that plants growing in the far northern Siskiyou County were superior to others. They realized that chilling made the plants more productive and vigorous, a response that is fairly typical of fruit-bearing plants of the Rosaceae family.[62] And so, in the 1960s, most of the major nurseries expanded to high-latitude, high-elevation locations (two to three thousand feet), where they remain today.[63] It is in these areas where they may propagate a fourth generation. The yields from these runners may be less, more like 30 to 1, rather than 100 to 1 (or about 350,000 plants per acre, rather than a million), but the timing of the chill serves the fruit industry well. The nurseries plant bare root (plants without soil or foliage) in these areas in April, as soon as the snow melts. The plants produce runners all summer when it is hot. Then, as the nighttime temperatures begin to drop below freezing in late September or early October, the chill encourages the plant to start conserving to survive the winter—in effect, to go dormant. Just as nighttime temperatures reach below 20 degrees Fahrenheit, sometime in late October, the nurseries begin their harvesting—at night, to keep the roots chilled, only switching to daytime harvest when the days are just as chilly. (Among other things, this makes for very harsh working conditions.) They'll refrigerate the harvested material for one to three days, then ship it to Southern California for immediate planting in mild temperatures. Having been chilled, the plant experiences early springtime and begins to grow and produce fruit immediately. Growers who propagate their own plants, as some still do, use cold storage to achieve the same effect. Plants destined for the Central Coast fruiting area may not be planted until late fall, and so can be propagated in the low-elevation nurseries that receive a late-fall chill. Either way, the chill the plant experiences tricks it into dormancy so it can experience springtime when it is planted on the coast.

Today nearly all of the six or so nurseries locate their high-elevation operations in an obscure valley on the northeast side of the Mount Shasta volcano, very near the Oregon border. Ranching once held sway in Butte Valley until the strawberry nurseries came along and found it useful for its chilly fall nights and well-drained volcanic soils. It's a brutal, harsh environment; I am told that probably few other valleys in the western United States are as cold in September. In any case, they would not have the same proximity to the rest of the industry in terms of minimizing transportation costs. The location, in other words, has several distinct natural advantages for California growers.

Despite its highly remote and undesirable location for residential living, the Butte Valley does not immunize growers from the land issues that haunt the industry. I was told that strawberry land around the outpost of Macdoel sells for $5,000 per acre, twice the cost of alfalfa land. Nursery growers primarily lease land, though, and just like on the coast have been known to compete for available land, leading to high lease prices. Adding to their grief, they pretty much have to grow on a three-year rotation to keep disease load down and avoid exhausting the soil. Therefore they have to maintain access to three times as much land as they use in any one season, effectively tripling their land payments. If they are growing foundation stock there, they have to find isolated pieces, as well. Yet these producers cannot work out the clever rotation arrangements that fruit growers do. If they rotate with other growers, they risk disease and losing their certification, and therefore the right to export plants. Even though they have the methyl bromide exemption for

FIGURE 14. High-elevation plant propagation, Macdoel. Photo by author.

nursery crops, it is never completely effective. Propagation, in short, creates another set of land challenges even if it takes place in cheaper land markets than the coast. Nevertheless, it saves the precious coastal land for fruit production.

FICTITIOUS CAPITAL GOES TO TOWN

Put together, all of the innovations that intensified the output of strawberry land on the coast—fumigation that eliminated rotations, plant breeding that extended the length of seasons, rotations with vegetable growers that eliminated remaining gaps in land use, and the displacement of plant propagation to other locations—effectively drove up land values in the fruit-growing regions. As I explained in my book *Agrarian Dreams: The Paradox of Organic Farming in California* (2004), agricultural land values generally derive from the capitalization of expected income (for example, rents) from any piece of land. Landowners will recognize the value of that land and demand higher prices to access it.[64] This can be traced historically. Land planted to vegetables became more expensive than land planted to wheat; land planted to vegetables that could generate multiple harvests in a year became more expensive than land that demanded rotations, and land planted to lengthy, annual rotations of strawberries became and remains more expensive than land planted to, say, broccoli. In practice that means it has become uneconomical for growers to pay rents on strawberry land and grow anything that produces less income than strawberries do, when planted year after year, on the same block, without rotations except for lettuce, and for as long a season as possible. And since fumigation has enabled this regime, the practice has effectively been capitalized into land values. A case in point: annual land rent has climbed from $150 per acre in 1969 to $2,700 in 2014 (with many growers paying more than that), more than doubling the percentage of total production costs.[65]

In strawberry land, alas, it is not only the capitalization of crop value that bears on agricultural land values. As it happens, suburbanites enjoy the same cool summer breezes coming off of the Pacific Ocean and the mild winters that strawberry fruit does. And so most of the fruit-growing regions have seen extensive urban development over the past several decades. So thick are the houses and shopping malls that abut berry fields in Oxnard, Salinas, and Watsonville that few would continue to describe them as rural environments. And, indeed, farmland loss has led to land conservation efforts in all three areas. In Ventura County, individual cities have passed a series of initiatives since 1995 that require a vote before agricultural land can be rezoned for development—the so-called SOAR initiatives (for Save Open Space and Agricultural Resources). In 1996 Monterey County nearly passed a similar initiative (Measure E). In 2002 Watsonville passed a fairly watered-down Measure U that limited how the city could annex farmland. Farming landowners have been heavily opposed to these measures, however, precisely because they restrict the value of that land for resale.[66]

Strawberry regions are not immune from the global land grab, either. While almonds are notorious in California for soaking up outside financial capital, strawberry land has also garnered interest from wealthy investors. Consider Gladstone Land Corporation, a real estate investment trust (REIT) listed on NASDAQ that specializes in purchasing farms and farm-related properties and leasing them to farmers. In 2017 it claimed to own forty-eight row crop properties in Arizona, California, Colorado, Florida, Michigan, Nebraska, and Oregon, totaling 23,857 acres.[67] Flipping through their portfolio on their website, I noted about a dozen properties in the prime strawberry-growing zones of California, all with ostensibly good water sources. Yet the prices they pay for this land suggest that something more is afoot. For instance, in 2014, Gladstone purchased sixty-eight acres of strawberry farmland for $6.9 million, or just over

FIGURE 15. Strawberry fields abutting housing, Oxnard. Photo by Sam Hodgson.

$100,000 per acre—an unheard-of amount for row-crop land, akin to prices for land with recognized and valued geographic indicators for wine production.[68] Regardless of whether this constitutes a good investment (I tend to think that REITs will go the way of other derivatives that have bankrupted many), these wouldn't even be salable without a (perceived) shortage of "good" strawberry land—land with the natural advantages of soil, climate, and good water. And growers I met with corroborated this perception, often speaking of land shortages and land prices (along with fumigation restrictions, drought, and labor shortages) as a major threat to the industry.

Crucially, efforts to address rising land prices and land shortages are exacerbating problems with soil disease. We already saw that rotation with lettuce growers worsens the pathogen load. With high prices and land shortages, growers are looking beyond the core strawberry areas for ground. But that means that they are "pushing the envelope," as put by a UC extension agent, in terms of natural advantages. For instance, growers are planting farther south in the Salinas Valley, where the hills block the ocean breezes. Or they are planting on the hillside benches in Salinas or Santa Maria. Or they are moving farther east of the coast in Camarillo. Some of these areas contain less-than-ideal soils: deep and poorly drained loamy sands, silty clay loams, or even the deathly calcareous soils.[69] Such soils worked well enough with methyl bromide, which could propel the active ingredients through just about any-thing, but chloropicrin is insufficient to the task of dealing with the disease load of poorly drained soil. Nor is clay soil conducive to the bed fumigation often necessary in hilly terrain. These areas are also more arid and hot, with less summer fog, inducing the sort of stress that *Macrophomina* preys upon. For that matter, these are completely different ecosystems in terms of soil microbiology, which could make them more—or less—suppressive to patho-gens. The upshot is that growers are using more chemicals in these areas and/or seeing more disease. This is precisely the kind of situation captured in geographer Jason Moore's conception of negative value. For him, negative value arises when the evolutionary pace of extra-human natures such as super-weeds exceeds human ability to control them. The fixes become more short term, especially without new frontiers (read: new land) to colonize, until they run out. At that point, costs make production no longer profitable, hence negative value.[70]

Another kind of envelope pushing involves transitions to organic produc-tion. Keen to get in on the high prices for organic strawberries, some growers have sought out already-certified land or gone through the standard process of transitioning land into organics. The latter requires three years without a disallowed material (that is, no fumigation) when berries must still be sold as conventional, garnering conventional prices. But some have found land previ-ously in pasture, what they call "virgin" ground, which has required no tran-sition period. Since most such growers have only been dabbling with organics for a short time, they are generally satisfied with the results. Not only do they see nice yields for lack of pathogens, but perhaps these soils are naturally suppressive because of microbial diversity. But this is a very short-term strat-egy for organics, especially among growers who have little apparent interest in learning or applying the best organic practices of rotating strawberries with low- or no-value crops for several years in a row to prevent pathogen buildup (see chapter 8). Many informants concede that these growers move on within a few years, leaving behind diseased soil.

It is the lease relationship that makes these variations of pushing the enve-lope workable, at least for the time being. Proponents of agrarianism have long claimed that landownership is the basis of land stewardship—that those who have a long-term interest in a piece of property will gain knowledge of its soils and treat them and all they contain to be fruitful over the long haul.[71] The converse is only implied—that tenant farmers will not obtain intimate knowl-edge of the land, and thus will abuse and degrade it. Tenancy, in this view, creates a tragedy, if you will, of not giving a damn. Although these suppositions have certainly been debated and contested, the land behavior of at least some in the strawberry industry seems to corroborate these relatively standard suppositions about tenants' lack of commitment to land stewardship.[72] Agricultural economists working on the problem of *Verticillium* wilt in lettuce have made that argument, albeit with assumptions that in some ways complicate the point I am trying to make here. Their models show that longer-term farmers, whom they dub "owners," are more likely to practice disease prevention than short-term farmers, whom they dub "renters."[73] However, in their study, both methyl bromide fumigation and rotating with pathogen-repellant broccoli were counted as equivalent strategies of disease prevention.[74] While it may well be the case that growers with longer-term leases are more attuned to disease prevention, my point is that it is the lease arrangement itself that has allowed growers to walk away when such strategies fail.

That some strawberry growers might contribute to and then walk away from diseased land should not be all that surprising. After all, that is what they have been doing from the get-go, indeed were *advised* to do for many years. But the "natural" advantage of abundant land no longer exists, and growers may find this approach too risky, especially with landowners all too willing to lease or sell for "highest and best" use.

OPPOSING MATERIALITIES

Theories of property assume the alignment of land quality and property value.[75] For example, classical political economist David Ricardo's theory of differential rents held that land that was more productive would earn more income for farmers, and this unearned income (a product not of the farmers' efforts but of the soil itself) would be siphoned off by landowners as payment for access to the better land.[76] While others have corrected that theory to suggest that landowners might want to allow farmers to share in the extra income, the basic tenet is still there: more productive land is valued higher in the market.[77] Geographers have also noted the advantages of certain locations, whether for their proximity to other natural resources, markets, or supporting industries. Land with locational advantages is also valued higher in the market because it can reduce transportation costs, among other things.[78] The converse then should hold as well: land that reduces profits should be valued less. In the perfect world of neoclassical economics, a decline in profits or production would manifest in a decline in land prices because growers would not be able to make payments for the land.

If these theories wholly explained agricultural land values, land prices in the strawberry regions would most certainly be on the decline. For one, fruit prices have tended to slip, largely from overproduction. According to USDA data, inflation-adjusted prices for fresh berries to growers went from 84.7 cents per pound in 1970 to 78.6 in 2012, with much variation in between.[79] Although more recent official data is not available, growers say that prices have worsened considerably, pushing some of them out of business, particularly in the Oxnard area. One consequence is that nursery producers in the Butte Valley had to let go of land leases in 2016–17 because fruit production was down, diminishing demand for what is the only game in town. A 2016 report published by UC researchers, based on production budgets in the much longer growing seasons of Monterey and Santa Cruz Counties, further suggests dwindling profits in the industry, especially with per-acre costs reaching over $68,000, largely due to increases in labor costs.[80]

Fumigation restrictions should also be affecting land prices—not only because yield is being lost to plant death and some of them make the process itself more costly, but also because the nature of many of these restrictions affects how much land can be used for production. Buffer zones take land out of production. Forty-acre treatment limitations make it more costly to fumigate. Bed fumigation forces a trade between smaller buffer zones and less efficacy. Township caps reduce the amount of Telone that can be used on a first-come, first-served basis, making availability unpredictable.

Significantly, increased urbanization has been the impetus for many of these restrictions, as the middle-class people moving into the areas abutting strawberry land are ready and able to complain if harmed by fumigants. Restrictions on fumigants are also much more likely to kick in in more densely populated areas, since they are regulated by the Clean Air Act, which limits emissions of volatile organic compounds (VOCs). Beginning in 2008, Ventura County, the home of Oxnard, had to restrict the use of all VOCs because it was a "non-attaining area" under the Clean Air Act; there was too much ground-level smog. The county was given an allocation for each fumigant, of which growers in turn had to request a portion. Since there were more acres in production than overall allocation, each grower received only a fraction of their request, about 40 percent. Having made the goals, the following years they needed only to track and cap fumigant use. The recent rise of the Santa Maria growing region relative to Oxnard is about escaping these regulatory restrictions along with the high land prices. In the fifteen years between 2001 and 2016, strawberry acres harvested increased from 3,092 to 8,055 in Santa Barbara County alone.[81]

Finally, land prices should be declining because the material conditions that real estate contains—the natural advantages—are threatened. Much of the soil in strawberry land has become diseased, so that extension agents are now suggesting that the best insurance is finding the rare piece of disease-free land.[82] While rising sea levels have not had any demonstrable effect (the sandy soils have not washed away), climate change is beginning to take its toll. The year 2017 saw record heat waves, with temperatures even on the coast rising above 100 degrees.[83] Previous to that, 2016 was the hottest year ever recorded in California since modern temperature records were first taken in 1895, with the previous two records set in 2015 and 2014.[84] Meanwhile, four years of epic drought between 2012 and 2016 diminished the aquifers in the strawberry regions. One grower had his well collapse during our interview! Some have suggested that over-drafted water resources in Mexico's strawberry regions have staved off competition with California growers, but that competitive edge could well diminish for the California strawberry industry.[85] And with less groundwater available, most of these low-lying areas near the coast see saltwater intrusion. Both heat and saline water stress berries, making them more susceptible to diseases like *Macrophomina*.

By my reckoning, an adjustment in land prices that incorporates the declining conditions for strawberry production has yet to materialize. Others have suggested otherwise, citing anecdotal evidence (a personal communication) that land values have dropped as much as 25 percent when it is discovered that acreage is contaminated with *V. dahliae* and that renters have asking for reduced rent because of *V. dahliae* contamination.[86] This evidence simply does not square with the stories of S Farms and Gladstone Land Corporation, nor my many discussions with growers that took place after this personal communication was obtained in 2013. What I saw instead was that the abstract value of land is diverging from the material value of land, that speculative capital is at odds with natural capital, and that the exchange value of land bears little relation to its use value. I thus found it particularly amusing when a scientist who found my speculations on pathogen virulence to be fantastical suggested that there ought to be a mechanism of valuing land that takes into account soil conditions such as organic matter composition and pathogen load. Such a proposition imagines that land functions only as a medium for crop production, never as fictitious capital (see opening epigraphs).

In practical terms, this divergence further challenges the strawberry industry to farm without fumigants. Not only are fumigation and breeding successes capitalized into agricultural land values, making them unusually high. Not only has the urbanization of berry-growing regions created an absolute shortage in land suitable for growing strawberries under its current intensive regime without the full array of fumigants and fumigation techniques which such urbanization has made illegal. Land subject to real estate speculation makes unimaginable systems in which growers pass up profitable years to allow disease-suppressive microbes to reinhabit the soil—that is without major price hikes in strawberries to compensate for the years without strawberries. In a funny way, then, the lease arrangements that have long roots in California strawberry production serve as perfect, if temporary, accommodation for grower and landowner alike: for now, growers can continue to add to the soil's pathogen load, often unknowingly, while landowners hold out for a higher and better use, also in ignorance of the land's decreasing quality for agriculture. It is not a lasting solution.

There is one possible escape from this nearly impossible position for the strawberry industry: growers could follow in the footsteps of their European counterparts and move to soilless systems. Growing in climate-controlled greenhouses and using non-fertile media like coconut coir, peat, or even water to grow strawberries would allow them to escape the ecological threats that increasingly bear down, albeit no doubt creating others, as I will discuss in chapter 8. But, as I will also address there, this is not where the California industry is headed, and for good reason, because moving in the direction of controlled-environment agriculture would undermine the biggest competitive advantages of California strawberry growers: a temperate, coastal climate and high-quality land. Land, that is, remains the industry's most important asset, even as it has become the source of many of its challenges. Oddly, though, it does not nearly grab the amount and degree of complaints that labor does—the topic of the next chapter.

SIX

Scarce Labor and Disposable Bodies

Distinctive features of the [strawberry] commodity, its technologies, and its production processes have restricted the strategic options of growers, with the result that control over harvest labor has become the key to profitability.

Anthropologist Miriam Wells,
Strawberry Fields: Politics, Class, and Work
in California Agriculture, *1996*[1]

As a farmworker employer, we optimize every day in the calendar to make money to bring home, money so we can pay our bills. Every day is crucial during the season. And when you have a field that is infected with some soil disease and the plants are wilting, when there is very limited fruit on the plant, it's an obvious choice. Many workers will come to me and say, "We can't make it here." They're in one crew that has more of that percentage of that infected area, and they're saying, "This isn't fair; we need fields that are producing to make money." So we have to make arrangements and split fields so they're more evenly distributed. We don't know where the disease is going to pop up each year. So clean soil is integral to the farmer, of course, but the worker is being forgotten here. They have to make a living.

Grower testimony in support of methyl iodide
registration at a February 14, 2012, hearing of the
Monterey County Board of Supervisors

Although we know the risks, we have to work. This is the life of an immigrant, and of the worker of the *campo*.

Farmworker interview, 2014

Capitalists, including especially growers whose production process is so discontinuous, want labor power; they want a factor of production. What they get are people.

Geographer Don Mitchell, "They Saved the Crops":
Labor, Landscape, and the Struggle over Industrial
Farming in Bracero-Era California, *2012*[2]

ON ANY GIVEN WORKDAY DURING the long harvest season, it is possible to witness a strawberry harvest just by driving through one of California's prime coastal strawberry-growing regions. Although you might see acres and acres of berry plants, you will spot the harvest by the large, new pickup trucks (those of the field supervisors, all men) and tired, dusty sedans (those of the workers) lining the roads. A row of portable toilets will likely sit at the side of the road, and one or two large hoop tents may be set up to provide shade during breaks. Certainly at least one open-bed truck will be parked and packed with boxes, with a scale placed to one side. One or two workers will be stationed by this truck, holding clipboards to mark off boxes of strawberries. Yet what will likely most catch your eye, especially in late spring and early summer, is the movements of pickers, sometimes several hundred in any given field. They will be bending over and swiftly using two hands to twist the berries off the bountiful plants and pack them into baskets already placed in market-ready boxes. The plants will be bountiful because they are high-yielding varieties, and the soil was likely fumigated before planting. The workers will be moving briskly because that is how they maximize their hourly wages, especially in the height of the season, when there are many berries in the field and wages are almost always calculated at least in part by the number of boxes they pick. When a box is full, the worker will run down the narrow rows between the raised beds of berries to have the box inspected and tabulated at the open-bed truck. They will then race back to where they left off and begin picking for the next box.[3] When berries are plentiful on the vine, an average worker can pack about six boxes an hour, so they say, earning about $11 for the hour in piece rates (in 2014). A highly skilled worker can make much more.[4]

Growers like to say that harvest workers don't mind running through the fields. They can make a lot of money that way; they will even set up competitions to see who can pick the fastest. Out of their employers' earshot, workers tell more complicated stories. Based on her research with melon workers in the hot Central Valley, anthropologist Sarah Horton confirms the practice of picking competitions and recognizes that they can be performances of machismo. But she emphasizes that such performances are ultimately attempts by workers to demonstrate their exceptionality in order to keep their jobs—and that working like that can often lead to illness and even death.[5] Writing on their research with strawberry workers in particular, anthropologists Seth Holmes and Miriam Wells further

FIGURE 16. Strawberry harvest, Salinas. Photo by author.

emphasize the brutality of work that often leads to injuries, illnesses, and chronic pain.[6] According to these anthropological accounts, working quickly uses bodies up.

The apparent disposability of harvest workers doesn't square with growers' complaints that they face a labor shortage even more threatening to their livelihoods than restrictions on chemicals, drought, low market prices, and even rising land values—the other standard complaints. Following years of manufactured labor surplus in order to create farmworker vulnerability— there were always people to replace those who complained too much— starting in the late 1980s the United States began to adopt ever more stringent border and immigration measures in an attempt to appease nativist constituencies. For a long time those restrictions worked well for growers, effectively deepening workers' vulnerability. But since about 2010, this policy of "prevention through deterrence," which as anthropologist Jason De León shows has been the source of unfathomable numbers of deeply cruel border deaths, seems to have finally caught up with growers.[7] Strawberry growers, especially, claim they can no longer find enough workers—at least at wages they're willing or even able to pay. And so they have attempted other approaches to attract workers to their fields, including attending to plant vigor and the conditions of strawberry harvesting to allow workers to move quickly and improve their hourly pay. Growers even justify the use of fumigants in this context since fumigants, applied before planting, appear to contribute to abundant and easier-to-pick berries. For their part, harvest workers aren't exposed to these harsh chemicals unless there are fumigations in nearby fields—although other fungicides and insecticides are used with regularity throughout the harvest season.

That the conditions inducing workers to race through fields are also intended to attract workers inverts and at the same time deepens Miriam Wells's groundbreaking observations of labor management in the California strawberry industry. Wells has written of how the industry's structure and the characteristics of the strawberry, in addition to the political construction of labor surplus and non-citizenship, coalesced to discipline workers and moderate harvest costs.[8] With the acute labor shortages to which strawberry growers attest, farmworkers seem to have obtained a modicum of leverage with respect to growers. Facing significant challenges for organized action, though, workers instead employ what James Scott has called "weapons of the weak."[9] These include walking off the job for greener pastures, or in this case redder ranches, with more abundant fruit on the vine. In a manner of speaking, workers discipline growers into creating the field conditions in which workers can maximize their still-limited pay.

It is nonetheless a glaring paradox that growers both need working bodies and seem to use them up. That they justify the use of chemicals, especially chemicals that potentially affect the lives of future farmworkers through intergenerational harm, to attract workers makes the paradox even more confounding. For that matter, it is curious that workers putatively in a position of strength feel compelled to work at a backbreaking pace. In this chapter I recount the origins of the labor shortage for harvest work and show ways in which harvest work, berries, and fumigation are entangled, so that the labor shortage, however manufactured, presents a clear and present threat to the assemblage. I then review how strawberry production is also a clear and present danger to workers. Insofar as the primary means growers have employed to address the shortage can only exacerbate that danger to workers is yet another example of the industry's neglect of ecologies that matter and, hence, its propensity for iatrogenic harm.

Agricultural labor markets in California have never operated by so-called laws of supply and demand, with wages clearly indicating the "price" of labor.[10] As Wells puts it, political forces have long constrained the way that labor markets operate, including "ideologies, laws and institutions that establish the legitimate entitlements of particular groups."[11] That strawberry growers are paying more attention to field conditions than wages as a way to attract workers must be seen through this filter and contextualized in a long history of farm labor in California in which agricultural industries actively recruited and imported racialized groups to do the work and created perennial conditions of surplus as mechanisms of labor control. Although California saw moments of farm labor radicalism, first during the early twentieth century and next during the Great Depression, industry invariably responded by replacing existing farmworkers with other, more politically marginal and hence more compliant, groups.[12] Codifying this marginality, agricultural workers were excluded from the US National Labor Relations Act of 1935 (the Wagner Act), which allowed trade unions to form, bargain, and strike. Due to that exclusion, farmworkers, unlike manufacturing workers, never enjoyed a stable period of strong organization that has since eroded.[13]

Geographer Don Mitchell traces California agriculture's reliance on unauthorized workers to the Bracero Program, the national guest worker program that operated between 1942 and 1964, during which growers effectively exercised their class power to set the "prevailing wage" so low that white citizen workers would never accept it.[14] Still, California farmworkers saw another blip of empowerment following the end of the Bracero period and into the early 1970s. Not only did the end of the program produce a new labor shortage, but with the support of progressive whites, California farmworkers were able to obtain new protective legislation in the form of the California Agricultural Labor Relations Act (ALRA). A wave of agricultural unionization led by Cesar Chavez and the United Farm Workers (UFW) ensued, raising wages far above previous levels.[15] It was precisely these gains that provoked major pushback from growers and their political allies. Under a new Republican leadership in the California government, many of the political gains of the 1970s, including the bite of the ALRA, were reversed or defanged.[16] By the 1980s growers were relying heavily on farm labor contractors both to subvert the farmworkers' union and to produce renewed conditions of surplus. An influx of new immigrants, due first to the Central American wars and then to neoliberal economic restructuring in Mexico and Central America, added to the surplus, cementing unauthorized workers as the labor force of choice.[17] Though unauthorized, these workers had little trouble getting in at the time; they just had few rights once they arrived.

The weak structural position of farmworkers was exacerbated by the strategic mistakes of the leadership of the once-triumphant UFW union. Chavez was always more a civil rights leader than a union leader, attuned to the support and excitement of the young urbanites who worked for the UFW during its famous boycotts. Consequently, few UFW resources went to organizing workers in the field. Plus, the union had made enemies of unauthorized workers, who had often served as strikebreakers, by explicitly excluding them from its organizing efforts. These decisions, among others, led to the union's decline as a major political force in California.[18] To this day, no organization has arisen to replace the union and regain the enthusiasm of California farmworkers for a union.[19]

Farmworkers were further disempowered by a major political and cultural revolt against the presence of unauthorized workers in the United States. In the context of neoliberal rollbacks in taxation, undocumented workers were cast as a drain on public resources, leading to the 1994 passage of Proposition 187, a voter-led initiative that would have curtailed services to undocumented workers in California.[20] (The new law was eventually thrown out when a federal court found it unconstitutional.) At the federal level, several measures to fortify the US-Mexico border were enacted during this same era, including Operation Gatekeeper in California. Coinciding with NAFTA and the waves of immigration it induced, these measures multiplied the budgets for border patrol agents, created fences and walls, and installed military surveillance devices to enable the apprehension of would-be crossers. Entrants were recast as "illegal."[21] Then, following September 11, 2001, interior immigration policing practices were extended, justified in the name of antiterrorism.[22] All of this has made it increasingly dangerous and costly to cross the border. Hundreds, sometimes thousands have died each year attempting to cross via desert routes and many, many more have been apprehended and sent back home. As De León argues throughout his gripping account of border deaths in *Land of Open Graves* (2015), these deaths have been part of a deliberate policy of channeling border crossers to the most inaccessible and dangerous parts of the Sonoran Desert and not apprehending people until their journey was far along in order to "deter" future crossers. Knowing the real possibility of death lay ahead, few would risk the crossing if they did not profoundly need the work.[23] They have also had to pay increasingly higher fees to coyotes to assist in their crossing.[24] Shoring up the border has therefore also shored up worker compliance; few would complain of wage theft, or safety or pesticide violations, in the face of imminent deportation.[25]

Although border and immigration policy mostly served growers well, by constructing highly vulnerable workers, it increasingly became a liability as the border tightened and enforcement went further afield, especially for labor-intensive fruit and vegetable sectors. As Mitchell suggests, wages to attract domestic workers have remained unthinkable, because of an implicit pact growers have long held not to budge too much.[26] And so, some say that more recent claims of labor shortage, amplified by the news media, are politically motivated and part of a long history of growers claiming labor shortages in order to gain access to more tractable workers.[27] To be sure, many growers with whom I spoke hoped for reinstatement of a guest worker program similar to the Bracero Program so they might re-create conditions of abundant access to politically vulnerable, unorganized workers. Such a program is certainly not receiving much traction in the late 2010s, though, given the wave of anti-immigration hysteria fanned by Trump and his promises to build a border wall.

In any case, there is good evidence that a lack of immigrant workers was affecting the farm sector even before the Trump election. A report released in 2015, developed by the Partnership for a New American Economy, a group explicitly seeking immigration reform, claimed that the decline in workers was reducing fruit and vegetable production by 9.5 percent, or $3.1 billion, per year due to a tightened border. The report noted that California had been hit particularly hard, with a decline of about eighty-five thousand field and crop workers between 2002 and 2014.[28] A 2014 Pew Research Center report found that the population of unauthorized immigrants living in the United States from Mexico fell by a half a million people between 2009 and 2012, with California seeing the biggest drop, of ninety thousand fewer people.[29] Trump's expansion of the program of deportations begun during the Obama era, along with his anti-immigration rhetoric, appeared to only worsen the labor shortage. Although no new studies have been conducted as of this writing, many major news sources reported on nationwide farm labor shortages in 2017.[30] Wine grape growers, for example, who offer much better conditions, have raised wages $4 more per hour and have not attracted the domestic white workers that Trump's border games are supposed to entice.[31]

Adding to growers' grief, in 2016 California passed new minimum wage and overtime laws, with specific provisions for farm work. Although the laws mandate that the hourly minimum wage rise to $15 by 2023, the laws' overall impact on agricultural wages are not entirely clear. Piece rates often exceed the state's current legal requirement, yet only for the fastest workers. The overtime provisions may prove even more challenging for growers. Previously fieldworkers could work up to ten hours per day or sixty hours per week without overtime wages; as the program is phased in, the maximum hours per week without overtime pay will drop to forty.[32] Growers also complain about increases in workers' compensation insurance and other obligatory labor costs. With increased obligatory pay, growers have little room to maneuver for incentive-oriented pay like piece rates—that is, without eating into the profits they have come to enjoy.

THE WORKER AND THE BERRY

In the beginning, the strawberry industry was able to skirt around California's notorious problems with farm labor, already well established in other specialty crops. Early strawberry farms tended to be small truck farms that relied on their own family labor until sharecropping became the dominant way to manage labor before World War II. Conveniently, Anglo farm owners were able to take advantage of the surplus of landless Japanese farmers, made available by the Alien Land Acts that prohibited noncitizens from owning land. They recruited these farmers, many with large families, to be sharecroppers, allocating two or three acres for each family. While landowners provided the land and equipment and helped prepare the land, sharecropping families provided the labor, from planting to harvest. Together, owner and sharecropper shared in the marketing of the plant, and sharecroppers mainly acted independently.[33] With the war and Japanese internment, strawberry growers had little choice but to turn to braceros for farm work.

After the war and the repeal of the Alien Land Acts, those of Japanese origin, most by then American born, returned to farming. Many obtained land with the aid of Anglos they knew before, allowing them to farm independently rather than as sharecroppers. Without their labor, white strawberry growers continued to rely on braceros. But the termination of the Bracero Program in the early 1960s and the increase in labor protections that followed left growers without a sufficiently subservient and cheap labor force.

Coinciding with this situation, many growers were adopting fumigation. The enormous expansion of acres that fumigation precipitated therefore also greatened the need for reliable and disciplined workers. And so at that point growers reverted to sharecropping.[34] But in this round, sharecropping operated differently. Unlike in previous years, farm owners enforced arrangements through a written contract stipulating how production should be carried out, without negotiations with the so-called sharecropper. Farm owners would provide the land and planting materials, and conduct all field preparation and marketing. Generally only maintenance and harvest labor was performed by the cropper. In return for their "services," the croppers received a portion of proceeds from sale, generally 50 to 55 percent, out of which they were obligated to pay for mandated expenses. And unlike the sharecroppers of yore, most of these new sharecroppers were Mexican immigrants.[35]

This particular sharecropping system was eventually put to rest through a 1975 lawsuit against Driscoll's, by then the state's most prominent strawberry shipper and user of the contracting system. The court decided that the owner-growers exerted too much control over production practices, making sharecroppers more akin to wage laborers than independent farmers. It was fairly clear that the sole purpose of these sharecropping arrangements was to exempt growers from applicable labor laws. The case was settled out of court in 1981, but essentially growers could no longer continue a practice that disguised wage laborers as independent contractors.[36] Subsequently, sharecropping did not entirely disappear, but instead assumed a form less likely to face legal challenges. Namely, grower-shippers and other intermediaries began to forge contractual "partnerships" with former farmworkers and ranch managers, providing financing and market access in exchange for a portion of the proceeds.[37] I will discuss these further in the chapter that follows. For now, note that this newer variation didn't obviate the need to hire workers. To the contrary, labor demand increased even more with the adoption of high-yielding cultivars and various practices that increased productivity, such as the use of plastic tarps and high-density planting. The institutions of repair that promulgated these practices were in no position to assist growers with growing labor needs—indeed, their interventions in yield-enhancing technologies exacerbated the problem.

Luckily for growers, both push and pull immigration conditions—the Central American wars and Mexican debt on the push side and farm labor contracting on the pull side—broadly solved the need for more bodies in this time of intensive expansion. But crucial for the industry's continued success were also forms of labor control that would ensure loyalty for harvest while keeping costs as low as possible, as sharecropping once did. The importance of such controls, as Wells argues, lay with the enormous capital outlays of strawberry production, including for fumigation. When Wells conducted her research, strawberry growers expected to invest $18,000 to $25,000 per acre.[38] Making sure those berries were picked thus became crucial to recouping their investments. By the same token, with the substantial amount of labor that the abundance of fruit required, controlling the costs of labor became the fulcrum of profitability.[39]

But working bodies needed to be more than simply available and cheap. Growers wanted workers to attend to the particularities of the strawberry. Despite long-standing efforts in varietal development, strawberries were still subject to bruising and molding, and the exacting market standards of buyers for berries of a certain size and shape required significant culling, sometimes up to 30 percent of the harvest. Strawberry picking thus required knowledge about which berries to discard and great care in packing the good ones so that they would appear attractive in a grocery store display. The berries' high perishability was an additional reason growers needed to cultivate reliability among their workers.[40] It wouldn't do to have berries rot on the vine.

In a general sense, the political construction of farm labor markets to create both labor surplus and noncitizen status was the central mechanism to control costs and ensure loyalty.[41] Growers also capitalized on workers' social networks to ensure commitment and minimize insubordination. Core workers would recruit family members and others from their region of origin, even those living within California. In effect the workforce became "self-recruiting."[42] These familial ties ensured a level of intra-group loyalty, especially when having such ties was the only way to land a job. In effect, these networks tied workers not only to each other but to their employers. Many sending families developed long-term relationships with their employers, returning to the same workplace year after year. In turn, growers developed paternalistic relations with their workers, providing favors that would also shore up loyalty.[43]

Industry success also rested on bringing the biological characteristics of the berry and field conditions into close alignment with the workforce and labor market. Trends in varietal development were in that way both helpful and challenging. As we have seen, both public and private plant breeders bred for yield, color, extended shelf life, and other qualities important to growers and retailers. Some of the qualities for which they bred also would help moderate labor costs—even if those were not the primary aims. For example, berries that are visible on the plant, easy to remove, and large (this quality is presumably also of interest to consumers) allow workers to pick quickly, thereby reducing per unit labor costs. At the same time, plants bred to produce berries over long periods of time, such as day-neutral varieties, or produce berries at the beginning and end of the season, such as short-day varieties, could aid with labor recruitment and retention.[44] And yet there was little growers did about field conditions other than experiment with row spacing. Both the advantages of well-drained soils as well as the importance of building up planting beds from fumigated soil meant that strawberry plants had to stay fairly close to the ground, meaning in turn that workers would have to bend or squat to harvest them.

And so growers landed on piece rates as the primary way to address their varying needs for care and loyalty, while keeping labor costs as low as possible. Paying workers by the box incentivized productivity amid otherwise uncomfortable field conditions. Piece rates thus accomplished reducing wages on a per-unit basis while putatively satisfying workers because they could make more money in a short time than they might with an hourly wage. According to Wells, unauthorized workers, especially, were fond of piece rates so they could make a day's wage before immigration officers would come around.[45] Piece rates lent themselves to quality-control problems, however, especially since piece-rate workers were not paid for the berries they had to cull. Growers thus implemented systems of close supervision. *Mayordomos* (field supervisors) walked the rows, scrutinizing the quality and rate of harvest. Checkers not only kept tabs on boxes brought to the truck (as a basis of piece-rate wages), but also spot-checked baskets for bad fruit. Supervisors would fire people on the spot if the quality was low.[46]

Things changed after Miriam Wells published *Strawberry Fields*. When she was conducting research in the early 1980s (the books spans research that took place over twenty-five years, or close to it), the Central Coast strawberry industry was enjoying exceptionally stable profits despite wider instabilities in the broader economy.[47] There were few bankruptcies of established growers, and labor markets were flush with workers. None of these conditions remained when I conducted my research, as the industry faced not only labor shortages but also pathogen reemergence, fumigant regulation, drought, high land values, and weak prices. Moreover, per-acre costs had risen tremendously. A 2016 cost study showed typical outlays to be about $68,000 per acre in the Central Coast region, having risen nearly $20,000 per acre just since 2010, much related to increased costs of labor.[48] Ironically, the same set of policies that had once made for caring and loyal workers was now undermining those very qualities, especially loyalty—at least as growers saw it. Those were the border and immigration policies that made workers "illegal" and thus valuable to growers for their docility but also made it increasingly difficult for them to get to the strawberry fields.

SIGNS OF LABOR SHORTAGES IN
THE STRAWBERRY FIELDS

I did not expect to find growers more challenged by a labor shortage than the loss of methyl bromide. Nevertheless, virtually every grower we interviewed spoke of labor shortages, even when the interview was very brief. Many expressed that the shortage was the biggest threat to the future of the industry. "A lot, a lot, a lot, about 80 percent [of growers] have problems with the lack of staff. There are not enough field pickers to cultivate," said one grower. Another estimated a 15 percent shortage in staff the previous year and said it was becoming worse. Employees and principals of the major strawberry buyers echoed these concerns, as did representatives of the California Strawberry Commission. Corroborating this verbal evidence were the many billboards advertising berry jobs that we passed when driving from one interview to another.

Not all growers had been personally affected by this shortage, although most had seen at least some impact. And quite a few had struggled mightily. Numerous growers spoke of having abandoned already-planted fields, sometimes for multiple years, because they did not have the workers to harvest them. Those who had not abandoned fields themselves knew someone who had. The acres they had abandoned were substantial, with growers reporting they had not harvested twenty, fifty, or even ninety acres in a given year. Other growers had not abandoned fields as much as ceased harvesting them for a period. In choosing what plots to pick in these situations, some opted to "sacrifice a little bit of the field to keep the rest of the field fresh," whereas others considered timing, returning to a field when prices were higher late in the season. Even though these growers might have gotten something back, they did not come close to recouping the substantial investments in planting and field preparation.

Experience of a labor shortage—or the prospect of higher labor costs— had driven some growers out of business. My review of county pesticide use reports, from which I obtained grower information, revealed significant turnover between 2010 and 2013, especially in Ventura County, which has seen significant declines in acreage.[49] Although defunct businesses were more difficult to contact, a few I reached by phone were quick to say that the labor shortage had done them in—as well as incoming laws requiring much more obligatory compensation. Growers who had not "lost their shirts" had reduced the size of their businesses or opted not to expand, even when other conditions were favorable, including access to land. "You really can't go ahead and make these huge investments if you don't have some sort of a guarantee that you can harvest your crop," said one.

In explaining the origins of the shortage, many growers noted their interest in a guest worker program, their way of acknowledging that the labor shortage was rooted in border and immigration policy. A few attributed it to the improved Mexican economy, including the expansion of strawberry production in Mexico, so that fewer workers were migrating to California. Some growers averred that government assistance had worsened the labor shortage, as if legal residents would otherwise be doing the work. Importantly, growers also spoke of the burgeoning blackberry and raspberry industries on the coast, with which they were directly competing for labor—unless they themselves had diversified into those crops. These "cane berries" are more attractive to workers because they can pick without bending over. Many strawberry harvesters had moved to raspberries when the opportunity arose. A ranch manager for a large company spoke about this at length:

> There are four berries: there's blueberry, blackberry, raspberry, and strawberry. The first three that I named, they're harvested standing up. Strawberries are harvested slunched over all day long. So we have seen that out of the four, everybody gravitates to the three. And then now, as of the last three years, with the borders really, really tightening up, we have seen about a 35 or 40 percent labor drop. It's affecting the strawberry more than any other commodity, veggie, berries, because nobody wants to pick them.... It's like the bottom of the job pool, because nobody wants to do it. Because you're working in mud, trenches, you're in the trenches.... And I feel for those people. I mean, they're all day long in there. You can make the most money on strawberries, but ... you have to put in [a lot of effort] to get [the money] out.

At the "bottom of the job" pool, the strawberry industry was necessarily more acutely affected by the shortage.

In clear distinction from the scenario that Wells painted, many growers complained of worker disloyalty. They told of workers who stayed for only a few weeks during peak production times, or who walked off the job and didn't finish out the week. They often contrasted this from how it used to be:

> One of the big issues is that even when we get done harvesting in time, they just say, "I don't want to do it," and then leave. But there's nothing you can do. Like Mondays: so many people don't show up. Five years ago they would have gotten written up, second time it's three days, third time you're fired. Now we just beg them to come on Tuesday.

This grower also discussed how the quality of the pack had gone down because he could no longer enforce strictness. Doing so, he posited, would induce crew members to walk off the job and go where quality standards were more lax. Another grower concurred: "Even workers that have been with you for a while, they'll say, 'Hey, I'm being offered a job over here at this much.' ... *You're almost forced to pay them whatever they want now*" (my emphasis).

Most striking were the stories that many growers told of potential workers who came to their ranches unannounced and drove or walked the perimeters to inspect the fields. According to these growers, workers were examining the health of their plants, the state of the rows, the evenness of the fields, the size of the berries, or the lack of culls, all conditions that would affect how many berries they could pick. Unimpressed workers would choose to work elsewhere, they said. Growers also talked of workers who used their cell phones to communicate with their friends and relatives about which ranches offered higher pay or grew bigger berries or had easier-to-navigate fields. Learning of a better situation elsewhere, some workers would apparently walk off the job that very day in what were previously unheard-of acts of insouciance.

It is not as if such "weapons of the weak" were not deployed during the time that Wells did her research. She wrote of workers packing hollow-centered baskets or including damaged fruit in their boxes. But she portrayed these as "petty acts of sabotage" in response to personal affronts from their superiors.[50] The behaviors I describe here appeared more widespread, and not particularly hostile or even personal. To the contrary, they seemed to simply reflect a change in workers' possibilities. As put by one grower, "There is nobody to replace the laborers who are working now. The laborers now don't work as hard because they know there is a labor shortage and they have the upper hand."

Growers in the plant propagation business also complained of labor shortages, but in less specific terms.[51] This is in large part because of the isolation of the nurseries; few farmworkers even travel to those regions to see what is on offer, and so the behaviors witnessed in the fruit growing regions would

FIGURE 17. Plant propagation workers picking blooms off new plants, Shastina. Photo by author.

be very unusual. But this isolation made these shortages both more severe and, hence, more potentially damaging. The propagation business is more labor intensive that one might imagine. Nurseries need fieldworkers to plant starts and tend to quickly running plants. Once plants are pulled, they need workers to inspect, cut, count, and place the shippable plants in boxes. If nurseries don't have enough workers, they can't deliver the plants on time to fruit growers. Without beginning fruit production on time, growers miss out on windows for the best possible production. Therefore, a late delivery might cause a grower to look elsewhere for plants in the following years. At the same time, nursery work is also very harsh, as it involves harvesting in below-freezing conditions, or nipping buds while physically contorted in strange positions. Perhaps that is why nurseries especially tend to rely on the more formal means of labor recruitment and retention discussed in what follows.

RESPONSES TO LABOR SHORTAGE

In the face of labor shortages, growers have attempted a range of approaches. One has been to revert to forms of recruitment that historically shored up the labor supply, such as the use of labor market intermediaries. Some growers reported returning to the use of farm labor contractors, a form of recruitment that may have fallen out of favor because some of its advantages eroded (for instance, growers must share responsibility for labor law compliance).[52] Although they claimed it was more expensive and that they did not get the best pickers, growers who had taken this option said it was far less expensive than losing a field. With no itinerant workers, nursery growers have had little choice but to use labor contractors. Yet, as we learned from talking to growers who doubled as farm labor contractors, they too have been experiencing a labor shortage owing to the tight border, not surprisingly since they tend to work with the newest migrants. Growers have additionally experimented with a sort of labor exchange, called Berry Central, but that wasn't very effective either. As a last resort, growers have been turning to H2A, a temporary guest worker program. Created after the end of the bracero era, for many years it was largely underutilized because the program requires that workers be in the United States only on short-term contracts and that growers provide housing and transportation for them, making them landlords as well as growers. But by the mid-2010s, H2A usage was on the rise, despite the expense and increased regulation.[53] Several growers with whom I spoke had begun using H2A workers, even though they were finding these workers less skilled. Locating housing for them was difficult as well, especially because residents of the middle-class suburbs that increasingly abut strawberry fields are averse to having farmworker housing nearby. Growers who rely on H2A were sometimes housing workers fairly far from the strawberry fields, perhaps an hour's bus ride away. Meanwhile, nursery growers were even renting motel rooms for workers.

The labor shortage thus has also given new impetus to move into robotics. Several companies, including Driscoll's, are collaborating with engineers to design a plant sorting machine to be used in the nurseries. Driscoll's has also been leading the charge with a harvesting machine. Although the mechanism is being kept under wraps, I understand that it will only be usable with waist-high tabletop systems where berries are planted in soilless substrate and hung off the edge.[54] Such planting systems are at the same time being promoted to address the labor shortage in a completely opposite way: rather than replacing workers with robots, the improved ergonomic conditions enabled by waist-high trays, matching those of other berry work, could attract more workers. But again, despite the advantages of planting in substrate, including the avoidance of soil disease, the California industry as a whole had thus far been highly reticent to adopt this approach for reasons I will discuss in chapter 8.

For the most part, then, grower efforts to address the shortage have been both inversions and extensions of how they managed labor under conditions of surplus. Rather than using familial networks to recruit workers and ensure loyalty, growers were employing more proactive strategies. To aid in recruitment, one mentioned beginning his season early, and another discussed diversifying his crops to provide employment during the off-season. From one grower we learned of cash incentives for recruiting others. This grower spoke of an industry practice of offering dollar-per-head bounties for bringing in someone from another crew. Such strategies seem a far cry from the "self-recruiting" labor force about which Wells wrote. Rather than rely on discipline to force weak workers out, growers reported on various incentives they used to retain workers throughout the entire season. Some adjusted the balance of piece rates and hourly wages over the course of the harvest period, paying more in hourly wages and less in piece rates when plants were thin of berries at the beginning and end. Some offered end-of-season per-box bonuses to encourage workers to stick around. One grower said he kept more people on during the slow season just to shore up loyalty. And some even paid for dangerous border crossings.[55]

To attract workers, some growers have even conceded the need to tinker with wage rates, adding to the overall costs of labor that had been stipulated by law. It appeared that the chief way of increasing pay was through increasing piece rates, generally calculated by the box. From interviews we learned that the basic expectation is that workers should pick no less than three boxes an hour, average workers will pick about six boxes an hour, while very fast workers may pick upward of nine boxes an hour. Increasing per-box payments by ten or fifteen cents, as several claimed to do, would therefore attract the fastest workers to their ranches, who could see increased wages of more than a dollar per hour. A grower who claimed to have had a loyal crew of workers since he began had implemented regular pay increases. Still, in keeping with the long-standing implicit pact to keep wages low, some growers expressed great disdain for these raises, as well as the aforementioned recruitment practices, suggesting they were effectively beggar-thy-neighbor strategies that did not resolve the basic problem. One further argued that such practices could be "counterproductive," as they encouraged an "in your face" disloyalty that could result in substantive wage gains.

Finally, in keeping with abiding attempts to bring berry characteristics and field conditions into alignment with the workforce and labor market, perhaps the most standard (or at least uncontroversial) approach to addressing the labor shortage has been to attend further to planting decisions and row conditions. But whereas in earlier times such attention was intended to save on labor costs, now the objective was cast as meeting the needs of laborers—responding directly to workers' attention to fields, plant health, and berry size. For growers it nevertheless presented a more tenable strategy than major overhauls in remuneration. Many growers thus spoke of selecting varieties that are high yielding and easy to pick. For example, plants that stand up and are tall rather than bushy can more easily be seen by the harvest crew and perhaps require less bending. Growers also discussed plant and row spacing, noting that decreased plant density enhances ease of picking. They mentioned the need for clean rows that would allow workers to move quickly when depositing filled boxes at the counting truck. Some installed "mobile picking aids" (such as trolley systems or conveyer belts) so that workers no longer had to carry their boxes. And in the interest of "making it better for the worker," some, again, even defended fumigation as a way of addressing the labor shortage. By controlling pathogens, that is, fumigation would contribute to the robust, high-yielding plants that workers putatively want—or, as it really is, need.[56] And that, I think, is the rub, as it suggests that even workers' wages have been tied to the presumption of fumigation, further entrenching the assemblage.

BODIES VERSUS LABOR

The very wording of a labor shortage is revealing. It is an abstraction connoting an insufficient quantity of a factor of production. Yet labor power comes from living, energetic, material bodies. They are themselves socioecological elements of an assemblage (as are owners, pesticide applicators, and buyers, for that matter). Indeed it is the bodies, not the abstractions, that growers need to do the work of strawberry cultivation and harvest.

The problem for growers and other capitalist employers is that these bodies must be replenished and reproduced on a daily, seasonal, and lifetime basis, across generations, requiring nourishment, housing, sleep, health care, relative freedom from harm, and much else in order to make future production, hence future exploitation and profits, possible. At the same time, if the fundamental tenet of classical political economics holds, profit stems from the extraction of labor power from human beings (and other working animals) at less value than what the workers produce. Therefore, profit also rests on the rift between the needs of capital for labor and the needs of workers for life, notwithstanding that such "needs" are everywhere and always socially and culturally constructed—as well as politically negotiated.[57] Still, when too much labor power is extracted, or when bodies suffer too much harm in the course of their work, they become literally exhausted. Karl Marx himself recognized that maximizing labor power would shorten its life, "in the same way that a farmer robs soil of fertility."[58] Ostensibly that would pose some limit to exploitation.[59]

These days, however, laboring bodies are used up with stunning regularity. So-called disposable workers are those whose labor power is highly valuable, yet whose lives are otherwise of little value for capitalism.[60] These "necropolitical subjects," as geographers Michael McIntrye and Heidi J. Nast call them, are generally racially marked subjects whose future is not protected because of the existence of surplus populations—those willing to step in and work when others are used up—as well as legacies of colonial racialization that have made such bodies ideologically less important or even threatening. As such, "Little heed is paid to the needs of labor to reproduce itself. Instead, the overriding concern is to subject workers to the most accelerated form of exploitation possible to optimize profit."[61] These laborers are even more valuable to capitalism precisely because they have been constructed as disposable and thus readily left behind when they become sick or less productive.[62]

California farmworkers are classic disposable workers, and Sarah Horton's research on melon workers in California's hot Central Valley illustrates this phenomenon well. She notes that male workers, especially, feel compelled to be exceptional to keep their jobs. They do not take breaks or speak up when feeling ill, and they routinely work to the point of exhaustion. They don't drink sufficient amounts of water, especially in heat, because they worry about the repercussions of taking breaks, and they are prohibited from bringing their own drinking water onto the fields for food safety reasons. As a result, chronic kidney disease is highly prevalent among these workers, and many die early deaths from it.[63] Berry workers may not be subject to heat exhaustion to the extent that melon workers are, working in conditions generally twenty or thirty degrees cooler, but they do experience chronic neck and back pain from bending over all day, for years on end.[64] And nearly all farmworkers are subject to chronic stress from fear of immigration enforcement and much else. Although the bodily impacts of stress are rarely mentioned as a health condition of farmworkers, Horton notes that that the continual release of adrenaline and cortisol, along with rapid heartbeat, in stressed bodies weakens organs, hardens arteries, and raises blood pressure, also contributing to early death.[65] This says nothing about exposures to pesticides. As discussed in a previous chapter, pesticide drift is a regular occurrence, and many workers complain of difficulty breathing, itching and burning, tearing, nausea, headaches, and more from acute exposure.[66] And not all exposure manifests as acute illness. The disproportionate adverse effects from chronic exposures on farmworkers is a story much less told, but no less pernicious. Least understood and recognized are the intergenerational effects of pesticides mentioned in that same chapter.

All of these health risks are compounded by the social and legal means that sustain farmworkers' vulnerability to illness. Horton discusses how workers are ineligible for income support until they are completely disabled, so they continue to work until they are very sick.[67] Seth Holmes writes not only of the inaccessibility of the health care system for migrant workers but also about the disregard farmworkers face when they do see doctors, with ailments dismissed as hangovers and common colds.[68] And Jill Harrison has written about the fear of complaining about pesticide drift or even violations of pesticide protocols for fear of firing and deportation.[69] All suggest that these social and biological insults are both concatenating and inseparable. And yet, as environmental historian Linda Nash has written, while histories of exposure, inadequate health care, and other forms of neglect have combined with external factors to exacerbate embodied vulnerabilities for farmworkers, the cumulative and compound nature of these concerns also makes it very difficult to prove illness with current modes of epidemiology.[70]

With all that said, disposability is theorized to stem from conditions of labor surplus. Again, it only makes sense to use up and exhaust laboring bodies when there are vast armies of the unemployed (as Marx would put it), many residing in the Global South, ready to take the place of damaged workers, or even those who complain.[71] But that was ostensibly not the case in California in the late 2010s. Indeed, the irony at hand is that the tightened border that had so deeply benefited the industry was creating a condition that would seem to make those who have successfully crossed, as well as their children, all the more essential.

Apparently it had not. In conducting my research during the time that growers were crying shortage, I saw strawberry workers picking rapidly in the same "slunched over" positions and then running through fields to have their boxes checked. My team interviewed workers who were well aware of the risks associated with pesticides but felt they had no choice but to work. We talked to women who worked while pregnant even though they had experienced miscarriage from what they believed to have been pesticide exposure, or knew someone who had. I saw workers trimming blossoms in impossibly uncomfortable positions while suspended from a rig that combed the plant propagation fields (see fig. 17 earlier in this chapter). I watched videos of workers bundled up in the below-freezing trimming sheds of the high-elevation nurseries. And I am not alone in having noticed the change in affect among farmworkers since Trump was elected. Their faces showed more fear than earnestness, and their stress levels can only be imagined. If this is insubordination, I'm not sure what compliance looks like.

In that light, I think it is worth returning for a moment to the primary mechanisms growers have been using to address labor shortage—mainly increased piece rates and changes in field conditions. For workers, robust, fruitful plants, with visible berries, along with clean rows and perhaps trolleys, no doubt make it easier to increase hourly yields. And higher piece rates do represent some increase in overall pay. If workers are prone to walk off the job to find greener pastures, with bigger berries or higher piece rates, it may be less an act of insubordination than one of great rationality in an industry where there are few opportunities for upward mobility (as study after study has shown), and the only way to make more money is to work faster or receive higher piece rates.

Yet it cannot be denied that even with these modest gains in pay precipitated by labor shortages, to earn that extra money, piece-rate workers must be complicit in their own exhaustion. Piece rates, that is, continue to encourage workers to exert their bodies to the fullest extent and even engage in practices that further put themselves at risk. Reports of workers eschewing the use of pesticide-protective gloves, since gloves impede the efficiency of harvesting, or eating in the fields without washing their hands to maximize their time, are indicators of this tendency.[72]

For growers, piece rates are probably the best option to keep labor costs controllable in a time of labor shortage, at least in the short term. By themselves, piecework wages do nothing to increase the overall wage of the worker; they improve productivity by increasing the hourly wage, while shortening the time that workers pick. Even when increases in piece rates effectively allow workers to exceed minimum wage, they may earn less overall because of shorter days. This, with seasonal hiatuses, means that most workers receive an annual wage significantly less than an annualized minimum wage.[73] By increasing piece-rate pay, growers are also assured that workers will move quickly and no money will be lost to downtime.

If labor shortages are here to stay, it makes little sense to use up workers' bodies, any more than it makes sense to increase the pathogen load of soil; they are both bad strategies for industry longevity, albeit highly predictable strategies when the political-economic dynamics of land and labor markets come into tension with the materiality of soil and bodies. But whereas evolving agronomic science is beginning to recognize the mistakes of disregarding soil health for the future of agriculture, the same cannot be said of workers' health. And it is unlikely to anytime soon given the various economic squeezes growers face.

STRAWBERRY SQUEEZE

In many respects the labor issues facing the strawberry industry today are nothing new under the sun. California agriculture was founded on cheap, available, compliant labor. That's what made labor-intensive, high-value fruit and vegetable production possible—and consumption of these crops affordable. When these conditions were threatened, growers fought back to restore the conditions, whether through bringing in more compliant workers or using political machinery to suppress labor organization. Growers were rarely acting as a matter of unbridled greed, though. They constantly faced a cost-price squeeze between what buyers were willing to pay and their own costs. The competitive nature of farming, especially, meant that few growers ever developed effective monopolies to protect themselves from such squeezes. And so, as Don Mitchell has put it, "The squeeze growers constantly feel in their struggles to grow a profit (and keep at least some of it out of the hands of the financiers) is managed, sometimes, by jiggering the labor process, and sometimes therefore transferring the effects of the squeeze—the pain—onto the shoulders (or perhaps the aching lower back) of workers."[74]

What makes today's situation with the strawberry industry remarkable, if not exceptional, is the degree to which this squeeze has intensified owing to the evolution of the strawberry assemblage. The imperative to control pathogens through fumigation has made for unusually steep up-front investments in land preparation. The benefits of sandy soil and temperate climates along the coast has put strawberry growers in competition for land with urban uses, at the same time that a production system grounded in fumigation has given landowners the expectation of high rents from strawberry production. And the short stature of the strawberry plant and low planting beds, along with the fragility of the berries, has made for highly arduous working conditions, so that in the political conditions that thwart border crossings, potential workers are likely to turn elsewhere.

A further curiosity is that the many problems that strawberry growers face—pathogen outbreaks and lost production, increased costs of cultivars due to the proprietary behavior of the strawberry breeding apparatus, loss of the most effective fumigation chemicals, unrelenting increases in land values, and rising labor costs—have not raised the price of berries. According to US agricultural census data, prices declined precipitously from a peak in 2014, just as the perfect storm was taking shape.[75] Are strawberry growers just incessant whiners (as some do say), or is there something more going on? As I discuss in the following chapter, many of these problems have been precipitated by efforts at repair. And yet since many growers can no longer survive these conditions, it would suggest that the repairs have become harmful, and the industry itself is becoming pathological.

SEVEN

Precarious Repairs and Growing Pathologies

I think there will be an industry shakeout and increasing industry consolidation. If regulations continue to tighten or certain critical fumigants are phased out, such as chloropicrin, small independent growers—say, growing ten to fifteen acres—will become less common, and more land will be farmed by ranch managers associated with large companies like Driscoll's. The industry might wind up just becoming more corporate owned—large companies that can afford it and foot the bill to do what you have to do to grow strawberries. That's where it may be headed.

Interview with grower, 2015

For many, many years, the strawberry industry was like a share-crop deal, sort of. Small growers, big shippers, right? These shippers taking thousands of dollars an acre off in profit before the income would ever hit the farms. . . . There's not enough room to do it anymore. So what's gonna have to happen is . . . the farmer needs to be the shipper. That's a weakness with the Driscoll's program, too, because Driscoll's is not a farmer. . . . Driscoll's is a shipper, and they want a big commission, and their quality standards are very difficult. So how are you gonna survive as a farmer if they're gonna grade you really hard and then take, I don't know what they're taking, that 18 percent commission? We have one farm, we charge 5 percent, at most 8 percent. The money has to go back to the farm. . . . So what I'm saying is . . . if it's Cal Giant, or whoever else, Cal Giant is gonna need to own that farm. Not just loan them the money and then take all this stuff off the top. . . . It's changed a lot in five years. . . . Cal Giant's a classic example because seven or eight years ago, they had no money in the ground . . . no investment in the farm. . . . Now I think they own 75 percent of their farms. They have to.

Interview with grower, 2015

I would bet . . . five years from now, you're gonna see problems in all crops due to no fumigants, and you're gonna see good markets and you're gonna see low yields, and so I don't know how the sharecroppers are paid, but if they're just a dollar per box or something like that, there's no trigger that gets them more money, they're gonna lose . . . and your shippers are gonna be the ones that make obscene profit off that.

Interview with grower, 2015

From an individual grower's standpoint, yes, it's tough. Whatever happens—and I don't want this to happen—the industry will manage to muddle along. Production will go down, acreage will go down, individual growers will fail. That's the tragedy that my colleagues and I are trying to avoid. But we'll continue to produce strawberries in this area. There will be major changes.

Interview with extension scientist, 2014

THE SENTIMENT WAS THE SAME everywhere we went. An industry shakeout was coming (or already happening), and those who would lose were low-resource growers, farming on marginal land, and highly dependent on fumigants, while those growers who would stay had access to land less prone to disease, capital to invest in innovations, and a willingness to work with complexity (more on this in the next chapter). The big winners, however, would be the shippers—especially those who took little to no production risks and essentially made their money farming farmers—as long as they didn't squeeze their farmers dry. Shakeout scenarios, of course, are not unique to the strawberry industry. They are dynamics predicted by agrarian political economy. Generally, though, these theories assume that those who don't adopt agrochemicals will lose; the strawberry industry faces a situation where intensified fumigant regulation, as it intersects with other elements of the assemblage, will favor those who can figure out how to grow strawberries for the mass market, *without* chemical fumigants, while addressing the other constraints around land, labor, and drought.[1]

This chapter discusses how certain industry players have become well positioned for the future while others have become vulnerable to loss. I will show, for example, how Driscoll's and other industry actors have been able to mediate access to crucial elements of the assemblage in ways that bear on the prospects of various kinds of growers going forward. I will give special emphasis to Driscoll's, with its eye for competitive advantage and its tendency to seize on opportunities that have led to industry dominance.[2] In particular, the company's development of capacities at both ends of the supply chain has indeed allowed it to farm farmers.[3] By leaving the risky business of farm production, moreover, it has been able to seize upon sustainability and social justice concerns and make them its own. As part of this discussion I will also attend to the changing demographic of farmers. Here I primarily wish to highlight that Latinx (non-gendered plural of "Latinos") have become major players at a time that the industry faces its most significant crisis.

Characterizing today's industry requires further attention to aspects of the strawberry assemblage that have yet to be discussed. The chapter thus begins with a brief synopsis of the institutional development of the industry. The objectives of presenting this largely human history are twofold. One is to show in a general way how many of these institutions arose to address both human and nonhuman challenges to strawberry production and, hence, to provide repair. This history will thus illustrate that the industry has always been fragile, dependent on fixes to keep it alive. The other is to bring special attention to how these institutions of repair, and university science in particular, were unable to prevent shakeouts among growers and others. To the contrary, many of the innovations promulgated by the university seemingly worsened things by creating too much supply. Here a reminder is in order that crop science is better at addressing underproduction than overproduction.[4] As it happens, other organizations emerged to address overproduction, primarily through attention to marketing. Their eventual success in creating a mass market for strawberries, however remarkable, encouraged yet more production. Together, then, these fixes produced high expectations that strawberries could continue to be widely available at cheap prices.

A HISTORY OF FRAGILITY

The institutional development of the California strawberry industry can be told as three phases, each roughly marked by differing engines of growth in which new growers, businesses, and supporting institutions came on the scene to address what in many ways were recurring (and typical) industry challenges in production, finance, and marketing. They were recurring because repairs in one arena often exacerbated problems in another.

Experimental Beginnings

In the very beginning, strawberry growing in California was largely inconsequential, albeit promising, as innovative growers, experimenting with their land and crop types, recognized the potential profit of strawberry production.[5] Generally, early commercial production took place in close proximity to markets. In the 1850s, growers from the erstwhile rural Bay Area counties of Alameda and Santa Clara brought their produce to rapidly urbanizing San Francisco. In the peak of the 1868 season, strawberry growers in the Santa Clara Valley were shipping strawberries to San Francisco, as well—up to eight tons a day. Los Angeles was another budding commercial agricultural region, and even Yosemite Valley, occupied by settlers in the 1850s, had its strawberry patches, as did other areas in the state that were never to become prime growing regions.[6]

Still, as a loosely coordinated group of growers, the industry took root in the 1860s in the Pajaro Valley, straddling Santa Cruz and Monterey Counties. It was there that conditions aligned to make a serious go of commercial production. Not only did the valley have the soils and climate so conducive to strawberry farming; Watsonville WaterWorks' installation of flumes from the Corralitos reservoir provided growers a steady supply of water for irrigation.[7] Yet despite the productivity of these early strawberry patches, selling the berries almost immediately proved difficult. It wasn't until the late-1870s extension of the railroad into the Pajaro Valley that these growers could access San Francisco markets, allowing commercial production to take hold.[8] Forty-two acres were planted in 1881. In 1883, John T. Porter, a major landholder and grain grower, planted fifty more acres of strawberries on his ranch, bringing the total acres planted in the area to 118. Acreage continued to increase, from 185 acres in 1884 to 840 acres in 1902.[9] In 1886 alone, approximately 150,000 pounds of strawberries were shipped to San Francisco daily, "a considerable portion of which came from Pajaro Valley."[10] So impressive was this nascent industry that in August 1902, the *San Francisco Chronicle* wrote that "although apples lead, and although there has been a great planting in this fruit during the past ten years, berries have, all things considered, the prominent place as a profitable crop. The yield of strawberries is enormous."[11]

Two of these early entrants were Joseph "Ed" Reiter and Richard "Dick" Driscoll, friends and brothers-in-law, who began growing strawberries in the Pajaro Valley in 1896. Reiter's sister's discovery of the delicious Sweet Briar variety while vacationing in Northern California created an early advantage.

In 1904, they formed a company and began production of what they renamed the Banner variety, effectively gaining a monopoly position on a berry far superior to others cultivated at the time.[12] In 1911 they moved to a 130-acre ranch in Alameda, and planted thirty to fifty new acres per year "until the entire ranch was used."[13] It was when other growers began to copy the Banner that the company pursued the legislation that culminated in the Plant Patent Law of 1930.[14]

As businessmen and -women, few of these primarily white farmers actually worked in the fields—typical of California growers.[15] Instead, they depended on Japanese migrants to work under contract or as sharecroppers.[16] Yet, as the Japanese discovered their "unusual skill" in working with strawberries, many wanted to retain more of the profits of production and become independent growers themselves.[17] Requiring financing, they would borrow money from commission houses in the San Francisco Produce Market or turn to local merchants. To gain access to land, they formed associations with other family members and friends to lease the ground together. These cooperatives would then take on other sharecropping families, with each family growing two to three acres.[18] Their success as growers, competing with white farmers, was a major impetus for the Alien Land Laws prohibiting *foreign-born* Japanese from leasing or owning land. And yet there were workarounds, including forming partnerships with whites or passing ownership to one's native-born children.

White growers had concerns about credit as well, since the commission houses absorbed much of their profits. In addition, with such rapid entry, all growers began to experience problems of overproduction and erratic markets. Like growers of other California crops, they saw more orderly marketing as the solution, including standardization in packing to eliminate competition based on a lower-quality product. Led by Richard Driscoll, growers of both European and Japanese descent joined to form the Central California Berry Grower Association (CCBGA) in 1917—one of the first institutions of repair. The goals of the organization were "to promote, foster, and improve growing, packing, and marketing of berries grown by members." One of the first orders of business was to back a 1917 standardization law that required "strawberries be packed in pint baskets of 12 ounce capacity and that the top layer of fruit be a fair sample of the contents of the entire basket."[19] At the suggestion of members of the Japanese-run California Central Farmers Association, the new association's constitution mandated the board of directors to be made up of equal numbers of Caucasians and Japanese. The first board of directors thus included a mix of prominent growers at the time, including Driscoll and Reiter. Run as a marketing cooperative in which growers would pool their berries and sell them at negotiated prices, the association gained membership of 248 growers by 1920, representing 95 percent of the berry growers on the Central Coast. It then adopted Naturipe as its official trademark in 1922. Much later, in 1958, the association changed its name to Naturipe Berry Growers, which for a long time remained one of the largest berry cooperatives in the world.[20]

Given the character of Naturipe, it may not be surprising that the first large strawberry ranch was a partnership of those of European and Japanese origin. In 1920, Unosuke Shikuma, Heitsuchi Yamamoto, O. O. Eaton, and Henry A. Hyde purchased the two-hundred-acre Oak Grove Ranch in Monterey.[21] For years it was the largest and "most productive strawberry field in the world," with an "output of over three million baskets per year." Success was largely attributable to the owners' innovative production techniques, as well as innovations in shipping, allowing the berries to be transported farther afield. Rather than shipping strawberries in large chests on the railroad to cities, they were transported by "motorized truck with a cooling device, in small wooden trays holding twelve-pint baskets."[22] After witnessing berries that had arrived in Chicago too soft, Joe Reiter came up with the idea of pre-cooling the berries (rapidly reducing the heat of the fruit post-harvest). This same company then developed a pre-cooling device for shipping.[23]

As production expanded to meet the growing market, some problems ensued. Growers used planting stock consisting of runners stripped from old fields and generally lacked knowledge of how to cultivate strawberries in arid climates. When in the 1920s the various diseases, including *Verticillium*, had begun to afflict strawberries, it was CCBGA director Ernst Haack who called on scientists at the University of California to help out. When UC did step in, much of the research was first devoted to identifying the problems, and their early advice, such as moving to new fields, wasn't all that wise for the long run. Blight worsened in the 1930s, causing yields to drop, yet prices, which generally rise when yields are low, crashed as well, no doubt reflecting the economic hardships of the time.[24]

Despite its promising start, strawberries thus remained a minor industry until after World War II.[25] The challenges of pathogens and other pests had yet to be well met, while perishability and seasonality, problems endemic to fresh fruit and vegetable production, limited the fruit's markets and positioned the strawberry as a luxury good.[26] Just before the war began there were five thousand acres in strawberry production in California, dropping to nine hundred acres immediately following the war.[27] No doubt agricultural resources needed to be directed to more important staple crops during the conflict, but the larger insult to the industry was the Japanese internment, which forcibly (and horrifically) relocated even citizens of Japanese origin into camps, removing a key source of expertise and labor. Indeed, postwar Japanese internment "nearly caused the demise of the strawberry industry" and no obvious winners emerged.[28]

Freezing Out Glut

The reentry of Japanese Americans (now many US born) reinvigorated a nearly dying industry. They were able to lease land, and in some cases regain ownership of land to become independent farmers. In addition, innovations in the areas of production and marketing, abiding if opposing problems, enabled the industry to take off. These innovations, in turn, provoked new proprietary behaviors, particularly on the part of Driscoll's.

By the end of the war UC investments in plant breeding were finally coming to fruition. The five new varieties released in 1945 were "far more vigorous and productive" than the Banner variety that had first changed the industry.[29] So, when Ned Driscoll and his brother Donald, along with Joe Reiter— all sons of the founders—broke away from the UC program to found the Strawberry Institute of California, dedicated to the development of new and better varieties for themselves and their associates, they effectively went into competition with UC and those whom UC served.[30] Yet they likely wouldn't have been able to compete, much less excel, without the appropriation of knowledge and likely germplasm from UC. According to economist, former Naturipe director, and industry historian Herbert Baum, the transfer of knowledge from former UC employees "was in all probability the cause of [the Driscoll's] competitive advantage and continued industry dominance."[31] To be sure, the institute was first to develop short-day varieties out of everbearing forebears, allowing Driscoll's growers to launch production in Southern California in advance of others.[32]

Meanwhile, the more productive and reliable plants that UC breeders had developed increased the industry's susceptibility to price crashes. When berries were plentiful on the vine during high season, prices would drop. At this point, repair came from the private sector, as innovations in freezing and the popularization of frozen fruits and vegetables—mainstays of 1950s diets—

came to save the day.[33] Growers could now contend with midseason gluts by sending excess fruit to the freezers, to be sold outside of seasonal windows. At the same time, the frozen food industry expanded the market for berries that could now be shipped to distant markets. By the 1950s more than 55 percent of the US strawberry crop went to processors, with California alone harvesting and processing half that amount.[34] Predictably, the frozen food market encouraged farmers to expand production. By 1956 there were 22,500 acres of strawberry production in California.[35] Much of that expansion occurred in the Salinas Valley, which became the largest strawberry-producing region in the world.[36]

With the increase in acreage, growers needed more plants. So it was during this period that plant growers separated from fruit growers—producing "plugs" that could be sold back to farmers.[37] First was Lassen Canyon Nursery, established in the 1950s to become one of the largest, capturing something like 35 percent of the plant business as a major cooperator with UC.[38] Other early entrants included Sierra Cascade and NorCal Nursery, the latter owned by the Sakuma Brothers of Washington state.[39]

The 1950s also saw a shift in the structure of fresh marketing, which for more than three decades had been operating cooperatively under one organization: Naturipe. Here again the Driscoll and Reiter group broke from the pack, taking many of the white growers with them, to establish Driscoll Strawberry Associates, incorporated as a for-profit company in 1953.[40] It was also in this decade that some Driscoll's growers began operating in the more southern Santa Maria region, taking advantage of the early-season varieties that had been developed at the Strawberry Institute.[41] A decade later the non-profit Strawberry Institute merged with Driscoll Strawberry Associates, effectively making Driscoll's the first berry company to work at both ends of the supply chain (breeding and marketing). At that point, Driscoll's left production farming altogether, seeing more money to be made by farming growers.[42] That left Naturipe as a collection of almost entirely Japanese-origin growers. Taking it as a sign that the organization needed to increase its market clout, Naturipe began its own expansion to other regions in the state. Rejecting such expansion, this move precipitated a group of growers calling themselves "traditionalists" to organize and create Watsonville Berry Cooperative in 1957. With the brand name Berry Bowl, this cooperative was constituted by small and medium Japanese growers in the immediate area of Pajaro Valley.[43]

Notwithstanding this strategic move on the part of Driscoll's, spinning off new marketing organizations was not going to solve problems that continued to plague the industry as a whole, boom or not. Growers continued to see poor yields from diseases and pests, while at the same time complained of poor prices related to excess supply. And so the industry, imagining it could kill two birds with one stone, approached the California legislature to establish a marketing order. Participation would be mandatory, and growers, processors, and shippers would have to pay assessments on all harvested berries. The funds would go toward research to address diseases and pests, including the development of new varieties—and marketing to enhance competitiveness. The legislation was a success. Another institution of repair, the California Strawberry Advisory Board (CSAB), was established in 1955 to administer the program. In 1993 it changed its name and expanded its purpose as the California Strawberry Commission (CSC).[44]

A decline in the frozen strawberry boom, as well as concern regarding the collusive practices of the processing companies, impelled growers to seek more support for that segment of the industry, too. A bumper crop in 1956–57 had led to another glut in the market owing to slowing of per-capita strawberry consumption. Prices for farmers had declined considerably.[45] Growers tried managing the surplus by exporting fruit to Canada, but the Canadian government responded with tariffs.[46] So they established a Processing Strawberry Advisory Board in 1960. Although it had a similar purview as the CSAB, it was far smaller and mainly worked to enforce quality standards.[47] Since the two organizations were somewhat redundant, they eventually merged.

As always, the contradictory missions of such organizations to both fix production problems and address gluts, first encased in CCBGA (Naturipe) and then enshrined in CSAB, didn't seem to give the industry much pause. One of the first orders of business of the CSAB was to ask that the Department of Plant Pathology at Berkeley amplify research to address soilborne diseases.[48] Out of that alliance came not only many of the technologies this book has discussed, including advances in breeding, fumigation, plasticulture, and nursery production, but also new miticides (to address perennially troublesome mites), fertilizers, and irrigation techniques. Though developed and promulgated by institutions of repair, adoption of these new technologies precipitated the first clear shakeout. Total acres planted with strawberries plummeted 61.4 percent in a decade—from 20,700 acres in 1957 to 8,000 acres in 1967—and many growers went bankrupt or sold their land. Notably, strawberry yields and total production did not decline nearly as much as acreage, as these technologies unleashed an unprecedented boom in per-acre productivity. Total production of strawberries only decreased 6.6 percent, compared to the 61.4 decrease in acreage planted, and fresh market berry production actually increased by 20 percent during that period.[49] Clearly markets were changing, as well as production practices. Even though the state would not again reach its 1957 acreage levels for more than three decades, some growers were better able to ride the wave by embracing these new technologies.[50]

A Fresh Boom

This third period saw both more growth and more fracturing. Acres planted, tons produced, and value of production fluctuated during the 1960s and 1970s. Into the 1970s, though, further developments in breeding, nursery production, bedding, plant spacing, use of polyethylene mulch, and irrigation (a shift to drip) allowed California to take its place as the world's leading strawberry producer, having outcompeted other regions.[51] Most notably, developments in breeding for short-day varieties allowed production to take hold firmly in the southern part of the state, which, along with Mexico, would become the source of late-fall and winter fruit. The southern regions, including Oxnard and Santa Maria, would come to produce up to 60 percent of berries in the state.[52]

In addition, the face of growing began to shift when the 1970s saw the entrance of Latinx as strawberry growers. While some had been farmworkers and some had been the so-called sharecroppers discussed in the previous chapter, notably some came into farming as production cooperative members. During the 1960s, public agencies and private charities had created such cooperatives as a means to fight rural poverty. Through these and sharecropping arrangements, the new farmers were able to gain some access to land, albeit often in more marginal areas. For example, in the Central Coast area they found plots in the Monterey Hills—an area that Japanese American farmers had all but abandoned for higher-quality land in the Pajaro Valley. Although some became independent with support from former employers who loaned them equipment and gave them advice and contacts, many didn't have the savings, experience, or legitimacy to obtain either capital or membership in marketing cooperatives. Moreover, the production cooperatives were being disestablished and the reinterpretation of sharecropping as disguised wage labor threatened to demote them back to wage laborers. But in the mid-1970s they got a break—as far as that goes. A wave of supply shortages prompted processors and shippers to finance these potential growers, up to 75 percent of up-front costs, in exchange for their agreement to sell to the buyers who financed them. This influx of new growers thus contributed to a new wave of industry growth.[53] Production as measured by strawberry tonnage almost doubled from 1970 to 1980. Yield per acre increased 38 percent, from 34,000 pounds to 47,000 pounds, and harvested acreage increased 29 percent, from 8,500 acres to 11,000 acres.[54]

Again, growth was a doubled-edged sword. These berries had to be sold. One set of fixes centered on fruit durability and transport. Breeders had been developing berries to last, but researchers also worked on environmental technologies that would keep them from rotting. In competition with earlier advances in pre-cooling by Driscoll's, UC researchers invested their own efforts in mechanisms that would slow down respiration. For example, they developed a process that involved injecting pallets with a carbon-dioxide-based product to extend shelf life. They supported changes in packaging, such as cardboard crates rather than wood, and plastic baskets rather than woven wood, to encourage more air circulation and hence less rot. Innovations in fruit durability paired well with new transportation technologies to get fresh fruit to distant markets: refrigerated trucking, interstate highways, and eventually air travel allowed California's strawberry growers to supply markets thousands of miles away.[55]

Such innovations might have kept berries from rotting but, again, they didn't solve the core problem of overproduction. Somehow, the industry had to generate demand. As a result of a long campaign led by smaller growers, this time repair came in the form of the CSAB substantially raising its assessments to enhance funding for both research and sales promotion. In 1974 the organization launched a major marketing campaign, modeling its advertising and merchandising after Sunkist Growers.[56] At first they invested in television ads, radio ads, and public relations campaigns. When organization leaders soon realized that these were fairly ineffective for the money spent, they then turned to working with retailers to enhance demand, a wise move in an increasingly consolidated retail sector. CSAB's provision of financial incentives for retail merchandising of strawberries was just the thing to encourage retailers to highlight strawberries in their sales displays. CSAB also focused efforts on catering and food service sectors, which proved to be important markets for strawberries. Restaurant usage increased 89 percent in the 1980s, and a 1989 survey of two hundred restaurant owners said the berries were the most profitable menu item.[57]

During the late 1990s, the CSAB—by then the CSC—began to put more effort into general public relations, using consumer magazines, food editor contacts, infomercials, and billboards to boast the benefits of strawberry consumption. By 2004 CSC had all but abandoned giving price incentives to retailers and had shifted its marketing budget entirely into research and public relations on the health benefits of strawberries. Such efforts clearly paid off in that strawberries became a favored food of parents who cajoled their children into eating "healthy food." More generally, dietary advice at the turn of the century emphasized fresh and less processed foods, further boosting consumer interest in strawberries. Strawberry consumption statistics were a testament to the payoff. While annual frozen berry consumption rose somewhat between 1962 and 2000, from 1.25 to 1.94 pounds per capita, fresh consumption rose considerably, from 1.6 pounds per capita in 1962 to 4.27 in 2002.[58] Even these statistics somewhat overstate the importance of frozen berries. Unlike in the previous era when consumers sought out frozen berries, most of the freezer market was destined for remanufacturing. Indeed, prices to growers became so low for frozen berries that for the most part growers used the freezer market only to offload low-quality or excess fruit when fresh market prices dropped too much.[59] Very few growers actually continued growing specifically for the processing market. In 2016, 98 percent of the value of production was sold in the fresh market.[60]

With growing markets came growing numbers of those in the distribution business. Rather than becoming more consolidated, that is, the industries surrounding the farm became more competitive, seeing opportunities that existing institutions did not provide. Shipping, for example, saw the proliferation of smaller independent operators, some of which were growers that developed their own brands, hoping to maintain control of prices and profits (for instance Rincon Fresh, Red Blossom). In addition, some shippers with multiple products began small strawberry operations, including the Salinas vegetable giant Tanimura and Antle. The main development in shipping, though, were the new shippers that came on the scene to compete directly with Driscoll's and Naturipe as big players in the retail market. WellPict, for example, founded in 1969, employed the Driscoll's model of developing its own patented varietals along with managing marketing and sales, working with experienced independent growers.[61] California Giant, begun in 1983 as a partnership of three growers, opted for a different model, one less extractive with respect to growers.[62] The founders remained growers and selected a few larger growers to work with, helping them with financing. Dole (of pineapple fame) set up shop in Oxnard, working on a similar model, to the extent I was able to learn. (Growers and buyers rarely reveal the terms of their contracts.)

Other shippers, such as Andrew & Williamson, came on the scene to develop private labels for large retailers such as Walmart and Costco. That competition became the name of the game is evident in the demise of the cooperatives. In 2007 Naturipe Berry Growers converted from a cooperative to a corporation, while Watsonville Berry Cooperative lost most of its affiliated growers and was out of business by 2017.[63]

During this period, breeding also became more competitive and proprietary, although Driscoll's had created its own breeding facility long before. Some say the problem was with UC, whose budgetary cutbacks in research and extension gave the appearance that it was not fully committed to breeding. Yet the trend began before those cutbacks were evident. WellPict started obtaining its private patented berries from Watsonville's Plant Sciences, incorporated in 1985, and its affiliate Berry Genetics, incorporated in 1993—another organization that was rumored to have obtained UC stock illegally. When Douglas Shaw and Kirk Larson were making moves to jump ship from UC, other shippers and growers began to take an interest in proprietary varietals as well. Some joined what are called breeding clubs, in which growers support private breeders with royalties in return for access to their varieties. Naturipe and Andrew & Williamson thus became breeding club members of Berry Genetics, although they did not require their affiliated growers to purchase these varietals. Even Giant, whose growers had always used university varieties, joined the breeding club of California Berry Cultivars, the organization that picked off Shaw and Larson from UC. That one of CBC's founding partners, A. G. Kawamura, was the former California secretary of agriculture as well as the owner of Orange County Produce while this deal was in the works says something about reduced public commitment to breeding. Or it may say something about Kawamura, who went on to become one of Trump's agricultural advisors.[64]

Not all this splintering took place in the private sector. CSC's concerns that UC had pulled back from supporting the industry induced the CSC to develop an additional partnership with Cal Poly San Luis Obispo, a tech-focused public university in California. Begun in 2013, the Cal Poly Strawberry Center boasts that it is the only organization in the United States dedicated solely to strawberry research and education.[65] According to announcements made at a 2018 field day there, more than one hundred students participate in its research program, and much of the programming is conducted in direct service to the industry. One of the areas of research is plant breeding.

With its close ties to the breeding business, the nursery business also changed. Lassen Canyon continued to mainly produce plant material of university cultivars, although it employed its own breeder. Newer entrants, such as Cedar Point and Crown, appeared to grow primarily university varieties as well. But Sierra Cascade, 50 percent owned by Driscoll's, and Plant Sciences, which established its own nursery, came to serve proprietary organizations only. In 2017 the Spanish company Planasa bought out NorCal, with the aim of introducing their proprietary varietals to the California market and becoming one of the largest nurseries in the world.[66]

Although most of these new businesses were formed to compete, and were often based on proprietary knowledge, in one case an organization was able to take advantage of regulatory changes to create a monopoly. Here I refer to TriCal, a business that provides fumigation services. Previous to their domination, many growers conducted their chemical pest management in-house or used TriCal's largest competitor, Crop Protection Services. But at some point Crop Protection Services ceased doing fumigations, allegedly because it was a publicly traded company and didn't want the exposure of handling frowned-upon chemicals. And then, it is also alleged, TriCal helped manipulate the significant regulations that developed around fumigation to secure a monopoly.[67]

Similarly, farm labor contractors became important in the 1980s. Primarily as a means of avoiding grower responsibility for employment conditions, as well as to ease recruitment across agricultural industries, they seemed to have found a particular niche in strawberry production, given its recruitment challenges.[68] A full discussion of the evolution of farm labor contracting is too far afield; the main thing I want to note is that the two large labor contractors I encountered during the research, RAMCO and Nor Cal, had growing operations that sold to Driscoll's.

. . .

In reviewing this history, we do not see an industry on a steady growth trajectory. To the contrary, 150 years of strawberry growing in California have been marked by both growth and decline in production and booms and busts in markets. As these problems waxed and waned, the industry created or called on other institutions that, somewhat incoherently, aimed both to improve productivity and address market excess.[69] Having introduced breeding, fumigation, and other technologies, UC served as the primary institution of repair for production problems until Cal Poly's more recent arrival. When improving upon production (and shipping) almost always ramified in excess supply and low prices, other institutions stepped in to try to help with marketing. Although it was a private-sector-led frozen food boom that mitigated problems of overproduction in the middle period, some of these institutions, like CCGBA and the CSAB/CSC, were not-for-profit organizations developed by the industry. In many respects, these industry investments paid off: UC's contributions to productivity-enhancing technologies were tremendously successful, as were CSC's investments in advertising and retail. As a result, markets continued to expand more or less in sync with ever-increasing productivity. Thanks to cooperation, in short, the California strawberry industry as a whole became exceptionally productive and also came to enjoy worldwide markets, allowing it to achieve its national dominance and a large international imprint.

But even as the industry cooperated in these endeavors, it also became more competitive, and saw a hiving off of businesses that would sell services back to growers in areas such as breeding, plant propagation, fumigation, and labor recruitment. Breeding was especially prone to privatization because of the money to be made in licenses for proprietary varietals. Marketing became more fractured as well, though, with different organizations opting for different business models, hoping to land on one that would sustain. Even the wholly public institutions of repair faced competition, as evident in Cal Poly's entrance into activities that had been the sole purview of UC, as a land grant institution. And so, the overall tendency was not toward monopolization per se, but more fracturing, born of imagined and real competitive advantages. Although market expansion has its limits even under conditions of cooperation, I want to suggest that these latter dynamics are more clearly precipitating a shakeout that may well result in monopoly. To see how that may be, a look at contemporary relationships between growers and shippers is in order.

GROWERS AND SHIPPERS

If size alone were an indication of growers' market power, things would look better than they did a generation ago, when Miriam Wells completed her research. Wells's research spanned a period that began in 1976 and ended in the early 1990s, and thus took place during the long boom that made the industry so prominent. And although it concerned only the Central Coast and not the southern regions, the study was widely applicable to California as a whole since so many of the organizing institutions of the industry were born on the Central Coast.

Wells noted that as a general trend, berry farms were relatively small, averaging thirty-four acres, and even in the highly capitalized Salinas area rarely exceeded one hundred acres. I did not set out to update her study, and so I cannot provide directly comparative statistics. Nevertheless, by the time I conducted my research in 2014–15, the size of farming operations had trended significantly upward. Although ranch sizes were delimited by available acreage, many growers were operating on multiple ranches, in multiple regions, and a significant number were farming more than five hundred acres per year, a huge capital undertaking given that per-acre cost, including labor, was nearing $68,000 at the time. My interview sample was not necessarily representative, given the difficulty of reaching some low-resource growers. Still, more than half of the seventy-four growers I interviewed were farming more than one hundred acres. Wells attributed relatively small farm size among strawberry growers to the robust per-acre profits they earned; growers could do very well with relatively small farms.[70] Today, profit margins appear much smaller. The latest cost study of strawberries, based on 2016 figures, predicted average returns of about $2,300 per acre on the Central Coast, a significant drop even from the $8,000 of 2010.[71]

Another change since Wells conducted her research is that Latinx have taken their place as the most dominant grower group throughout the state, particularly as growers of other ethnicities have retired and raised children not interested in farming. That Japanese-origin growers who once struggled to maintain a toehold in strawberry farming, from the Alien Land Laws to the internment, have largely left the business is particularly striking. According to figures posted on the CSC website in 2017, Japanese and other Asian-origin growers comprise only 20 percent of the state's growers—a figure that includes the recently immigrated Hmong, who have taken up strawberry farming on small plots in the Central Valley—and whites comprise only 15 percent of growers. Latinx, meanwhile, comprise 65 percent of those in the state, 25 percent of whom, CSC estimates, started off as fieldworkers.[72] These figures should not be understood to mean that low-resource growers have largely taken the place of well-capitalized growers, as it would be incorrect to equate categorically Latinx growers and low-resource growers. I noticed vast differences between the capitalization and social networks of second- and third-generation Latinx growers versus those who had recently come up from the immigrant farmworker ranks. I met Latinx growers who were well off and had significant acreage under their management, with one having more than one thousand acres in production. At the same time, I met quite a few more Latinx growers who were on the brink of financial ruin, many of them new farmers with few resources, and many others who had already gone out of business.

The fate of all growers stems from the nature of the contractual relationships between growers and shippers, especially as related to pricing, financing, and risk. In theory, these contract arrangements devolve varying degrees of upside reward and downside risk to growers. Operating nearly risk free are "custom growers" who are paid to produce strawberries regardless of outcome. The shippers finance these operations in their entirety and sometimes even control the ground lease. In return, custom growers receive a flat management fee or are paid by the acre. Such contracts are offered only to highly experienced growers and, as far as I could tell, are fairly rare. For whatever it is worth, the custom growers I met were white, and they would not say to whom they sell. Another arrangement is the one that Driscoll's has with its historical partner in the strawberry business, the Reiter Company. Owned by grandsons of the original Reiter, Reiter Affiliated Companies (RAC) does all its marketing through Driscoll's. Some RACs are wholly owned production operations, managed by internal employees and ranch managers. In effect this is a partnership of Driscoll's and Reiter, and the potential for this arrangement to become more widespread is alluded to in the first epigraph of this chapter.[73]

A much more common arrangement is when growers sell to shippers on marketing contracts. Although contracts are set in advance, these so-called independent growers are generally paid by the box at a fluctuating market rate. The viability of independent growers undoubtedly is connected to how they contend with the biological risks of production, including soil pathogens. Still, their profit potential is also affected by the shippers with which they work. Independent growers who sell to Driscoll's and WellPict putatively obtain higher prices. These prices owe in part to the branding of these berries as higher quality. This branding derives not only from the private breeding arrangements that allegedly makes tastier, more attractive berries. Driscoll's, at least, maintains quality standards that require growers to cull up to 30 percent of the berries they grow. So higher prices are possibly offset by higher cull rates. Moreover, growers' potential for high profits is limited by the various fees they owe the company. Driscoll's and WellPict growers must pay these companies hefty licensing fees for proprietary varieties, even though they still must pay their assessments to the CSC, some of which goes to support the development of university varieties. At the other end of the supply chain, these growers also must pay steep sales commissions. According to one grower, Driscoll's was charging an 18 percent commission for cooling and sales. In addition, growers have to purchase the marketing material, such as baskets and packing boxes. Growers who work with Giant or Naturipe, in contrast, apparently do not pay such high fees and have the option of using lower-cost university varietals or other proprietary varietals coming out of the breeding clubs. Nor are their cull rates as high. But these varietals tend not to command the same prices in the market. Given the various contingencies in these arrangements, it is difficult to say categorically how independent growers fare with different companies.

That is not the case for the low-resource growers, involved in what are called partnerships. These partnerships are the latest iteration of the much-disparaged sharecropping arrangements that were deemed illegal, and many in the industry still refer to them as sharecropping. Yet the partnerships really are a different model, especially because partners hire outside labor rather than rely on family labor. The way they work is that intermediary companies look for particularly entrepreneurial and competent farmworkers, ranch managers, or other hired hands and tap them to go into business together. The two parties set up a separate company, generally a limited liability company (LLC), in which the managing partner puts up something like 70 percent of the capital and the grower partners put up 30 percent. The grower generally begins with about twenty acres in production, which, given typical *capital* investments of no less than $25,000 per acre (for lease costs, soil preparation, irrigation, and plants), is a significant investment. (Labor costs are what bring the costs per acre to $70,000 or more.) To raise this capital, growers either take second mortgages on homes they acquired as successful ranch managers, or they go to the bank. Either way, the managing partner helps them secure the loan, for which the grower is generally not otherwise eligible. Often the managing partner also secures access to land and then controls the land they secure, holding the master lease and subleasing it to the partner growers. Managing partners were rumored both to allocate the best land to their favored growers and even to engage in arbitrage by leasing the farm back to the growers at amounts higher than they were paying. If and when the grower begins to make money, they expand acreage and gradually take more equity in the company. Although some growers have become well established this way, many have not.

Driscoll's has been the primary shipper that works with the intermediaries, one of which is Reiter Affiliated Companies and the other Ocean Breeze Ag Management. There may be others as well, although one mentioned by

Eric Schlosser in a 1994 exposé has since become defunct.[74] For many reasons, these growers are prone to fail, especially those who work with Ocean Breeze Ag management. For one, Ocean Breeze operates solely in the Oxnard area of Ventura County, which has become a particularly challenging area for strawberry production ecologically. Second, their debt loads are substantial and always growing. Even those who do not lose money on the operation borrow money to keep their accounts open and maintain credit. Third, like other Driscoll's growers the partners have to pay for Driscoll's varietals, sales, and shipping, and likewise be subject to Driscoll's grading and pricing mechanisms. In addition, partner growers rent equipment from the companies as well as pay management fees. Although all Driscoll's growers pay many of those fees and expenses, partner growers have to pay fees to the intermediary and shipper before residual revenue is calculated and divided, providing almost no opportunity for upside reward. Hence many of them do fail. It is the status of these growers that has led some to question the Driscoll's model of farming farmers. As put by one observer outside the company, "Driscoll's goes through growers like my kid goes through socks."

MAKING WINNERS AND LOSERS

Thus far, I've discussed the varying contractual relationships growers have with shippers, which impinge on their prices, costs, and debt load and therefore shape growers' abilities to stay in business. Here it is important to recognize that relationships between growers and the rest of the industry exceed the formal stipulations of contracts. Shippers not only mediate access to credit and markets; they also, albeit informally, mediate access to other elements of the strawberry assemblage, including varieties, chemicals, land, and possibly even labor. Other industry actors mediate access to needed inputs as well.[75] Since these are the factors so deeply entwined with growers' abilities to manage pathogens, the structure of the industry bears on the ability of individual growers to contend with the challenges moving forward, just as individual growers' ability to contend with the challenges moving forward will shape the structure of the industry.

Consider breeding. As we have seen, the breeding apparatus has become more proprietary—and has benefited from the taking of germplasm from the public domain. The growth of private breeding in turn has meant an increase in growers who rely on these berries. In 1955, 95 percent of growers were using university varieties. By 2005, 38 percent of berries grown in the state were proprietary varieties, and by 2014, nearly one half were such.[76] We know that those who contract with Driscoll's and WellPict necessarily must pay licenses for those companies' proprietary varietals, but if the low-cost shippers have turned to breeding clubs, that means their growers are looking to proprietary varietals as well. Notwithstanding UC's renewed commitment to breeding, the risk is that the breeding apparatus will continue to become more privatized. That means that growers who might want to plant pathogen-resistant cultivars in lieu of fumigation, should those cultivars be successfully developed, will have to pay more for them. Low-resource growers may then have to suffice with varietals more vulnerable to disease.

Tighter regulations on fumigation presumably affect all growers, yet some are better positioned for workarounds. For one, those who are located in more marginal, hilly land have not ever been able to flat fumigate. Although bed fumigation may be desirable from a public health and environmental standpoint, as we saw, novel pathogens have recolonized some fields that have been bed fumigated. Better-endowed growers not only have had access to land amenable to flat fumigation, but additionally have been able to access land that doesn't require large buffer zones. The role of TriCal in delivering fumigation has been a less obvious source of grower differentiation. As many growers mentioned in interviews, TriCal seemed to have favored its oldest, most loyal customers in getting them access to the remaining inventories of methyl bromide as they dwindled. TriCal's monopoly position also allowed them to charge extraordinarily high prices for the chemical. It is impossible to know what favors TriCal may do for its loyal customers in a fumigant-restricted future—or if TriCal itself will be shaken out—but such evident favoritism surely affects grower viability.

As for access to labor at a time of shortage, I have no direct evidence that the two major farm labor contractors favor other Driscoll's growers in helping to locate workers. Yet, since their principals also grow for Driscoll's, the possibility is not negligible.

Land may well be the most significant source of grower differentiation and eventual failure. We have seen that land well suited for strawberry production has become scarce, and that land prices have risen even with the ecological threats that have been bearing on the industry. At times, the difficulty of obtaining good land has kept many established growers from expanding. In this context, Driscoll's, some have alleged, has been employing land scouts to locate available, good, relatively disease-free land, obtain the master leases, and sublease it to their favored growers, while more marginal growers, such as partner growers, have been doled out land with low-quality or insufficient water and high disease load. As we saw, drought conditions stress the plants and make them more susceptible to soil pathogens. These growers have thus found themselves having to spend a great deal more money on pesticides and fertilizers than they wish. Their sentiments about the future are captured in the following quotes from two different growers, translated from Spanish:

> Yes, [the future] looks challenging. We don't know ... how things are as of now; we don't know how long we'll last. Hopefully it rains and the weather will get better because each day we are spending more. The cost of water and chemicals spikes up and pathogens grow because mountains no longer have weeds, so they come down to the valley. All of the pathogens are on the plants because there hasn't been any rain; there isn't any place else where they can grow, so they come down here, where we are. So, it's difficult to combat all the pathogens. Furthermore, the salt in the soil has risen into the springs, making salt water. So now the plants are suffering even more. We are spending more money on fertilizers to give the plants nutrition, much more than in the past.

> Now it is more difficult. Now we get plagues with fumigants that had always kept things under control. Now, the pests have become immune [to the fumigants] and do not die with anything. The chemicals are not strong enough, so now we have to invest more money [for more chemicals].

Growers, in other words, are hardly on an even playing field in getting access to critical industry resources, especially those that allow them to contend with plant pathogens. This, along with the squeezes between the prices received and the costs of inputs experienced by Driscoll's growers in particular, have driven some growers entirely out of business, while others are merely wilting. To harken back to Steve Hinchcliffe et al., these growers are also experiencing disease situations, with pathogeneity arising from the growing intensities of the ecological and political-economic conditions they face.[77]

What is clear in these arrangements is that the parties who have to contend with the least risk, both economically and ecologically, are in the best position to survive and thrive. And it appears that the Driscoll's business model puts the company exactly in that position. Consider the various fees that go to Driscoll's for proprietary plants, marketing materials, sales, and management. Notwithstanding the significant investments in developing new cultivars, these fees are nearly risk-free sources of revenue for the company that accrue before any crops are sold. Having extracted the company from actual production, except for their alliance with RAC, Driscoll's faces none of the more-than-human risks of disease, weeds, drought, poor-quality water, and more, nor does the company pay the immediate and increasing costs of regulatory compliance, such as major losses of yield in buffer zones.

This casts in a different light the sizable investments in social justice and sustainability that Driscoll's has made. As I've discussed in previous chapters, the Driscoll's breeding program was the first in the modern era to allegedly emphasize pathogen resistance. Driscoll's has also invested significant human resources into developing an organic program, through breeding varieties conducive to organic production and providing technical support to would-be organic growers. Scouting for suitable land for these growers also supports these transitions. Overall, Driscoll's seeks to be a leader in reducing the strawberry industry's dependence on fumigants, and in the chapter that follows I'll say more about these efforts, including their research into eliminating fumigants at the final stage of plant propagation.[78] In addition, the company claims to be working on ways to improve the working conditions, remuneration, and respect for farmworkers associated with the Driscoll's brand. To that end, they have implemented a system of worker welfare standards and audits to apply to their independent growers.[79] These initiatives are mostly to the good and can't be categorically dismissed (although their concern with workers could be questioned, given their significant investments in robotics as well, which would displace many workers from their jobs). But it still bears considering what allows Driscoll's to do this. The enabling factor is that their contracting growers, not the company, take the ecological and economic risks of farming without fumigants and increasing the wages of harvest workers.

The competitive advantages of these initiatives for Driscoll's are considerable. Not only do these investments burnish the company's image in international markets, but they put them in the lead of developing workable solutions should fumigants be phased out altogether. Solutions will then be passed on to their growers—and their growers alone. Those who stay in business will continue to pay the various fees to Driscoll's, and possibly expand from their success, providing yet more revenue and potentially cementing the corporation's position as the unquestioned leader in the field.

If you were to travel through strawberry country you might find it difficult to imagine an industry in crisis. Unless you knew where to spot diseased fields, or knew where berry plantations once existed, you would see vast acres of lush green plants and many workers speeding through the rows, picking berries. For that matter, if you were to go to a large midrange or upscale supermarket, you'd be equally convinced of the health of the industry. Most times of the year you would likely see a large display of strawberries, or maybe two—one organic and one conventional—piled high on wooden crates, enticing you with their big redness and sweet smells. But you wouldn't see the statistics, nor hear the gossip that courses through the industry.

Thanks to institutions of repair, the industry saw huge growth between 1960 and 2014—when acreage more than tripled, production increased tenfold, and the value of production in *real dollars* increased by 424 percent in Monterey and 593 percent in Santa Cruz Counties, the original centers of strawberry production.[80] Equally striking was that net returns for growers were exceptional as well, no doubt a reflection of the long season that growers in these two counties enjoy. Although net returns vary considerably, according to UC Davis agricultural economists, in 2010 conventional growers were typically making profits of nearly $8,000 per acre, while farms with productive soils, experienced managers, optimal production plans, and robust marketing plans were doubling that, even while others, with less experience and endowments, were losing money.[81]

Even the much remarked-upon threat of the strawberry industry's move to Mexico hadn't come to pass. The initial basis of this threat was a different phaseout date for methyl bromide for developing countries like Mexico. Developing countries had until 2015 to cease its use, while others were first mandated to phase out by 2005.[82] Once the United States was able to extend its phaseout to 2015 as well, the conversation focused on the low labor costs in Mexico. As it happens, strawberry production has indeed been on the rise in Mexico. Driscoll's has massive operations, and all the other major strawberry shipping companies have a foothold, too. By 2018, nearly twenty-eight thousand acres of strawberries were planted in Mexico.[83] That said, Mexico largely has served as a site of winter production. Therefore, it mostly only competes with Florida and to some extent the far reaches of Southern California, which has never been a major production area.

Zeroing in on the last few years of that trend, things appear less rosy. Acreage surveys from the CSC show a gradual decline in statewide planted acreage between 2013 and 2018, from 40,816 to 33,792. Notably, these changes are uneven across regions, and the much-challenged Oxnard region saw more than three thousand acres lost, although Santa Maria made up for some of those acres.[84] News stories corroborate that trend, as long-established farms closed in Oxnard, and Dole left the area, while Santa Maria expanded.[85] Perhaps more damning is the last cost study of strawberries, which again showed a significant decline in per-acre profits since 2010, from about $8,000 to $2,000.[86] Yet land values continue to rise despite this dip—a trend that would not be predicted by neoclassical economics, but is completely in keeping with the financialization of land markets. It's of course hard to know from these figures if these trends simply signify a blip from which the industry will soon recover, a restructuring of growers and regions—with both Oxnard growers and low-resource growers leaving in droves—or an indicator of a more severe decline to come.

My ear to the ground suggests that it is more than a blip; too many growers I interviewed expressed doubt and anxiety about what the future might bring and foresaw a major shakeout of growers. When I tried contacting growers in late 2018 for a related study I encountered another wave of bounced emails and nonworking phone numbers. By the same token, during this earlier research I spoke to some who were ready to welcome a shakeout and reconfigured industry. A shakeout, they said, would solve the problem of overproduction and falling prices, and would solve the labor problem, too. A shakeout would also allow those willing to experiment with new technologies to redefine strawberry growing and make the laggards go away. No one disagreed that there would be winners and losers and a reshaping of the strawberry industry as we know it.

While this chapter has identified some of the more obvious winners and losers, much still depends on who will develop successful technologies to contend with pathogens, what the technologies will be, who will (be able to) adopt them, and what those technologies will mean for consumer prices and availability. More proprietary technologies, that is, could make strawberries considerably more expensive than they currently are, while more agro-ecological approaches would reduce their availability. Therefore, much also depends on what consumers are willing to pay and what the public is willing to live with.

EIGHT

Imperfect Alternatives and Tenuous Futures

There are no kind and gentle ways to kill things.

Interview with Brian Leahy, director of California Department of Pesticide Regulation, 2015

Systems need to be integrated but simple. Growers can't deal with twenty-five moving parts—and I can't sell that.

Interview with a farm advisor, 2013

While forms of biosecurity seek to manage circulations to build resilience in the face of imminent but unavoidable disaster, probiotic approaches figure the present as already disastrous. They seek to reverse, restore or otherwise address deleterious existing transitions.

Geographer Jamie Lorimer, "Probiotic Environmentalities: Rewilding with Wolves and Worms," 2017[1]

THIS LAST ANECDOTE IS PERSONAL, kind of gross, and may at first glance seem off topic. It's about our family dog, Bernie. Before Bernie had seen much of the world, she was diagnosed with giardiasis, following a routine test for parasites. Giardiasis is spurred by an intestinal parasite called *Giardia lamblia* that survives by colonizing an animal's gut. It is transmitted through contaminated water and food, and can cause diarrhea and other gastrointestinal problems. It is widely prevalent in untreated water. Since the test itself is of fairly recent provenance, and not all dog owners pay for such tests, it is safe to assume that many cases of giardiasis, like pathogen wilt, have been undetected and likely attributed to random gastrointestinal upset. Not so Bernie's, and the vet put her on a course of metronidazole, a strong drug associated with neurological problems and not entirely effective at treating giardiasis. After her first treatment and an additional test, the infectious cysts were no longer visible, but before long Bernie was sick again and had another positive test. I will you spare you the remaining details, some scatological, but suffice to say that over the course of the next nine months Bernie underwent a range of treatments, involving less-toxic drugs, stronger and longer doses of metronidazole, commercial herbal remedies, custom herbal remedies, dietary changes, probiotic supplements, and sometimes a combination of several. Some made her symptoms worse and some curbed them, but none were able to eliminate the parasite from her system for more than a few weeks—that is, according to the tests. Nevertheless, through dietary changes, her symptoms improved considerably, and she finally filled out after having become nearly emaciated. And while she never lacked for energy, her energy levels became even more pronounced. Once she seemed very strong, her vet recommended one more round of metronidazole to clear the infection for good. Yet by then we had come to realize that the parasite was ubiquitous in her environment, and she would likely be reinfected after treatment. Given that her symptoms were manageable, we opted for a course of "living with." Although "living with" was not always easy, it worked well enough, and Bernie continued to thrive, if her energy levels were any indication.

Bernie's maladies occurred when this book was well under way, so I couldn't help but think about her treatments in the context of my research on soil pathogens. What struck me about the various treatments she received is that they rested on entirely different and often opposing ideas. Some were antibiotic, designed to eradicate the parasite, while others were probiotic, designed to encourage suppressive flora.[2] Of the antibiotic treatments, some involved harsh, synthetic drugs, while others used putatively less-toxic formulations or natural, plant-based remedies. Some were accompanied by instructions for strict sanitation in order to eliminate the parasite from the environment, while others were based on the presumptions that the parasite was ubiquitous in the environment and that the dog needed to build resilience to fight it off. And while none favored the life of the parasite, some aimed to eliminate the parasite to make the dog healthy, whereas others aimed to make the dog healthy in order to eliminate, or at least manage, the parasite.

The treatments Bernie received have clear parallels to the varying approaches to soil pathogen management currently being developed and assayed by the strawberry industry to replace fumigants. The harsh chemical fumigants are the obvious analogs to metronidazole, not all of which were entirely effective at managing the pathogens even before they were threatened with elimination. Of the remaining options, like the alternative treatments for giardiasis, some involve less-toxic methods of eradication yet are still antibiotic, while others involve adding organisms to the environment and thus are probiotic. Some involve trying to control the agricultural environment—and even to minimize the role of biology altogether—while others assume such control an impossibility and attempt to build immunity and resilience.[3] Notably, agricultural advisors believe that a combination of approaches, along the lines of integrated pest management, is the most optimal course, even if that means combining antibiotic and probiotic approaches, as we did with Bernie.

These approaches to soil pathogen management also vary in respect to their ability to fit into the current assemblage of strawberry production: some are near substitutes for existing treatments, while others would lead to entirely different configurations. The treatments under discussion also vary in regard to their susceptibility to proprietary behaviors. Each therefore would make for very different human winners and losers, were they to be widely adopted.

After a brief discussion of why alternatives were so long in coming, this chapter reviews the main approaches currently on the table: the kinds of elements they assemble, their biotic underpinnings, and what they portend for the character of the industry. I exclude plant breeding for disease resistance in this discussion since that has been covered elsewhere in the book. In considering the remaining options, readers should not assume the existence of an optimal pathway. Solutions that are efficacious, reasonably harmless, and economically viable remain elusive. In that light, the last part of the chapter draws on recent work on emergent ecologies to raise the possibility of letting the industry go to ruin. While putatively hopeful, such an approach, I argue, would have significant consequences for many of the people involved in strawberry assemblages, including eaters, and so in no way would minimize harm. Industrial agricultures involve social entanglements, too.[4]

FROM HAND-SITTING TO ACTION

The Montreal Protocol on Substances That Deplete the Ozone Layer first mandated the phaseout of methyl bromide in 1991. Although an international conference has been held annually since 1994 to report on emerging alternatives to methyl bromide, for a long time most strawberry growers sat on their hands rather than experiment with feasible replacements. The industry had been built around effective fumigants, and it appeared that most growers preferred that it expend its efforts on avoiding the ban. It is in that context that the industry successfully lobbied the US government for a protracted phaseout that was supposed to end by 2005, then sought the critical use exemptions (CUEs) that successfully extended the official phaseout until 2015, and secured the continuation of the quarantine exemption for nurseries. Privately, some have argued that the US success at obtaining these exemptions made many growers too complacent. But by sitting on their hands, these complacent growers allowed a few forward-looking growers and shippers to obtain a head start on figuring out how to farm without fumigants. Anticipatory development of sustainability programming on the part of Driscoll's was in that way yet another shrewd move.

Even as the writing on the wall became clear, many pinned their hopes on the discovery and registration of a suitable "drop-in" replacement—a chemical that would work much like methyl bromide and cause the least disruption to the entire production system. (And many of the alternatives discussed at the annual international conference were of that kind.) Long waiting in the wings, methyl iodide seemed to fit the bill. Moreover, in keeping with the logic of chemical substitution as discussed in chapter 4, it was a different compound that could be regulated entirely independently.[5] But activists were not complacent, and methyl iodide met its demise in 2013. At about the same time, DPR published its nonfumigant production plan that threatened the eventual phaseout of all fumigants. It was meant to be a wake-up call.

Even though most growers continued to use allowable and available fumigants for as long as they could, by the 2010s the industry had become more serious about investing in research on alternatives to methyl bromide. Supported by grower assessments, by 2013 the California Strawberry Commission (CSC) had funneled more than $13 million into such research.[6] More followed after that. Through annual grant programs, the USDA and the DPR provided substantial support for research as well, including two large collaborative grants funded in 2017 through the USDA's Specialty Crop Research Initiative.[7] And University of California extension agents were experimenting with various replacements. Much of what follows is drawn from documents created during this period, one from DPR and one from the CSC.

For the most part the industry has looked to alternatives to chemical fumigation that have antibiotic qualities. Designed to kill, albeit sometimes more gently, they do not fundamentally alter the existing strawberry assemblage. Some extension agents have thus vigorously promoted them, presumably in their ongoing efforts to please their clientele.

Less Toxic Drop-in Replacements

An area of great aspiration is the development of less-toxic fumigants.[8] Throughout the industry there is keen interest in chemicals with biological origins, or that are synthetic analogs of substances found in nature—qualities that presumably make them safer and more publicly acceptable. Consider Paladin, with the active ingredient dimethyl disulfide (DMDS). DMDS is a synthetic version of a naturally occurring compound found in garlic and onions, as well as plants in the *Brassica* genus, with their known fumigant-like properties. Paladin appears not to be carcinogenic or disruptive to endocrine and neurological systems, although condensed into a chemical formulation, it is highly irritating to the skin, nose, and lungs. In addition, the substance apparently emits a very strong and unpleasant smell of garlic and decaying fish, which has caused headaches, dizziness, and nausea among those exposed.[9] Dominus, another methyl bromide replacement candidate, has similar qualities. Made from purified mustard oil, its active ingredient is allyl isothiocyanate. Allyl isothiocyanate is also found in cabbage, kale, horseradish, and wasabi. Like DMDS, it has low toxicity, but in concentrated form can cause eye damage, skin burn, and respiratory irritation.[10] As such, one extension agent's claim that it is safe because it is the same as wasabi, and we eat wasabi, seems somewhat disingenuous. Another product under development carries the brand name Muscodor. Developed by researchers at Montana State University, it is made from a fungus that produces volatile compounds that can kill nematodes, insects, and plant pathogens.[11] During my interviews I heard of one or two other possible replacements, and one grower even sought my advice about how to get it approved. Following a short article I published on the near impossibility of growing strawberries without fumigants, my email inbox was peppered with plugs for additional biological alternatives. (Purveyors are clearly keen to get in on this market.)

One, called Ennoble, derives from a fungus found under the bark of the cinnamon tree, and according to the purveyor "produces a cocktail of gases" that kill disease-causing fungi.

But drop-in replacements are no sure thing. Although these putatively less-toxic compounds are least disruptive to existing practices, they are not generally as efficacious as the legacy fumigants. As an oil, Dominus, for example, does not volatilize. For it to be effective it must be mixed with water and spread throughout the entire field, which is not an easy task. Another concern is that virtually all such replacements are being developed, patented, and sold by small companies that intend to charge for their use as long as intellectual property rights hold. If they turn out to be effective and not too costly, they would not categorically eliminate low-resource growers, since those growers already pay for fumigation. But if they are costly, low-resource growers might not be able to adopt them, risking failure.

The primary obstacle to their adoption, however, is safety. Notwithstanding their biological origins, obtaining approval by California DPR has proven challenging, given its increased concern with community health. The US EPA has registered both Paladin and Dominus for use, albeit with required mitigation measures, and both are already used in Florida. But neither have been approved for use in California, and my discussions with the head of DPR suggested that neither would be anytime soon. For that reason, the makers of Paladin eventually withdrew it from consideration. The fact that these substances are structurally similar and functionally identical to naturally occurring counterparts does not guarantee safety any more than was the case for methyl bromide. It also bears noting that their antibiotic character may work at cross-purposes to managing pathogens in the long run, insofar as they also kill potentially pathogen-suppressive microbes and otherwise inexorably alter ecologies. What makes them least disruptive to existing assemblages, in other words, also makes them contribute to the fragility of existing assemblages.

Nonchemical Disinfestation

Recognizing the limitations of these less-toxic alternatives and usually not guided by profits, public researchers have put greater efforts into developing nonchemical means of eliminating pathogens. These are generally antibiotic to the pathogen although not necessarily to other life in the soil. Since they are not pesticides per se, they do not require regulatory approval.

Solarization involves the use of extensive plastic tarping to trap solar energy, with the intent to kill soilborne organisms with heat. Theoretically, it can be widely accessible to growers, as long as tarping is not more expensive than existing treatments. The problem is that researchers have not found it suitable for California's relatively cool and often foggy coasts. In cooler climates, the heat reaches only the top several inches of soil, while many of the target organisms reside deeper beneath the surface.[12]

More promising in terms of effectiveness has been steam treatment, which, like solarization, applies heat to kill soilborne organisms, but the heat is generated by machines rather than the sun. Steam has been proven effective at controlling both weeds and pathogens, but it has significant drawbacks. One is that steam must be applied to a given patch for several minutes to be effective. So with a machine the size of, say, a tractor, only a small part of the field can be treated in a day. A second is that the machinery is expensive, which may well put the technology out of reach for low-capitalized growers. Even were custom applicators to take over the business of steam disinfestation, as they did for fumigation, the costs might be prohibitive. A third is that the technology relies on large amounts of water and fossil fuels, resources that are neither cheap nor plentiful in California, and whose heavy use in agriculture is already subject to much controversy.[13] Researchers developing steam technology see it as an interim solution.

Some researchers and growers have pinned their hopes on anaerobic soil disinfestation (ASD). ASD involves creating anaerobic (oxygen-free) soil conditions by adding a major carbon source such as rice bran or molasses to fields and then flooding them with water. In addition to depriving microbes of oxygen, like the other two methods, ASD volatilizes organic compounds present in soil that are toxic to soil microorganisms.[14] But apparently it is not wholly antibiotic. ASD seems to promote soil bacteria and fungi, which, through competition, can be suppressive to pathogens.[15] Notably these microbial communities differ from those found in fumigated or nonfumigated soils.[16] Bio-fumigation with substances like mustard seed meal is similar in intent. Unlike a chemical analog to mustard seed, which significantly concentrates the active ingredient, this approach involves incorporating the biological material into the soil and then covering it with tarps to create an unfavorable environment for pathogens and weeds.[17]

There is nothing particularly proprietary about either of these latter technologies. Conceivably, they could be widely adapted. Yet there is concern about whether ASD could be a widespread solution, given the limited avail-

FIGURE 18. Anaerobic soil disinfestation. Copyright © 2016 The Regents of the University of California. Used by permission. Photo by Anna Howell.

ability of carbon sources within the region and dependence on California's ever more precious water. So far, those experimenting with ASD generally import rice bran from the northern Central Valley, where rice is extensively grown, but with wide adoption, those sources would not be adequate. Both also require large amounts of plastic tarping, as do solarization and conventional fumigation, another sustainability concern. As far as growers are concerned, the issue has been efficacy. UC researchers have conducted field trials on many different farms, and many growers have adopted ASD on portions of their farms. Yet thus far the results for ASD have been mixed in terms of controlling pathogens and weeds—and few growers have attempted it with more than a few acres at a time. Meanwhile, the results from mustard meal fumigation have been simply poor.[18] Besides the resource concerns, it is worth considering that the efficacy of either is contingent on the disease load of existing soils. It is difficult to completely eradicate disease with a nonchemical solution. For that matter, soils that have been heavily fumigated may be challenging as well, bereft as they often are of diverse biota. There's a bit of a paradox at work: returning soils to a disease-free state seems to require fumigation, yet fumigation sets back the development of suppressive soils.

SOILLESS SYSTEMS: THE ABIOTIC ALTERNATIVE

An entirely different approach is premised less on killing than on containment, somewhat akin to a cordon sanitaire. At minimum, this approach involves growing strawberry plants without soil, thereby avoiding the material that is the ostensible source of disease. It may additionally involve elaborate greenhouse operations, a version of controlled environment agriculture (CEA). Such systems have been widely adopted in South Korea, China, and throughout Europe, where prices for strawberries are significantly higher than in the United States.

In California, a few growers have experimented with the more minimal approach, in what they call field-scale hydroponics. This involves cutting troughs into traditional strawberry beds, lining them with landscape fabric, and filling them with a substrate such as peat moss, coconut coir, or mixtures of these and other materials. Some have also experimented with substrate in waist-high trays. Growers experimenting with substrate have thus far achieved only fair results. According to one I interviewed, the volume produced was close to 90 percent of normal, but the quality of the strawberry suffered greatly, lacking both sweetness and firmness. Suspecting that the berries had accessed too much nitrogen, he dialed back the nitrogen, but then lost volume. This grower fed his plants all of the required nutrients, but was not satisfied. Others corroborate that the quality of berries grown in substrate does not come close to field grown. The problem seems to lie with a reductionist approach to nutrition and the soil itself; soil, many now argue, has qualities that are neither quantifiable nor extractable, and so adding back "essential" nutrients misses the point.[19]

In full-on hydroponic operations, plants are brought entirely indoors and grown in tightly controlled conditions so to limit biological activity to only the most essential—it is a deliberate *simplification* of environments. Because

FIGURE 19. Strawberry plants in soilless substrate, experimental field.
Photo by Sandy Brown.

of these near-"sterile" conditions, growers are often able to use beneficial insects rather than pesticides in controlling insect pests. Whole companies are moving in this direction, and it was estimated in 2017 that already three or four hundred acres of California-grown strawberries were under high-tech greenhouses. The biggest operation is Windset Farms, which is reputed to have figured out how to grow tasty berries in substrate. Windset is owned by a Canadian company that, like others that have developed a competitive advantage, is highly secretive. What is most curious about Windset is that they are growing in Santa Maria, one of the top fruit-growing regions in the state because of its soils and climate.

On that note, the quality of the strawberry is just one of the reasons most California strawberry growers are reluctant to move forward with soilless systems. The approach is highly disruptive to the existing strawberry assemblage and its embeddedness in the place of California. These systems require major capital investments in infrastructure. The infrastructure includes trays, irrigation, and nutrition systems, and sometimes large amounts of plastic materials (although I understand there has also been experimentation with recycled materials). Since roots cannot grow deeply in these systems, plants also require daily monitoring for water and nutrition. So these systems tend to involve automation, as well.[20] Secondly, the substrate itself is costly and, like rice bran, not limitless. The peat moss comes from Canada, and thus far the coconut coir from Sri Lanka. Even the most hardened growers recognize the irony of importing materials from elsewhere—some of which would eventually have to be produced, not just extracted—to create an ostensibly more sustainable system in California. Third, they are not fail-proof: one grower experienced a *Fusarium* infection using substrate. Owing to these issues, some estimate that these systems double or triple annual per-acre investments, and were they to be widely adopted, the cost of strawberries would be out of reach for many consumers.

Mainly, though, indoor growing ignores the California industry's most central advantage: its location on the sandy-soiled, temperate coast. While using substrate outdoors maintains the climatic advantages of California, California growers are understandably concerned that buyers and others will set up greenhouse operations much closer to urban markets in the east, putting them out of business. They have reason to be fearful, as states like New Jersey have become sites of massive greenhouse operations. For now, then, the industry as a collectivity is not moving forward with this option. To the extent that soils become more unmanageable from disease and the climate continues to warm, this calculus could well change. But then California growers would be at a disadvantage in having to pay more for land highly subject to real estate development and speculation, without accessing its most primary agricultural benefits.

PROBIOTIC APPROACHES

In theory, probiotic approaches have vastly different underpinnings than these others. They are premised on the idea that controlling, containing, or simplifying the agro-environment, from eliminating the hedgerows and weeds that might be harboring pests to full-on CEA, are folly, if not utterly incoherent. Not only might containment leave out beneficial organisms; models of disease control that focus on preventing a pathogen's entry into a population are based on a conception of health as being free of disease rather than resilient and able to coexist. A containment or eradication approach renders the organisms of importance (in this case the plants) particularly vulnerable when efforts at containment fail.[21]

A probiotic approach is also premised on the idea that the old ways of killing life—antibiotic in an expansive sense—are no longer tenable, as eradication continues to undermine the function of ecological systems.[22] As put by geographer Jamie Lorimer, antibiotic approaches have become "excessive" to the extent that "obsessions with purity, division, simplicity and control are driving the emergence of new pathologies."[23] I'm inclined to agree, as I've tried to make a case that the strawberry industry's technologies of containment (for example plastic) and eradication (for example fumigation) may have drawn out the novel pathogens. In contrast, then, to these modern, antibiotic approaches, probiotic approaches take absences rather than "excessive presences" as foundational to ecological problems.[24] In a cropping system, for example, the problem may lie with insufficient and highly simplified soil biota rather than excessive pathogenic species. The aim is to encourage beneficial biota that will outcompete the more destructive kind. In that way, probiotic approaches do not avoid killing. Instead they involve killing in more creative and less violent ways, often by deliberately creating or allowing complexity in environments.[25] Yet there is a world of difference in what this means in actual farming practice.

Probiotic Inputs

With increased awareness of the importance of soil and soil microbiology, many pest control advisors and extension agents are recommending against strictly antibiotic approaches in favor of probiotic inputs. This can mean as little as adding soil amendments after fumigation, an approach equivalent to eating yogurt after a course of medical antibiotics to restore gut flora. Others are giving more credence to wholly probiotic approaches that work with the microorganisms that can suppress disease and stimulate plant growth. This can involve inoculation, where natural enemies of key pathogens are released directly into the soil, or amendments that alter microbial communities.[26] Researchers find these by locating soils where growers have reported no disease and mining that soil to isolate beneficial organisms. No doubt influenced by similar trends in human nutrition, researchers are also experimenting with feeding beneficial microbes—learning what causes them to grow and adding that material to the soil so these microbes have a better chance of surviving and multiplying. So far the interactions between soil microbial ecology and the cropping environment are poorly understood, however, as even suitable methods to assess disease presence are lacking.[27]

At this point it is unclear whether an input-based probiotic approach would ever be a sufficient fix, given that the amendments are incomplete substitutes for soil ecosystems. In addition, microbial amendments, like biologically derived fumigants, are commodities that growers must purchase at potentially high prices. In both ways, they are much like probiotic dietary treatments: costly and necessarily reductionist.[28] And again, such a reductionist approach to probiosis may miss the point.[29]

Agro-Ecological Methods

In contrast to input-based probiotic systems are ones designed to create biodiversity in situ, based on the supposition that a complex system keeps diseases in check. One aspect involves multi-cropping and multiple crop rotations, making mono-cropping anathema to such systems. Practitioners of agro-ecology also aim to produce all inputs on the farm. Agro-ecological methods may additionally include applications of brassica seed meals or other antimicrobial crop residues, the planting of so-called trap crops to attract the pathogens away from valued crops (in marked contrast to advice to keep weeds clear as harbors of pathogens), cover cropping, and composting. In strawberry production, rotations of broccoli, which has mild fumigation properties, are especially recommended.

According to many long-term organic strawberry growers and Eric Brennan, a research scientist for the USDA who has been conducting such research in California, agro-ecological systems are the only nonchemical alternative that seems efficacious for the long haul. But the requisite crop rotations are most effective when strawberries are planted on the same block only every four to five years, making them a minor crop in such integrated systems and particularly challenging economically in areas with high land values. And in such systems they are never grown at the scale most strawberry production has reached. Even when agro-ecological methods are assiduously adhered to, they are not foolproof. They require a good dose of tolerance for the presence of pests and diseases—and a willingness to engage with complexity. It is also worth mentioning that in practice even integrated operations use large amounts of plastic for weed control and solarization—they rarely meet agro-ecological ideals.

ORGANICS: THE NONDISTINCTIVE ALTERNATIVE

By now readers may be wondering where organics fit within the menu of possible fixes. Organic strawberry production has grown significantly. In 2000 there were 509 acres in organic production; by 2012 it had grown five-fold to 2,681 acres. The growth in organic strawberry sales was even more robust, from about $9.7 million in 2000 to $93.6 million in 2012.[30] By 2018, organic strawberry acreage comprised 12 percent of all planted strawberries, and even that was not enough for extant market demand.[31] Not surprisingly, among the major shippers, Driscoll's has taken a lead in organics, and in 2018 had more than 66 percent of market share for berries as a category.[32] They have been able to do this by providing some technical assistance to their contract growers, as well as cultivars designed for organic systems. But ultimately it is the growers' choice whether to embark on an organic program. So why and how do growers make that decision?

The rise in organics is only partially related to the increasingly restrictive regulatory environment for fumigants, especially because an organic program requires more than eliminating chemical fumigation. Through interviews I learned that only a small percentage of growers were transitioning to organics to learn how to grow without fumigants as their primary reason, while another small percentage were moved by the health risks of pesticides in general—for their family, nearby residents, or workers. The vast majority of transitioning growers claimed it was for market considerations. For some that meant they could make more money in organics. Cost studies have shown that production costs for organics are fairly comparable on a per-acre basis to conventional, but revenues are strikingly higher—so that profits can differ by more than $12,000 per acre.[33] For others this meant that buyers, whether intermediaries like Driscoll's or Giant, or farmers' market consumers, had shown interest in organic strawberries.[34] These two considerations amount to the same thing: some consumers are asking for and willing to pay more for berries grown organically, as public knowledge grows about the heavy pesticide regime for strawberries. The market for organic strawberries might not be so robust without the same conversations that are also leading to enhanced regulation.

What may be surprising is that all of the approaches above could be included in an organic program, making organics a less than distinct alternative, especially given exemptions on the use of fumigated starts. This, of course, has everything to do with what organics is—a voluntary system of regulation in which growers agree to abide by a set of standards in return for the label and the higher prices that ensue. As I wrote in *Agrarian Dreams: The Paradox of Organic Farming in California* (2014), the devil is in the standards, which never mandated a truly agro-ecological approach and thereby allowed input substitution approaches. Based on review by standard-setting entities, a less-toxic alternative might qualify as organic if it meets the basic criterion of being non-synthetic. Exceptions are occasionally made to that criterion when synthetics are virtual analogs to their natural counterparts and no feasible alternative exists. Exemptions are also made if organic material is simply not available.[35]

As it happens, the makers of Dominus petitioned the National Organic Standards Board (NOSB) to have it included as an organic material. As of 2015 the petition was unsuccessful, in part because as a synthetic it couldn't meet that high bar; the board also noted that the material's non-selectivity would have negative impact on soil microbial health and ecosystem biodiversity.[36] Still, other bio-fumigants coming on line could well be considered acceptable. Even the use of soilless substrate is allowed in organic production systems, albeit not without having been the subject of a major fight. Although hydroponic production systems have always been allowed in US organic production, they have never been favored by old-school organic growers. Following years of debate, in 2017 there was a vote at the NOSB to ban soilless systems. The ban was defeated by a small margin.[37] For the old-schoolers, organic production is philosophically rooted in ideas about building healthy soil, so that any soilless production is fundamentally at odds with the original meaning of "organically grown." They thus saw this decision as the final insult to already watered-down standards. Those championing soilless systems pointed to the meaning that "organic" has come to have for the vast majority of the public: grown without toxic pesticides. For now, strawberries grown in substrate can continue to receive organic certification, as long as other disallowed substances are avoided.

In practice, those growing organic strawberries, including Driscoll's growers, have varying degrees of commitment to its fundamental (probiotic) precepts, and some are unlikely to last under the organic rubric given the challenges unique to organic production. Many newly transitioned are working with ASD—indeed, experimenting with ASD is allowing them to dabble in organics. Success has been largely predicated on finding land that is pathogen free or was already certified or certifiable organic, allowing these growers to skirt three-year transitions.[38] And so some have begun organic production on land that had not been in crop production, such as pasture (a more popular option in relatively rural Santa Maria), while others have begun on land that was already certified organic and tested for pathogens. The lack of availability of suitable organic land alone is an obstacle to organic production.[39] It would be folly to begin an organic program on land that was already heavily diseased when none of the fumigation alternatives are effective at wholly eradicating pathogenic fungi.

The more fundamental obstacle is the need to rotate land to remain disease free—especially in the absence of effective antibiotic approaches that are allowable for organic programs. Growers hoped that ASD would obviate this need, but it is not clear that it has. As we saw early on, planting strawberries in back-to-back years is a surefire recipe for disease infestation, and so organic growers are advised to plant strawberries on a given block no more than every three to four years and to use probiotic methods to reduce disease. Again, planting broccoli or cauliflower before a rotation of strawberries is especially recommended because of their mild fumigation properties, even if these crops are plowed into the ground. Yet current land markets make three- to four-year rotations economically unviable, especially in the face of a generalized land shortage and the high land prices on the coasts. Some newly minted organic growers have tried to reduce or eliminate rotations, and word on the ground is that the results have been terrible. While they may have success at first, by year three they see major pathogen outbreaks. These growers may not much care. They are entering into organics when opportunities arise—when a piece of land comes along—and they are not loath to exit when it doesn't work. The tragedy of such flash-in-the-pan entry is that it adds to the overall pathogen load of soil.

Those who have been most successful at growing organic strawberries for the long haul follow agro-ecological precepts, using on-farm probiotic methods, including rotation with brassicas, compost, and cover crops. But, again, strawberries are a minor crop in these systems, making this a very different model economically. Those who have successfully employed agro-ecological methods and remained viable tend to have access to relatively cheap land. These growers, moreover, are able to obtain high price premiums for strawberries since they sell in direct markets (not through the major shippers) where consumers are willing to pay—and purchase the crops they rotate with strawberries, such as broccoli. Yet even these growers must make compromises owing to the absence of organic starts.

As I discussed in an earlier chapter, nurseries have thus far been exempt from the methyl bromide ban because of the requirements of California's clean plant program to produce disease-free plants. Nurseries use large amounts of chloropicrin as well, and all nurseries at this point fumigate all of their starts. One company attempted to propagate plants organically, but failed miserably because of disease. Since organic starts are nowhere available, organic fruit growers are exempt from having to use them, and so they purchase their starts from the nurseries, whether they like to or not.

That said, the nurseries themselves are concerned that the methyl bromide exemption may eventually be removed and know that chloropicrin alone is not entirely effective. So they, too, are experimenting with alternatives. One nursery I visited had been experimenting with both steam and substrate. The nursery that works with Driscoll's is also devising a way for the last year of plant propagation to be done organically, I believe with substrate, even though the details, not surprisingly, are proprietary knowledge. All agree that should the nurseries be forced to forgo fumigation, especially for all three to four years of plant propagation, growers' expenses for plants alone would increase so to make the fruit prohibitively expensive.[40] The yet-to-be-overcome obstacles to producing organic starts suggest that the outlook is not particularly good for producing strawberries entirely without fumigants.

. . .

To sum up so far, there is no singular solution on the horizon for managing soil pathogens that would allow California strawberry growers to continue operating at existing spatial scales and temporalities without greatly increasing the costs of production. The most promising model ecologically, one based on agro-ecology, is highly challenging economically given current land markets, where rents are based on the production of high-value strawberries every year. With commercial development pressure bearing down, it is not as though these land prices would decline if the industry as a whole shifted to agro-ecological systems.

Some solutions would also radically change the constituent elements of the strawberry assemblage, even extending it to other continents, in the case of soilless systems using coconut coir. Still, even the costliest solutions would not be as reliably effective as the combination of methyl bromide and chloropicrin once was, nor as easy to implement. To the contrary, at this point, farm advisors are conceding that growers need to employ a combination of approaches—a patchwork of chemical and nonchemical technologies, along the lines of integrated pest management, somewhat begging the question of whether antibiotic and probiotic approaches can even work together coherently.[41]

Clearly, not all farmers can survive as farmers in this changing environment. Not only are some of these approaches highly capital intensive, while others are just costly; many of these technologies require new knowledges and changed sensibilities. Whereas the silver-bullet solutions of the past both produced and allowed ignorance, the integrated approaches now being advised call for knowledge, willingness, and patience to deal with many moving parts—along with an acceptance that some plant loss is inevitable and must be lived with. These qualities, too, may be cause for industry shakeout. Growers with social capital as well as financial capital may be better positioned to obtain support and acquire new skills—especially as grower support itself becomes more proprietary.[42]

Which brings me to another problem with an increasingly proprietary and competitive industry, somewhat captured in this conversation I had with a farm advisor in June 2017:

> FARM ADVISOR: It seemed like you wanted to ask about the future of the industry, like where we're going or what we could improve. The best thing people could do is tone down the politics and quit suing each other and quit being so arrogant and greedy and become a team. . . .
>
> JG: What should they pay more attention to?
>
> FARM ADVISOR: We should organize a series of meetings to deal with all of the issues. Create a roadmap of what we need. Disease resistance, what characteristics do we need? We've got vertical gardening, are they going to be growing strawberries outside of New York City? What's the future look like and how do we prepare for it? Everyone needs to take a breath, discuss it, and come up with a plan.
>
> JG: Isn't part of the problem that . . . there are differing interests within the industry?
>
> FARM ADVISOR: And these guys can't control that. . . . That's my recommendation, but it's probably not realistic. The [industry] is its own worst enemy.

Despite the concerns of those in the work of repair—those who in their roles are obliged to secure the industry as a whole—there are clearly industry players who are not particularly interested in developing alternatives for the public domain so as to keep existing growers and shippers in business. Cutbacks in university cooperative extension have provided additional ecological niches for private-sector interests, beyond private breeding.[43] Many prospective drop-in replacements are being promoted by entrepreneurial innovators who seek a return on their research-and-development investments. The problem is not only that these proprietary technologies can squeeze growers economically. It is that they are being developed and disseminated behind closed doors, and the support growers need to adopt them is therefore being issued in proprietary ways as well. Again, Driscoll's may well be in the lead of farming without fumigants, but the knowledge they are developing and disseminating is not for public consumption.

In short, given the state and character of most alternatives, even the most optimistic prognosis for the industry is that the costs of production will rise tremendously, squeezing out low-resource growers. That many of the alternatives are being developed and disseminated by highly proprietary companies seems to point in addition to a highly consolidated industry with Driscoll's at the center.

A more pessimistic prognosis, given both the profound ecological and political-economic challenges of almost all options, is that the industry will fail. Is ruin something we should wish for?

ON THE (BIO)POLITICS OF
POST-PLANTATIONOCENE RUIN

The emergence of new pathogens in California's strawberry fields, with no obvious cure, is arguably indicative of a landscape in ruin. Echoing a point made by Lorimer that probiotic environmentalities take the present as already disastrous and requiring transformation, several scholars working in the register of multispecies ethnography write of landscapes "blasted by militarism and capitalism" and characterized by the "detritus of industrial food production."[44]

They call attention to the simplifications of what Donna Haraway first coined the "plantationocene," referring to an era of human-driven agricultural destruction, characterized by scaled-up industrial agricultures, guided by the exigencies of capital, which have depended on both highly alienated labor and highly toxic substances to manage the pests created by such simplification.[45] They note that such efforts at control have created novel ecologies beyond human control, in the form of pathogens, superweeds, or simply "dead land" that can no longer sustain much life at all.[46] For geographer Jason Moore, the increasing costs of making highly degraded nature suitable for capital accumulation signals the "end of cheap nature" and putatively a serious threat to capitalist accumulation writ large.[47]

But rather than imagining such landscapes as hopeless, at least some of these scholars see parasites, pathogens, and other emergent natures as "forces that generate diversity and thus open up new opportunities for flourishing."[48] Anthropologist Eben Kirksey, for example, writes that "such emergent dynamics may destroy existing orders but they also figure into collective hopes," and that we should pay attention to species that are thriving amid ruin as they beckon new futures, even if they hinder growth of the plants we desire.[49] Anthropologist Anna Tsing writes of the more modest possibilities opened up by the detritus of once mono-cropped forests. The delicious (and prized) matsutake mushroom thrives in this ruin, allowing precarious ecologies to connect up with the precarious people who forage these mushrooms and sell them for their livelihoods.[50]

These scholars also call for a different ethic, one that is less instrumental and more collaborative with other species. Kirksey calls for an ethic of working in collaboration with other species without intention to dominate, an "allowance of other life forms to flourish without utilitarian calculations," and a recognition of the degree of entanglement humans have with other species, perhaps leading to a deeper form of living-with.[51] Feminist soil scientist Kristina Lyons likewise extols those who "pursue sustainability with the understanding that growing food for humans necessarily implies nurturing an array of organisms that collaborate with (and oblige) farmers to regain a relative autonomy from chemical companies, aid packages, and other imperatives of capital."[52]

I share these scholars' concerns with climate change, degraded soils, toxic environments, and all other manner of ecological concern, but I grasp at trying to imagine how this ethic might apply to the strawberry fields of today. First of all, the strawberry fields may be highly troubled, even wilting as it were, but they are not landscapes in utter ruin.[53] They are surely not beyond reproach, but they are not dead yet. They remain working landscapes. And were they go the way of ruin, say, by the work of uncontrollable pathogen outbreaks—or increasing costs of pathogen management that would put so many growers out of business—they would not revert to the patches of life and refuge that are sometimes imagined in these accounts.[54] Let's be clear: in coastal California, what would flourish in the interstices would be sprawling suburbs.[55]

But I also wonder if it is desirable, much less possible, to shed the instrumental anthropocentrism of food production. It is axiomatic that agriculture entails collaborating with nonhumans and sometimes killing them to secure the health of the plant or animal that then becomes human food. Although industrial production may attempt to simplify these collaborations and relies more heavily on deliberate killing, probiotic approaches to agriculture do not hold a monopoly on working with and depending on other species and biophysical elements—as this book has surely made evident. That human life counts more in these interactions also seems evident.[56] Even probiotic interventions are almost always about making other life flourish in the interest of human life.[57] It is undoubtedly true that plantation economies have privileged capital accumulation over human life, in the sense that the structures of such economies have produced human hunger, as well as subjected certain subpopulations to arduous and toxic working conditions so that others may have affordable and abundant food.[58] Still, with so many of the planet's human inhabitants remaining food insecure, it seems untenable to dismiss the importance of food production, notwithstanding that food insecurity has rarely been driven by insufficient output. And with food insecurity in the United States increasingly recognized as a problem of poor nutrition and not so much insufficient food intake, it seems unreasonable as well to discount the importance of producing fresh fruit.

Finally, it seems that these somewhat hopeful accounts of ruin sometimes fail to take into account the many lives that are still intimately connected to the plantationocene—for better or for worse.[59] Plantations, though oppressive and exploitative, are not absent of people and social relations.[60] Here I'm reminded of Elspeth Probyn's *Eating the Ocean* (2016).[61] Though Probyn writes of loss, degradation, and poisoning of the sea and its inhabitants, she does not see a coherent politics of opting out of eating fish. Not only is fish in places we don't see (for instance the fish meal used in crop fertilizers), but fish are entangled with humans through and through, and an ethical food politics ought not to ignore the millions whose livelihoods depend on fish. Our politics of entanglement, Probyn suggests, must take account of which humans will get hurt by our actions. Letting go of a troubled landscape (or ocean) to let life live is more (dis)utopian than it admits to being.

With these arguments in mind, a brief consideration is in order of what the various options discussed above would mean for humans in the current strawberry assemblage. There is nothing to defend about the use of highly toxic substances, which is why the industry is now facing its present situation. Yet even "less toxic" chemicals would involve exposing fumigation workers and neighbors to toxins, so that eaters could continue to have affordable, healthy strawberries on a regular basis. Nonchemical soil disinfestation programs such as steam, solarization, and anaerobic conditions reduce immediate exposure to toxic chemicals but rely heavily on water in a drought-scourged state and carbon sources such as rice bran that may well be produced under chemical conditions. Such alternatives thus displace toxicity and harm. So might the use of soilless substrate.

There's also nothing to defend about workers bending over all day to pick strawberries. The move to soilless substrate, when used as a medium in waist-high trays, could improve working conditions and in that way might help resolve growers' labor shortages. But it could also relocate the industry to less high-value land markets and closer to East Coast urban centers. Or, just as likely, it could usher in the robotics that could displace so many workers. Here it must be remembered that the strawberry industry is extremely labor intensive, employing tens of thousands of people. Workers labor in strawberries because they have no other options, labor shortage notwithstanding, and so to relocate or, worse, eradicate those jobs would harm many lives, too.

The probiotic, integrative approaches of agro-ecology reduce farmworker exposure to toxic substances, and can create more employment for cultivation. But with strawberries as a minor crop, eaters would not see abundant, affordable, healthy strawberries year-round, not to mention that parents would have to cajole their children to eat more broccoli than strawberries. Indeed, except for drop-in replacements, all options seem to point to a reduction in supply and much higher prices for strawberries.

For some readers, this last point may seem a non-problem. You may ask, as many have asked me, Why do we need affordable strawberries year-round? They were once a luxury, so why can't they be again? Why *shouldn't* people pay more? Besides the usual objections to the inherently elitist solution of making high-quality food only available for the few, I see an infinite regress here. Eliminate strawberries, and the next-most-problematic fruit or vegetable becomes our object of concern. Eliminate fruits and vegetables—which all tend to be produced under at least somewhat toxic chemical and working conditions—and we are left with staples (think corn and soy) and livestock— as if they haven't been subject to criticism! At some point we have to grapple with how we produce food, and the one-by-one elimination from diets of those we deem most toxic will not do. It may be too glib to write off strawberries tout court. And it may be simply unfair to make responsibility lie with consumers, especially since the excess of affordable strawberries was the industry's and supporting science's own doing—a product of an assemblage bent on high productivity, whose excesses were only mitigated by attention to marketing.

All of this is to say that there are no easy outs—and no optimal solutions. Not even close. Antibiotic approaches are toxic and may be creating ecologies that further deteriorate the conditions of strawberry production, while continuing to allow ignorance about how those conditions are transforming. Probiotic approaches face formidable political and economic challenges around land costs, affordability, and input availability—especially inasmuch as they require a changed sensibility toward living-with. But abandonment of the industry would entail widespread loss of livelihoods, and arguably worsen the political vulnerability of those whose presence in California is unauthorized, while enabling strawberry fields to become suburbs. Short of a major political-economic transformation that would allow at least some of this to give (hope springs eternal!), rehabilitation may be the better ethic. But that, I suggest, would take a different kind of science. I take up this last concern in the conclusion that follows.

Conclusion

THE PROBLEM WITH THE SOLUTION

Solutions have become problems, putting biopower out of joint.

Historian of science Hannah Landecker, "Antibiotic
Resistance and the Biology of History," 2015[1]

One of the recommendations from our nonfumigant work group is we need to understand soil, if we're ever going to break this.... It's kind of basic and we haven't.

Interview with Brian Leahy, director of California
Department of Pesticide Regulation, 2015

IN THE 1920S, WHEN THE EARLY STRAWBERRY GROWERS first encountered wilted plants, abandonment may have crossed their minds. No doubt they had invested significant personal resources into those plants, yet strawberry production was still experimental—and certainly consumers hadn't come to expect to purchase strawberries whenever they wanted. Strawberries were indeed a luxury crop. Yet instead of abandonment, these growers, no doubt recollecting years of high profits, turned to the University of California, seeking repair for the conditions they were encountering. And UC complied, first with help in identifying the problem, next with providing advice, not all of which was sound, then with establishing a breeding program, and finally with fumigation and the various infrastructures that accompanied it, including plasticulture and drip irrigation. On-farm productivity went through the roof.

The solution of fumigation, a technology of eradication of, rather than cooperation with, some of the other biota that might have been found in strawberry fields, reverberated loudly throughout the rest of the production system. As other elements evolved in concert with fumigation, a new assemblage came into being. Fumigation, for one, allowed annual planting of strawberries on the same blocks of land, and rents and mortgages adjusted in expectation of yearly planting—and negligible plant death. Breeders were able to relinquish concerns with pathogen resistance and turn their gaze to productivity. Breeding for long seasons further increased yield (and land values), and breeding for durability allowed strawberries to be shipped to distant markets. New markets further incentivized innovations in productivity. Learning that plants would produce runners better in hot weather and that a subsequent cold snap would trick strawberry plants into dormancy, the industry moved its plant propagation activities to areas beyond the increasingly expensive coast, further extending the season and making room for more fruit production on the ideally temperate and sandy-soiled coasts. Separation of plant production from fruit production, along with fumigation, also helped keep seedlings and starts disease free. The high costs of these up-front investments in plants, fumigation, and other ground preparation, provided by hived-off businesses, made cheap, reliable labor an imperative for growers. So they settled on piece rates and close supervision as a way to control labor costs and ensure that the fragile berries were handled carefully. Fumigation also required that planting beds be made with the fumigated soil, keeping them close to the ground. But with politically constructed labor surpluses, growers worried less about making workers comfortable and chose instead cultivars and field conditions conducive to high productivity.[2] Since fumigation was key to productivity, piece rates were effectively set with the presumption of continued fumigation. The repair of fumigation, in short, produced a particular political economy of strawberry production that depended on the continued availability and efficacy of fumigation—as much as the discounting of workers' bodies and livelihoods.

Over time, advantageous aspects of that political economy began to falter: land suitable for strawberry production became scarce and more costly, particularly with the suburbanization of the coastal fruit-growing regions and farmland financialization; enforcement of the US-Mexico border—and better jobs elsewhere—made labor more scarce and somewhat more costly; the suburbanization of the coasts also brought residents less willing to live with the possibility of pesticide drift, adding to the regulatory scrutiny of fumigants. In addition, the institutions of repair themselves became more proprietary—no longer available for everyone, and often more costly. Not atypically, growers were squeezed between the low prices offered by shippers and the high costs of inputs and land demanded by providers.

Most crucially, the repair of fumigation, as it intersected with both local and global ecologies as well as that political economy, unleashed new ecological conditions or aggravated old ones, none well understood nor easily repairable. New pathogens emerged and were attributed to the restrictions on and inefficacy of the remaining chemical treatments. But it is just as possible that they emerged as intensified, mono-cropped strawberry production reached its tipping point, as years of drought and hotter-than-normal temperatures made the plants more vulnerable to opportunistic fungi. Other elements of the assemblage also might have been transformed by years of repair. Plant traits possibly useful to contend with changing climatic and regulatory conditions might have been buried or lost in a breeding mechanism that, in emphasizing productivity, used newer cultivars as parent plants. Breeding and propagating in fumigated soil might have selected for plants that were less resilient. Fumigation might have altered the character of the pathogens to make them more virulent and promiscuous. And fumigation most likely transformed the bodies of those who had become the labor force of choice. The repair of fumigation, in short, created novel ecologies that in all likelihood have denigrated the futurity of this particular assemblage. Fumigation, that is, made the strawberry industry a major contributor to the Anthropocene as well as a victim of it.

It is not as though fumigation was the sole pivot of the industry's success and its incipient decline. I have also suggested that the entire production system was capitalized into land values, bred into the strawberry, and emerged from the pathogen. And I have alluded to Miriam Wells's argument that the fixed costs of land, plants, and fumigation made the variable cost of labor the driving force of grower practice. That's the nature of an assemblage— elements act together and on one another, and no one node is the source or recipient of all transformations. One of the reasons that there is no simple and singular solution to replace fumigation is that all elements, including soils, plants, land values, wage rates, markets, climate, science, and more, have all coevolved at different temporalities and different spatial scales. Nevertheless, fumigation played a critical role in bringing the strawberry assemblage to the state it is in today, and the threat of no fumigation is what is driving industry imperatives to rework the assemblage.

One largely unrecognized problem is a lack of knowledge about exactly how and to what extent these transformations and coevolutions have taken place. Writing on the rise of antibiotic resistance, owing to widespread, more-than-prophylactic use of antibiotics in livestock feed, historian of science Hannah Landecker warns of the dangers of ignorance—and the danger of a science that doesn't take seriously its abilities to change the material world. As she discusses, theories of antibiotic resistance have shifted from ideas of mutation and "vertical inheritance" to those of horizontal gene transfer, in which genes coding resistance and virulence travel much more promiscuously than previously thought. Some scientists, she points out, have gone so far as to say that "the exchange of genes is so pervasive that the entire bacterial world can be thought of as one huge multicellular organism in which cells exchange their genes with ease."[3] There is a key difference between antibiotic treatments and chemical fumigants. In antibiotic formulations the biotic material, prone to gene transfer, is an active ingredient. That is not generally the case with synthetic chemical fumigants. But fumigants still do change the ecologies in which they are embedded. How can they not?

Those in the business of repair are therefore trying to fix something that is changing because of what they thought they had fixed. The assemblage, in other words, has evolved as a result of both the problem and the cure, and even the cures to the cure (the alternatives discussed in the last chapter). That makes pesticide reduction more than a political-economic problem of growers being squeezed or an ecological problem of pathogen resistance to fumigation. It makes it a knowledge problem, too.

The solution of fumigation was developed in institutions geared toward supporting farmers with their production problems: the land grant institutions. Narrowly disciplined and not integrated, the knowledge produced in these institutions was generally productivist and instrumental, with research directed at knowing pathogens, plants, chemicals, and soil only as they were detrimental or useful to the strawberry industry, and then isolating solutions to specific problems. Scientists thus paid little attention to how pathogenicity arose in the first place, how pathogenic fungi coevolved with the berry, whether ancient varieties were hosts to the pathogens, or even how various solutions transformed the pathogens and others in the assemblage. No doubt some of these questions were more difficult to research than others, and some questions were simply not formulated. That is the nature of undone science, and some questions had to await new scientific developments. Knowledge of how epigenetic and microbiomatic processes may affect the organisms of concern is still inchoate, for example. Nevertheless, fumigation as a technology of eradication was particularly conducive to ignoring complexity as a practical matter. (It's not that scientists were incapable or incurious!) As repairs became business opportunities and sources of competitive advantage within an increasingly proprietary industry, a more integrative science became an even more remote possibility. As with other applied agricultural sciences, efficacy and grower satisfaction, rather than understanding, became the primary measures of success.

To summarize the situation of today, the strawberry industry and the institutions that support it have created highly constrained conditions of possibility because of the way adopted solutions have reverberated throughout the assemblage. Compounding the problem, knowledge about how to address these conditions has itself been constrained by the solutions the science produced. To the extent that the solution of fumigation made the alternatives less viable economically, ecologically, and scientifically, the solution created the problem. Fumigation, that is, has become a source of iatrogenic harm—in multiple ways.

There are lessons here for other agricultural industries besides the deeply interesting strawberry industry. What this story tells us is that we should worry about chemical-intensive agriculture (of the manufactured kind, because, yes, chemicals are everywhere, as those in the industry like to remind us) not only for its toxicity to humans and others, and not even only for its capacity to unleash novel pathogens and other undesirable and often indeterminable environmental changes. We should worry about it because it gives rise to entire assemblages that are so entrenched that when toxicity becomes an issue or conditions of production are compromised, there is little room for maneuver. Agrochemicals, that is, allow us to avoid the hard work of figuring out how we're going to grow food in the future—for the many and not the few—and create agro-ecosystems more resilient to perturbations both ecological and economic.

. . .

I have now come to the what-is-to-be done moment that invariably concludes books like this. For those not attracted to the post-ruin imaginary discussed in the previous chapter, perhaps a more practical model of a more sustainable and just way to produce strawberries is what you're looking for. I have read more than one critique of agriculture that ends with just the example I am about to provide. It is Swanton Berry Farm, founded by longtime organic grower and quasi-visionary Jim Cochrane. Swanton Berry Farm became the first organic strawberry farm at a time when others said it couldn't be done. It became the only union strawberry farm in California when other growers said it had better *not* be done! Swanton also offers ownership opportunities in the form of stock bonuses to career-oriented employees, and is food-justice certified.[4] Cochrane was an avid spokesperson against the registration of methyl iodide. And I can say, without equivocation, that the strawberries that Swanton produces are delicious. There's much to admire about Swanton Berry Farm.

But it is also the exception that proves almost every rule. Swanton, the farm, was originally financed by a community credit union, not a bank, and has been able to obtain funds when others might not have—and with fewer stipulations. Swanton grows the Chandler variety, a university variety whose patent is expired and thus can be obtained at little cost. The variety is delicious and attractive but does not last long; it is thus not conducive to shipping too far. Swanton is located north of Santa Cruz city, about twenty-five miles away from the main strawberry-growing regions of the county and thus far from the centers of pestilence. Much of the land that Swanton leases is on conservation easement land, in a state park, or owned by a company partner, and thus is obtained at relatively low cost. Because the land is relatively inexpensive, the farm is able to rotate strawberries with brassicas or let fields lie fallow some years. Many of Swanton's customers recognize the importance of these rotations and make a point of buying their broccoli from the farm, as well as the delicious strawberries. And they buy the other farm commodities during the four to five months when strawberries are not available in that region. Swanton sells its crops on the farm, in local supermarkets, and in regional farmers' markets. The farm has attracted a loyal customer base that is willing and able to pay premium prices for an organic, union berry. Swanton's progressive labor arrangements have shored up a loyal and skilled labor force even though the farm is off the beaten track, in an area where farmworkers generally do not go looking for work. Swanton has seen challenges with both crop loss and water shortages, but has thus far weathered the storm. There is a good amount of living-with, made possible by a business that is more than a profit-making venture. The rule it doesn't break, because it can't, is the use of fumigated starts, a problem that Cochrane has protested. Nevertheless, it swims against the tide in most all other ways. For now it is not a replicable model for the larger strawberry industry—not in this ecology, not in this economy.

The challenges to a sustainable and just strawberry system are inseparably ecological, political-economic, and scientific. Again, there are no easy solutions—and certainly no perfection to be had. Even returning to the strawberry patches of yesteryear, and treating strawberries as a spring delight, would cause harm. But a lack of easy solutions should not prevent efforts to improve upon the toxicity, grower viability, and working conditions of current modes of strawberry production. Some of these improvements, most notably in regard to working conditions, can only come from political organizing and changed public policy. Technical approaches are simply not sufficient to improving wages and working conditions, even as many in certain science and engineering fields would like to pretend otherwise. But technical approaches can reduce the use of toxic materials, and we should continue to pursue these approaches, recognizing that none are perfect. But doing so, I think, will require a different kind of science as well as public investment, making for more than technical solutions.

Now that agricultural science is being called on to "repair the repair," as Christopher Henke puts it, a more integrative science can no longer be simply a scholarly pursuit.[5] Disciplinary siloing has contributed to the problem and remains an obstacle to knowledge that can contend with a rapidly changing environment. Radical interdisciplinarity is a practical imperative, and Department of Pesticide Regulation director Leahy's comment about how little attention has been paid to understanding soil is telling.[6] The strawberry assemblage of today is by definition an emergent one and requires a socioecological approach that can both bring understanding to assembled connections and then work with intention to do more good and less harm in producing what has come to be California's fruit. It thus requires a scientific sensibility that is cognizant of and deliberately incorporates knowledge of the political economy of agrarian production in addition to integrative biology and ecology and their post-genomic developments. Still, the context of science is as important as its epistemological foundations. To shift the assemblage to be less toxic yet still produce affordable strawberries requires a vibrant and well-funded public science—a university system that can pursue knowledge and technology on behalf of the broader good and not be beholden to whatever donors or "partners" can foot the bill. Finally, it requires that this knowledge be widely shared so that all have access to technologies and programs that encourage more sustainable practices. The monopolization of sustainability as a competitive advantage simply will not do.

Now, go eat your broccoli!

NOTES

PROLOGUE. THE BATTLE AGAINST METHYL IODIDE

1. For a fuller account of the methyl iodide story see Julie Guthman and Sandy Brown, "Midas' Not-So-Golden Touch: On the Demise of Methyl Iodide as a Soil Fumigant in California," *Journal of Environmental Policy and Planning* 18, no. 3 (2016): 324–41.

2. John Froines, Susan Kegley, Timothy Malloy, and Sarah Kobylewski, "Risk and Decision: Evaluating Pesticide Approval in California: Review of the Methyl Iodide Registration Process," Sustainable Technology and Policy Program, University of California at Los Angeles, 2013, https://www.pesticideresearch.com/site/wp-content/uploads/2012/05/Risk_and_Decision_Report_2013.pdf.

3. Julie Guthman and Sandy Brown, "I Will Never Eat Another Strawberry Again: The Biopolitics of Consumer-Citizenship in the Fight against Methyl Iodide in California," *Agriculture and Human Values* 33, no. 3 (2016): 575–85. Surprisingly, many of the public comments made by individuals stemmed from their positionality as consumers, even occasionally invoking a "not in my body" politics. This notion originally comes from E. Melanie DuPuis, "Not in My Body: BGH and the Rise of Organic Milk," *Agriculture and Human Values* 17, no. 3 (2000): 285–95.

4. Alison Hope Alkon, "Food Justice and the Challenge to Neoliberalism," *Gastronomica: The Journal of Food and Culture* 14, no. 2 (2014): 27–40; Dvera I. Saxton, "Strawberry Fields as Extreme Environments: The Ecobiopolitics of Farmworker Health," *Medical Anthropology* 34, no. 2 (2015): 166–83.

5. For discussions on food movement tactics see Patricia Allen, Margaret FitzSimmons, Michael Goodman, and Keith Warner, "Shifting Plates in the Agrifood Landscape: The Tectonics of Alternative Agrifood Initiatives in California," *Journal of Rural Studies* 19, no. 1 (2003): 61–75; Alison Alkon and Julie Guthman, *The New Food Activism: Opposition, Cooperation, and Collective Action* (Oakland: University of California Press, 2017).

1. Stephen Wilhelm, Richard C. Storkan, and John M. Wilhelm, "Preplant Soil Fumigation with Methyl Bromide-Chloropicrin Mixtures for Control of Soil-Borne Diseases of Strawberries: A Summary of Fifteen Years of Development," *Agriculture and Environment* 1 (1974): 234–35, my emphasis. Stephen Wilhelm was a plant pathologist at the University of California at Berkeley; Richard C. Storkan was the president of the fumigation company TriCal; and John Wilhelm was president of Niklor Chemical Company.

2. Heather Swanson, Anna Tsing, Elaine Gan, and Nils Bubandt, "Introduction: Bodies Tumbled into Bodies," in *Arts of Living on a Damaged Planet | Monsters of the Anthropocene*, ed. Anna Tsing, Heather Swanson, Elaine Gan, and Nils Bubandt (Minneapolis: University of Minnesota Press, 2017), M6, my emphasis.

3. California Department of Pesticide Regulation, "Our Director and Chief Deputy Director," accessed February 16, 2018, http://www.cdpr.ca.gov/dprbios.htm.

4. Department of Pesticide Regulation, "Nonfumigant Strawberry Production Working Group Action Plan," April 2013, http://www.cdpr.ca.gov/docs/pestmgt /strawberry/work_group/action_plan.pdf.

5. On iceberg lettuce see Susanne Freidberg, *Fresh: A Perishable History* (Cambridge, MA: Harvard University Press, 2009), 157–96. For broader discussions on the evolution of specialty crop production in California see Julie Guthman, *Agrarian Dreams: The Paradox of Organic Farming in California*, 2nd ed. (Berkeley: University of California Press, 2014); George L. Henderson, *California and the Fictions of Capital* (New York: Oxford University Press, 1998).

6. California Strawberry Commission, "California Strawberry Farming," January 2018, http://www.calstrawberry.com/Portals/2/Reports/Industry%20Reports /Industry%20Fact%20Sheets/California%20Strawberry%20Farming%20Fact%20 Sheet%202018.pdf; Dune Lawrence, "How Driscoll's Is Hacking the Strawberry of the Future," *Bloomberg Businessweek*, July 29, 2015, https://www.bloomberg .com/news/features/2015-07-29/how-driscoll-s-is-hacking-the-strawberry-of-the- future.

7. Laura Tourte, Mark Bolda, and Karen Klonsky, "The Evolving Fresh Market Berry Industry in Santa Cruz and Monterey Counties," *California Agriculture* 70, no. 3 (2016): 107–15.

8. Tourte, Bolda, and Klonsky, "The Evolving Fresh Market Berry Industry in Santa Cruz and Monterey Counties," 110.

9. Herbert Baum, *Quest for the Perfect Strawberry* (New York: iUniverse, 2005), 37–39.

10. Tourte, Bolda, and Klonsky, "The Evolving Fresh Market Berry Industry in Santa Cruz and Monterey Counties," 111.

11. Bernice Yeung, Kendall Taggart, and Andrew Donohue, "California's Strawberry Industry Is Hooked on Dangerous Pesticides," *Reveal News* (Center for Investigative Reporting), November 10, 2014, https://www.revealnews.org/article /californias-strawberry-industry-is-hooked-on-dangerous-pesticides/.

12. Environmental Working Group, "Dirty Dozen: EWG's 2018 Shopper's Guide to Pesticides in Produce," accessed February 16, 2018, https://www.ewg.org/foodnews/dirty_dozen_list.php.

13. Susan E. Kegley, Stephan Orme, and Lars Neumeister, "Hooked on Poison: Pesticide Use in California, 1991–1998" (San Francisco: Pesticide Action Network, 2000), accessed July 2, 2014, http://www.panna.org/issues/publication/hooked-poison (URL no longer available). This report is admittedly dated, but the fact that strawberries still are at the top of the Environmental Working Group's "Dirty Dozen" list, which is built on current data, suggests that the basic findings are not.

14. Eric Schlosser, "In the Strawberry Fields," *Atlantic Monthly* 276, no. 5 (1995): 80–108. Here Schlosser was riffing on a phenomenon that anthropologist Miriam Wells discussed in depth in a number of articles and also in her book released about that time. I will return to Wells's more scholarly account in due course.

15. Sameerah Ahmad, "Strawberries and Solidarity: Farmworkers Build Unity around Driscoll's Berries Boycott," *In These Times*, February 4, 2017, http://inthesetimes.com/working/entry/19927/farmworkers_build_unity_around_driscolls_berries_boycott.

16. Freidberg, *Fresh*, 157–96.

17. Dale Kasler, "Long-Standing Marriage Goes Sour for UC Davis, Strawberry Industry," *Sacramento Bee*, August 16, 2014, https://www.sacbee.com/news/business/article2606860.html.

18. Fumigation in that way seems to exemplify what Nils Bubandt terms the "necropolitics of the anthropocene," such that "human, animals, plants, fungi and bacteria live and die under conditions that might have been critically shaped by human activity but that are also increasingly outside of human control." Nils Bubandt, "Haunted Geologies: Spirits, Stones, and the Necropolitics of the Anthropocene," in *Arts of Living on a Damaged Planet | Monsters of the Anthropocene*, G124. Given the ubiquity of toxic chemicals, others define toxicity as a quintessential feature of the Anthropocene. On this point see Soraya Boudia, Angela N. H. Creager, Scott Frickel, Emmanuel Henry, Nathalie Jas, Carsten Reinhardt, and Jody A. Roberts, "Residues: Rethinking Chemical Environments," *Engaging Science, Technology and Society* 4 (2018): 165–78; Max Liboiron, Manuel Tironi, and Nerea Calvillo, "Toxic Politics: Acting in a Permanently Polluted World," *Social Studies of Science* 48, no. 3 (2018): 331–49.

19. On the nature of plantation agriculture see Donna Haraway, "Anthropocene, Capitalocene, Plantationocene, Chthulucene: Making Kin," *Environmental Humanities* 6, no. 1 (2015): 162n5. Note that Haraway's definition does not distinguish between plantations worked by slave labor and those worked by wage labor. Generally I do not refer to strawberry farms as plantations, for precisely that reason. I use this terminology here for its emphasis on ecological relations and not the social relations of production.

20. Haraway, "Anthropocene, Capitalocene, Plantationocene, Chthulucene," 159–65; Donna Haraway, Noboru Ishikawa, Scott F. Gilbert, Kenneth Olwig, Anna L. Tsing, and Nils Bubandt, "Anthropologists Are Talking—About the Anthropocene," *Ethnos* 81, no. 3 (2016): 535–64.

21. Jennifer Clapp, Peter Newell, and Zoe W. Brent, "The Global Political Economy of Climate Change, Agriculture and Food Systems," *Journal of Peasant Studies* 45, no. 1 (2018): 80–88.

22. On the possibilities of de-scaled, more heterogeneous modes of living see Eben Kirksey, *Emergent Ecologies* (Durham, NC: Duke University Press, 2015); and to a lesser extent Anna Lowenhaupt Tsing, *The Mushroom at the End of the World: On the Possibility of Life in Capitalist Ruins* (Princeton, NJ: Princeton University Press, 2015). For a critique of works that celebrate highly degraded nature or ruination as signaling the end of capitalism see Cynthia Morinville and Nicole Van Lier, "'Dead' Land and the End of Capitalism? Interrogating the Politics of Toxic Environments and Capitalist Accumulation," *Capitalism Nature Socialism* (in review). They argue that such works don't account for the violence that ruination entails for human lives still attached to capitalist natures. As a side note, Tsing implies that plantations are absent of sociality; workers are "self-contained and interchangeable" units with no attachments (38–40). Her characterization is notably different than that of anthropologist Sarah Besky, who has written of efforts to bring fair trade to Darjeeling tea plantations. Besky shows that tea workers, while clearly exploited, have certain attachments to the plantations on which they reside, to the degree that they regard fair trade efforts as having actually undermined their social security. Sarah Besky, *The Darjeeling Distinction: Labor and Justice on Fair-Trade Tea Plantations in India* (Berkeley: University of California Press, 2013). I would note in addition that the migrant wage laborers on strawberry "plantations" have strong social attachments, arriving and working through social networks, but also investments in the continuation of such plantations as a source of livelihood (see chapter 6).

23. See Rachel Schurman and William A. Munro, *Fighting for the Future of Food: Activists versus Agribusiness in the Struggle over Biotechnology* (Minneapolis: University of Minnesota Press, 2010) for their observations of the utterly distinct "lifeworlds" of the biotechnology industry and anti-biotech activists.

24. H. Vincent Moses, "'The Orange-Grower Is Not a Farmer': G. Harold Powell, Riverside Orchardists, and the Coming of Industrial Agriculture, 1893–1930," *California History* 74, no. 1 (1995): 22–37. Steven Stoll also documents those who saw fruit growing as "high-class farming" (46) and who saw themselves as businessmen, not toilers, in *The Fruits of Natural Advantage: Making the Industrial Countryside in California* (Berkeley: University of California Press, 1998).

25. Richard Walker, *The Conquest of Bread: 150 Years of Agribusiness in California* (New York: New Press, 2004), 10.

26. This thesis originates with the writings of Karl Kautsky, *The Agrarian Question* (London: Zwan, 1988). Writing in 1899, in anticipation of the Russian Revolution, Kautsky was concerned that farming had not even reached the stage of capitalism necessary for a socialist transition. Seeking to explain arrested development, he argued that "the peculiarities of agriculture, its biological character and rhythms coupled with the capacity for family farms to survive through self-exploitation (i.e., working longer and harder to effectively depress 'wage levels'), might hinder some tendencies, namely, the development of classical agrarian capitalism." Quoted in

David Goodman and Michael Watts, "Agrarian Questions, Global Appetite, Local Metabolism: Nature, Culture, and Industry in Fin-De-Siècle Food Systems," in *Globalising Food: Agrarian Questions and Global Restructuring*, ed. David Goodman and Michael Watts (London and New York: Routledge, 1997), 9. Kautsky therefore predicted that capitalism would take hold around the farm, but that farms themselves would mostly remain as noncapitalist enterprises. He also gave much emphasis to the land-based aspects of agricultural production as obstacles to capitalism. For variations on these ideas see also Ben Fine, "Towards a Political Economy of Food," *Review of International Political Economy* 1, no. 3 (1994): 519–45; Jack Kloppenburg, *First the Seed: The Political Economy of Plant Biotechnology* (Madison: University of Wisconsin Press, 2005); Susan A. Mann, *Agrarian Capitalism in Theory and Practice* (Chapel Hill: University of North Carolina Press, 1989).

27. Susan A. Mann and James M. Dickinson, "Obstacles to the Development of a Capitalist Agriculture," *Journal of Peasant Studies* 5, no. 4 (1978): 466–81.

28. This idea is based in the concept of "differential rent" first discussed by David Ricardo, *Principles of Political Economy and Taxation* (London: G. Bell, 1891). Ricardo theorized that land rents would vary in accordance with the yield that the land was able to produce.

29. Ted Benton, "Marxism and Natural Limits: An Ecological Critique and Reconstruction," *New Left Review* 178 (1989): 51–86; Mann and Dickinson, "Obstacles to the Development of a Capitalist Agriculture," 466–81.

30. David Goodman, Bernardo Sorj, and John Wilkinson, *From Farming to Biotechnology* (Oxford: Basil Blackwell, 1987).

31. On the history of shippers in the fruit business see Stoll, *The Fruits of Natural Advantage*.

32. Don Mitchell, *"They Saved the Crops": Labor, Landscape, and the Struggle over Industrial Farming in Bracero-Era California* (Athens: University of Georgia Press, 2012), 198–202.

33. Jill Lindsey Harrison, *Pesticide Drift and the Pursuit of Environmental Justice* (Cambridge, MA: MIT Press, 2011), 70.

34. Mitchell, *"They Saved the Crops,"* 198–202; Michael J. Watts, "Life under Contract: Contract Farming, Agrarian Restructuring, and Flexible Accumulation," in *Living under Contract*, ed. Michael J. Watts and Peter Little (Madison: University of Wisconsin Press, 1993), 21–78.

35. William Boyd, Scott Prudham, and Rachel Schurman, "Industrial Dynamics and the Problem of Nature," *Society and Natural Resources* 14, no. 7 (2001): 555–70.

36. Willard W. Cochrane, *The Development of American Agriculture: A Historical Analysis* (Minneapolis: University of Minnesota Press, 1979), 417–36.

37. Warren Dean, *Brazil and the Struggle for Rubber: A Study in Environmental History* (Cambridge, England: Cambridge University Press, 1987), 4, 163.

38. John Soluri, *Banana Cultures: Agriculture, Consumption, and Environmental Change in Honduras and the United States* (Austin: University of Texas Press, 2005).

39. Fine, "Towards a Political Economy of Food," 519–45.

40. On scientific practice see Bruno Latour and Steve Woolgar, *Laboratory Life: The Social Construction of Scientific Facts* (Beverly Hills, CA: Sage, 1979); Steven Shapin, *A Social History of Truth: Civility and Science in Seventeenth-Century England* (Chicago: University of Chicago Press, 1994).

41. W. Scott Prudham, *Knock on Wood: Nature as Commodity in Douglas-Fir Country* (New York: Routledge, 2005), 114.

42. Land grant colleges were a product of the 1862 Morrill Act, which mandated the US government to use the proceeds of public land sales to help states and territories create colleges to teach agriculture and the mechanical arts. The Hatch Act of 1887 built on the Morrill Act by establishing experiment stations connected to the land grants to conduct applied agricultural research. The Smith-Lever Act of 1914 then set up cooperative extension services, connected to the land-grant colleges, which would disseminate this research to farmers. Finally, the Bankhead-Jones Act of 1935 increased federal funding of the land-grant institutions. See Alan P. Rudy, Dawn Coppin, Jason Konefal, Bradley T. Shaw, Toby Van Eyck, Craig Harris, and Lawrence Busch, *Universities in the Age of Corporate Science: The UC Berkeley–Novartis Controversy* (Philadelphia: Temple University Press, 2007).

43. Christopher R. Henke, *Cultivating Science, Harvesting Power* (Cambridge, MA: MIT Press, 2008).

44. Henke, *Cultivating Science, Harvesting Power*, 10. Stoll makes a similar point, accentuating that crop specialization produced all sorts of problems and threats to which these institutions were set up to respond, in *The Fruits of Natural Advantage*.

45. Richard C. Sawyer, *To Make a Spotless Orange: Biological Control in California* (West Lafayette, IN: Purdue University Press, 2002); Stoll, *The Fruits of Natural Advantage*.

46. Timothy Mitchell, "Can the Mosquito Speak?," in *Rule of Experts: Egypt, Techno-Politics, Modernity* (Berkeley: University of California Press, 2002), 19–53. The essay title is a riff on Gayatri Chakravorty Spivak's essay about marginalized peoples who do not have the opportunity to write their own histories: "Can the Subaltern Speak?," in *Can the Subaltern Speak? Reflections on the History of an Idea*, ed. Rosalind Morris (New York: Columbia University Press, 1988), 21–78. Mitchell was thus implying that neither do nonhuman actors.

47. Mitchell, "Can the Mosquito Speak?," 24–25. In narrating a similar story of the outbreaks of dengue that followed the DDT era, Alex Nading hints at the contingencies by which such diseases move from minor problems to something more. Dengue, for example, has four different serotypes, which once were endemic only to distinct geographic regions. Noting that it takes the infection of multiple serotypes to produce a fatal case of the disease, these serotypes had to be brought together in one place to produce fatal illness. Yet even then, he says, the disease thrived on "underdevelopment of technology for housing, feeding and watering people." Alex M. Nading, *Mosquito Trails: Ecology, Health, and the Politics of Entanglement* (Berkeley: University of California Press, 2014), 131.

48. See Ben Anderson and Colin McFarlane, "Assemblage and Geography," *Area* 43, no. 2 (2011): 124–27; Bruce Braun, "Environmental Issues: Global Natures in the Space of Assemblage," *Progress in Human Geography* 30, no. 5 (2006): 644–54; Paul Robbins and Brian Marks, "Assemblage Geographies," in *The Sage Handbook of Social Geographies*, ed. S. Smith, S. Smith, R. Pain, S. Marston and J. P. Jones (Beverly Hills, CA: Sage, 2010), 176–94. The idea of assemblage has a much deeper provenance, and its usage is debated. Since I'm using the concept primarily as method, I do not elaborate those debates.

49. Scholars trying to address the agency of nonhuman actors in contributing to various phenomena have often drawn on the work of Bruno Latour and his influential actor network theory (ANT). Bruno Latour, *The Pasteurization of France*, trans. Alan Sheridan and John Law (Cambridge, MA: Harvard University Press, 1993). There are volumes written on this, and a particularly good discussion is presented by Edwin Sayes, "Actor-Network Theory and Methodology: Just What Does It Mean to Say That Nonhumans Have Agency?," *Social Studies of Science* 44, no. 1 (2014): 134–49.

50. For a useful discussion of distributed agency see Hannah Appel, Arthur Mason, and Michael Watts, "Introduction: Oil Talk," in *Subterranean Estates: Life Worlds of Oil and Gas*, ed. Hannah Appel, Arthur Mason, and Michael Watts (Ithaca, NY: Cornell University Press, 2015), 1–26.

51. The idea of intra-action is drawn from the work of feminist science studies scholar Karen Barad to connote phenomena observable only by relation, as things that act together (as opposed to interact, which connotes phenomena mutually shaping each other). Karen Barad, *Meeting the Universe Halfway: Quantum Physics and the Entanglement of Matter and Meaning* (Durham, NC: Duke University Press, 2007). Interest in relational ontologies has become quite pervasive in academic geography and elsewhere. As described by geographer Becky Mansfield, in a relational ontology, the relationship precedes the element being constituted "because it is through these relationships that objects themselves come to be defined, no one element can be determinant of the others. It is only in their constellation that any particular element (be it 'social' or 'natural') gains its significance." Becky Mansfield, "Fish, Factory Trawlers, and Imitation Crab: The Nature of Quality in the Seafood Industry," *Journal of Rural Studies* 19, no. 1 (2003): 13. See also Dawn Biehler and John-Henry Pitas, "From Mosquitoes in the System to Transformative Entanglements with Ecologies of Injustice," unpublished paper presented at the annual meeting of the American Association of Geographers, Chicago, 2015; Abigail Neely and Thokozile Nguse, "Entanglements, Intra-Actions, and Diffraction," in *The Routledge Handbook of Political Ecology*, ed. Tom Perreault, Gavin Bridge, and James McCarthy (Abingdon-on-Thames, England: Routledge, 2015), 140–49. While I find the idea of relational ontologies attractive, they seem more easily theorized than written. It is difficult to find language to discuss events and things whose ontological status is created by their intra-action. Sometimes our language doesn't allow us to make the theoretical claims we wish to make—and we ought to wonder why. I think it may be sufficient to say that all elements potentially shape and

transform one another. Marianne Elisabeth Lien, *Becoming Salmon: Aquaculture and the Domestication of a Fish* (Berkeley: University of California Press, 2015), 107.

52. Laura A. Ogden, Billy Hall, and Kimiko Tanita, "Animals, Plants, People, and Things: A Review of Multispecies Ethnography," *Environment and Society* 4, no. 1 (2013): 7.

53. For an introduction to critical realism see Andrew Sayer, *Method in Social Science*, 2nd ed. (London: Routledge, 1992). For an articulation of actor network theory see Latour, *The Pasteurization of France*. Some have attempted to reconcile Latourian perspectives with political economy. Harold Perkins, for example, locates the tension in the status of objects. Objects in Marxian terms cloak the commodity fetish and cannot have status similar to human labor. For Perkins, Marx's view of machinery as "dead labor" provides a starting point for thinking about how nonhumans might be conceptualized in a Marxian framework. Harold A. Perkins, "Ecologies of Actor-Networks and (Non)social Labor within the Urban Political Economies of Nature," *Geoforum* 38, no. 6 (2007): 1152–62. For another discussion of reconcilement see Brian J. Gareau, "We Have Never Been Human: Agential Nature, Ant, and Marxist Political Ecology," *Capitalism Nature Socialism* 16, no. 4 (2005): 127–40. Others say that it is simply not possible to combine these theories coherently. Rebecca Lave, "Reassembling the Structural: Political Ecology and Actor-Network Theory," in *The Routledge Handbook of Political Ecology*, 213–23.

54. Malini Ranganathan, "Storm Drains as Assemblages: The Political Ecology of Flood Risk in Post-Colonial Bangalore," *Antipode* 47, no. 5 (2015): 1300–1320; Robbins and Marks, "Assemblage Geographies," 176–94.

55. Appel, Mason, and Watts are similarly ecumenical in their introduction to *Subterranean Estates*. Rebutting accounts of oil in which oil the substance is singularly determinative, they call for a "critical topology" of oil in which "states, companies, universities, insurgents, local geographies, representations, and the materiality of oil itself cannot be approached as discrete, but must be understood as co-constitutive." Appel, Mason, and Watts, "Introduction: Oil Talk," 17. John Allen and Stephanie Lavau make a similar move in forwarding a relational economy of disease: "Rather than reify corporate power, we seek a more distributed and non-deterministic account of the various human and non-human agencies that come together in industrial animal production, and through which disease comes into being, multiplies and spreads." John Allen and Stephanie Lavau, "'Just-in-Time' Disease: Biosecurity, Poultry and Power," *Journal of Cultural Economy* 8, no. 3 (2015): 343.

56. Brian King and Eric Carter draw on similar analytics in explaining HIV positivity in South Africa and malaria in Argentina, respectively. Brian King, *States of Disease: Political Environments and Human Health* (Oakland: University of California Press, 2017); Eric D. Carter, *Enemy in the Blood: Malaria, Environment, and Development in Argentina* (Tuscaloosa: University of Alabama Press, 2012).

57. Bruce Braun, "Thinking the City through SARS: Bodies, Topologies, Politics," in *Networked Disease: Emerging Infections in the Global City*, ed. S. Harris Ali and Roger Keil (Oxford: Wiley-Blackwell, 2008), 250–66; Steve Hinchliffe, John Allen, Stephanie Lavau, Nick Bingham, and Simon Carter, "Biosecurity and the

Topologies of Infected Life: From Borderlines to Borderlands," *Transactions of the Institute of British Geographers* 38, no. 4 (2013): 531–43; Jamie Lorimer, "Probiotic Environmentalities: Rewilding with Wolves and Worms," *Theory, Culture and Society* 34, no. 4 (2017): 27–48.

58. Steve Hinchliffe, Nick Bingham, John Allen, and Simon Carter, *Pathological Lives: Disease, Space and Biopolitics* (Chichester, England: John Wiley and Sons, 2016). See also Allen and Lavau, "'Just-in-Time' Disease," 342–60; Lorimer, "Probiotic Environmentalities," 27–48. The metaphor of topology draws from geometric notions, in which constituent parts stretch and fold onto each other rather than tear apart, intensifying the relationships.

59. Hinchliffe et al., "Biosecurity and the Topologies of Infected Life," 539. Lien makes a similar point about the problem of sea lice in *Becoming Salmon*, 30.

60. The notion of chemicals as actors is increasingly gaining attention in both geography and anthropology. Importantly, this work is not necessarily anti-chemical but recognizes the ubiquity of chemicals in shaping socioecological assemblages, and therefore the impossibility of governing chemicals as a way to keep them at bay. Andrew Barry, "Manifesto for a Chemical Geography," lecture, University College of London, 2017, http://www.academia.edu/32374031/Manifesto_for_a_Chemical_Geography_2017_; Boudia et al., "Residues," 165–78; Liboiron, Tironi, and Calvillo, "Toxic Politics," 331–49; Adam M. Romero, Julie Guthman, Ryan E. Galt, Matt Huber, Becky Mansfield, and Suzana Sawyer, "Chemical Geographies," *GeoHumanities* 3, no. 1 (2017): 158–77; Nicholas Shapiro and Eben Kirksey, "Chemo-Ethnography: An Introduction," *Cultural Anthropology* 32, no. 4 (2017): 481–93.

61. Mitchell, "Can the Mosquito Speak?," 47, 48.

62. Nading, *Mosquito Trails*, 20. In the assemblage Nading describes, underinvestment in infrastructure is the source of problems, but in the strawberry assemblage, overengineering is causing the problem.

63. Alex M. Nading, "Local Biologies, Leaky Things, and the Chemical Infrastructure of Global Health," *Medical Anthropology* 36, no. 2 (2017): 141. Anthropologist Alex Blanchette's speculations about the constant exposure of livestock workers' bodies to porcine fecal dust presents a particularly disquieting case of such harm. Given that hog diets are laced heavily with antibiotics, hog workers are essentially receiving daily doses from inhaling the desiccated shit that is ubiquitous in these operations. Blanchette thus speculates that not only may such exposures exacerbate workers' own resistance to the powerful drugs, they may transform workers bodies morphologically, in ways similar to the intended effects of prophylactic antibiotic use of such drugs on hog bodies. Alex Blanchette, "Living Dust, Toxic Health, and the Labor of Antibiotic Resistance on Factory Farms," *Medical Anthropology Quarterly* (forthcoming).

64. Ryan E. Galt, *Food Systems in an Unequal World: Pesticides, Vegetables, and Agrarian Capitalism in Costa Rica* (Tucson: University of Arizona Press, 2014), 210.

65. Shapiro and Kirksey, "Chemo-Ethnography: An Introduction," 484.

66. Anderson and McFarlane, "Assemblage and Geography," 125.

67. Lien, *Becoming Salmon*, 73, 74.

68. Clevo Wilson and Clem Tisdell, "Why Farmers Continue to Use Pesticides Despite Environmental, Health and Sustainability Costs," *Ecological Economics* 39, no. 3 (2001): 449–62; Harrison, *Pesticide Drift and the Pursuit of Environmental Justice*, 26; Galt, *Food Systems in an Unequal World*, 65.

69. Galt, *Food Systems in an Unequal World*, 81.

70. Lien makes a similar point in arguing against standard accounts of agricultural industrialization. Even as farmed salmon, the object of her inquiry, are the most domesticated of fish, to tell the story as one of (political-economic) industrialization is to "gloss over heterogenic specificity as well as instances of noncoherence." Lien, *Becoming Salmon*, 6.

71. Henke, *Cultivating Science, Harvesting Power*, 144.

72. Henke, *Cultivating Science, Harvesting Power*, 51.

73. For an account of hybridization and the role of public universities see Kloppenburg, *First the Seed*.

74. Rebecca Lave, Philip Mirowski, and Samuel Randalls, "Introduction: STS and Neoliberal Science," *Social Studies of Science* 40, no. 5 (2010): 659–75; Philip Mirowski, *Science-Mart: Privatizing American Science* (Cambridge, MA: Harvard University Press, 2011); Rudy et al., *Universities in the Age of Corporate Science*.

75. On cigarettes see Robert Proctor and Londa L. Schiebinger, eds., *Agnotology: The Making and Unmaking of Ignorance* (Palo Alto, CA: Stanford University Press, 2008). Harrison and Robbins imply this about agrochemicals: Harrison, *Pesticide Drift and the Pursuit of Environmental Justice*, 85–144; Paul Robbins, *Lawn People: How Grasses, Weeds, and Chemicals Make Us Who We Are* (Philadelphia: Temple University Press, 2007), 45–71.

76. Scott Frickel, Sahra Gibbon, Jeff Howard, Joanna Kempner, Gwen Ottinger, and David J. Hess, "Undone Science: Charting Social Movement and Civil Society Challenges to Research Agenda Setting," *Science, Technology and Human Values* 35, no. 4 (2010): 444–73; Daniel Lee Kleinman and Sainath Suryanarayanan, "Dying Bees and the Social Production of Ignorance," *Science, Technology, and Human Values* 38, no. 4 (2013): 492–517.

77. Frank Uekötter and Uwe Lübken, "Introduction: The Social Functions of Ignorance," in *Managing the Unknown: Essays on Environmental Ignorance*, ed. Frank Uekötter and Uwe Lübken (New York: Berghahn, 2014), 1–11.

78. Henke, *Cultivating Science, Harvesting Power*, 51–57.

79. Uekötter and Lübken, "Introduction: The Social Functions of Ignorance," 1–11; Henke, *Cultivating Science, Harvesting Power*.

80. Frickel et al., "Undone Science," 448.

81. Frank Uekötter, "Ignorance Is Strength: Science-Based Agriculture and the Merits of Incomplete Knowledge," in *Managing the Unknown*, 130.

82. Frank Uekötter, "Farming and Not Knowing: Agnotology Meets Environmental History," in *New Natures: Joining Environmental History with Science and Technology Studies*, ed. Dolly Jorgensen, Finn Arne Jorgensen, and Sara Pritchard (Pittsburgh: University of Pittsburgh Press, 2013), 37–50.

83. Sawyer, *To Make a Spotless Orange*, xxiv.

84. Mitchell, "Can the Mosquito Speak?," 46.

85. Uekötter, "Farming and Not Knowing," 37–50.

86. Uekötter, "Ignorance Is Strength," 122–39.

87. Hannah Landecker, "Antibiotic Resistance and the Biology of History," *Body and Society* 22, no. 4 (2015): 1–34.

88. Brian J. Gareau, "Dangerous Holes in Global Environmental Governance: The Roles of Neoliberal Discourse, Science, and California Agriculture in the Montreal Protocol," *Antipode* 40, no. 1 (2008): 108.

89. Department of Pesticide Regulation, "Nonfumigant Strawberry Production Working Group Action Plan."

90. See Madison Barbour and Julie Guthman, "(En)gendering Exposure: Pregnant Farmworkers and the Inadequacy of Pesticide Notification," *Journal of Political Ecology* 25, no. 1 (2018): 332–49.

91. Scott F. Gilbert, "Holobiont by Birth: Multilineage Individuals as the Concretion of Cooperative Processes," in *Arts of Living on a Damaged Planet | Monsters of the Anthropocene*, M73–M90; Margaret McFall-Ngai, "Noticing Microbial Worlds: The Postmodern Synthesis in Biology," in *Arts of Living on a Damaged Planet | Monsters of the Anthropocene*, M51–M72.

CHAPTER TWO. EMERGENT PATHOGENS

1. Steve Hinchliffe, John Allen, Stephanie Lavau, Nick Bingham, and Simon Carter, "Biosecurity and the Topologies of Infected Life: From Borderlines to Borderlands," *Transactions of the Institute of British Geographers* 38, no. 4 (2013): 537.

2. Sandy Brown was an early collaborator on the project.

3. Sometime after that interview took place, DPR proposed to expand buffer zones around schools, generating an intense debate. For a provocative analysis see Keli Benko, "People Need To Know! Notification and the Regulation of Pesticide Use near Public Schools in California," in review, *Environmental and Planning E: Nature and Space.*

4. I. Berlanger and M. L. Powelson, "*Verticillium* Wilt," American Phytopathological Society website, accessed November 21, 2000, http://www.apsnet.org /edcenter/intropp/Lessons/fungi/ascomycetes/Pages/VerticilliumWilt.aspx.

5. Steven J. Klosterman, Zahi K. Atallah, Gary E. Vallad, and Krishna V. Subbarao, "Diversity, Pathogenicity, and Management of *Verticillium* Species," *Annual Review of Phytopathology* 47 (2009): 39–62; S. Koike, K. Subbarao, R. Michael Davis, and T. Turini, "Vegetable Diseases Caused by Soilborne Pathogens," Report 8099, University of California Division of Agriculture and Natural Resources, 2003, https://anrcatalog.ucanr.edu/pdf/8099.pdf; Albert O. Paulus, "Fungal Diseases of Strawberry," *Horticultural Science* 25, no. 8 (1990): 885–89; G. F. Pegg, "The Impact of *Verticillium* Diseases in Agriculture," *Phytopathologia Mediterranea* 23, nos. 2/3 (1984): 176–92.

6. Klosterman et al., "Diversity, Pathogenicity, and Management of *Verticillium* Species," 39–62; Koike et al., "Vegetable Diseases Caused by Soilborne Pathogens."

7. Pegg, "The Impact of *Verticillium* Diseases in Agriculture," 189.

8. Steve Hinchliffe, Nick Bingham, John Allen, and Simon Carter, *Pathological Lives: Disease, Space and Biopolitics* (Chichester, England: John Wiley and Sons, 2016).

9. Also see Berlanger and Powelson, "*Verticillium* Wilt"; Koike et al., "Vegetable Diseases Caused by Soilborne Pathogens."

10. Ingrid M. Parker and Gregory S. Gilbert, "The Evolutionary Ecology of Novel Plant-Pathogen Interactions," *Annual Review of Ecology Evolution and Systematics* 35 (2004): 675–700; Surinder Kaur, Gurpreet Singh Dhillon, Satinder Kaur Brar, Gary Edward Vallad, Ramesh Chand, and Vijay Bahadur Chauhan, "Emerging Phytopathogen *Macrophomina phaseolina*: Biology, Economic Importance and Current Diagnostic Trends," *Critical Reviews in Microbiology* 38, no. 2 (2012): 136–51. On metacommunity theory as it relates to pathogenicity, Borer et al. write, "The proliferation of pathogenic microbes depends on single-species dynamics and multispecies interactions occurring within and among host cells, the spatial organization and genetic landscape of hosts, the frequency and mode of transmission among hosts and host populations, and the abiotic environmental context." Elizabeth T. Borer, Anna-Liisa Laine, and Eric W. Seabloom, "A Multiscale Approach to Plant Disease Using the Metacommunity Concept," *Annual Review of Phytopathology* 54 (2016): 397.

11. Hinchliffe et al., *Pathological Lives.*

12. James C. Scott, *Seeing Like a State* (New Haven, CT: Yale University Press, 1998), 268–69. As noted by Scott, fields populated by genetically identical individuals are vulnerable to a given pathogen in exactly the same way.

13. Timothy Mitchell, "Can the Mosquito Speak?," in *Rule of Experts: Egypt, Techno-Politics, Modernity* (Berkeley: University of California Press, 2002), 19–53. Also see Marianne Elisabeth Lien, *Becoming Salmon: Aquaculture and the Domestication of a Fish* (Berkeley: University of California Press, 2015); Marion W. Dixon, "Biosecurity and the Multiplication of Crises in the Egyptian Agri-Food Industry," *Geoforum* 61, no. supplement C (2015): 90–100.

14. Ideas about changing this sensibility to one of "living with" can be found in Anna Lowenhaupt Tsing, *The Mushroom at the End of the World: On the Possibility of Life in Capitalist Ruins* (Princeton, NJ: Princeton University Press, 2015); Eben Kirksey, *Emergent Ecologies* (Durham, NC: Duke University Press, 2015). On the difficulties of living with see Jamie Lorimer and Clemens Driessen, "Bovine Biopolitics and the Promise of Monsters in the Rewilding of Heck Cattle," *Geoforum* 48 (2013): 249–59; Sebastian Abrahamsson and Filippo Bertoni, "Compost Politics: Experimenting with Togetherness in Vermicomposting," *Environmental Humanities* 4, no. 1 (2014): 125–48.

15. The science I refer to here challenges the individuality of organisms and the determinacy of Mendelian genetics. Research in microbiomatics, for example, vindicates ideas first forwarded by maverick evolutionary biologist Lynn Margulis

about the role of symbiotic bacteria in evolution. Lynn Margulis, *Symbiotic Planet: A New Look at Evolution* (London: Phoenix, 1999). Apparently bacteria and other microbes are able to evolve very rapidly and quite promiscuously from rampant horizontal gene transfer. See three entries in *Arts of Living on a Damaged Planet | Monsters of the Anthropocene*, ed. Anna Tsing, Heather Swanson, Elaine Gan, and Nils Bubandt (Minneapolis: University of Minnesota, 2017): Scott F. Gilbert, "Holobiont by Birth: Multilineage Individuals as the Concretion of Cooperative Processes," M73–M90; Donna Haraway, "Symbiogenesis, Sympoiesis, and Art Science Activisms for Staying with the Trouble," M25–M50; and Margaret McFall-Ngai, "Noticing Microbial Worlds: The Postmodern Synthesis in Biology," M51–M72. Through this lens the millions of bacteria that live within and around key organisms such as, say, a strawberry plant shape its propensity to survive under various conditions. Materials such as agrochemicals that eradicate these symbionts and otherwise simplify those conditions can therefore radically affect those potentialities.

16. Hans Jenny, *The Soil Resource: Origin and Behaviour* (New York: Springer-Verlag, 1980).

17. Amanda Hodson and Edwin E. Lewis, "Managing for Soil Health Can Suppress Pests," *California Agriculture* 70, no. 3 (2016): 137–41; Ron Nichols, "Soil Health Campaign Turns Two: Seeks to Unlock Benefits on and off the Farm," Natural Resources Conservation Service, USDA, n.d., https://www.nrcs.usda.gov/wps/portal/nrcs/detail/sd/home/?cid=stelprdb1261962.

18. James B. Anderson, "The Fungus *Armillaria bulbosa* Is among the Largest and Oldest Living Organisms," *Nature* 356 (1992): 428–31.

19. Maria Puig de la Bellacasa, "Making Time for Soil: Technoscientific Futurity and the Pace of Care," *Social Studies of Science* 45, no. 5 (2015): 691–716.

20. Kristina Marie Lyons, "Soil Science, Development, and the 'Elusive Nature' of Colombia's Amazonian Plains," *Journal of Latin American and Caribbean Anthropology* 19, no. 2 (2014): 231.

21. Tsing, *The Mushroom at the End of the World*, 138–39. Even mycorrhizae can be growth retarding, however. N. C. Johnson, J. H. Graham, and F. A. Smith, "Functioning of Mycorrhizal Associations Along the Mutualism–Parasitism Continuum," *New Phytologist* 135, no. 4 (1997): 575–85.

22. Tsing, *The Mushroom at the End of the World*, 170. On the ecological function of fungi more generally see Paul Stamets, *Mycelium Running: How Mushrooms Can Help Save the World* (Random House Digital, 2005).

23. Stephen Wilhelm and James E. Sagen, *History of the Strawberry: From Ancient Gardens to Modern Markets* (Berkeley: University of California Press, 1974), 6–7.

24. Hodson and Lewis, "Managing for Soil Health Can Suppress Pests," 137–41.

25. Klosterman et al., "Diversity, Pathogenicity, and Management of *Verticillium* Species," 39–62; Pegg, "The Impact of *Verticillium* Diseases in Agriculture," 176–92.

26. Pegg, "The Impact of *Verticillium* Diseases in Agriculture," 190–91.

27. Pegg, "The Impact of *Verticillium* Diseases in Agriculture," 190–91.

28. Pegg, "The Impact of *Verticillium* Diseases in Agriculture," 190.

29. Clive Brasier, "Plant Pathology: The Rise of the Hybrid Fungi," *Nature* 405, no. 6783 (2000): 134–35; Pegg, "The Impact of *Verticillium* Diseases in Agriculture," 176–92.

30. L. J. Ashworth and G. Zimmerman, "*Verticillium* Wilt of the Pistachio Nut Tree: Occurrence in California and Control by Fumigation," *Phytopathology* 66 (1976): 1449–51. The article by Ashworth and Zimmerman just mentions that *Verticillium* was found in California "native" soils. I learned it was endemic from an interview with an evolutionary ecologist who specializes in plant disease.

31. R. James Cook and Kenneth Frank Baker, *The Nature and Practice of Biological Control of Plant Pathogens* (St. Paul: American Phytopathological Society, 1983); A. H. C. Van Bruggen and M. R. Finckh, "Plant Diseases and Management Approaches in Organic Farming Systems," *Annual Review of Phytopathology* 54 (2016): 25–54; Jos M. Raaijmakers and Mark Mazzola, "Soil Immune Responses," *Science* 352, no. 6292 (2016): 1392–93; Mark Mazzola, "Mechanisms of Natural Soil Suppressiveness to Soilborne Diseases," *Antonie van Leeuwenhoek* 81, nos. 1–4 (2002): 557–64.

32. Pegg, "The Impact of *Verticillium* Diseases in Agriculture," 190–91.

33. Pegg, "The Impact of *Verticillium* Diseases in Agriculture," 190–91. The practice of tillage in particular can have the effect of spreading the pathogen to previously noninfested areas since it fragments, moves, and buries plant residues. At the same time, through heat, cold, and drying, mechanical tillage can kill pathogens that have been brought to surface exposure. See also Koike et al., "Vegetable Diseases Caused by Soilborne Pathogens."

34. Pegg, "The Impact of *Verticillium* Diseases in Agriculture," 178.

35. Wilhelm and Sagen, *History of the Strawberry*, 133.

36. Steven T. Koike, Thomas R. Gordon, Oleg Daugovish, Husein Ajwa, Mark Bolda, and Krishna Subbarao, "Recent Developments on Strawberry Plant Collapse Problems in California Caused by *Fusarium* and *Macrophomina*," *International Journal of Fruit Science* 13, nos. 1/2 (2013): 76–83.

37. Wilhelm and Sagen, *History of the Strawberry*, 193, 213.

38. Wilhelm and Sagen, *History of the Strawberry*, 213–14, 211.

39. Pegg, "The Impact of *Verticillium* Diseases in Agriculture," 205.

40. Wilhelm and Sagen, *History of the Strawberry*, 205.

41. George M. Darrow, *The Strawberry: History, Breeding, and Physiology* (New York: Holt, Rinehart, and Winston, 1966), 367–68.

42. Wilhelm and Sagen, *History of the Strawberry*, 211.

43. Herbert Baum, *Quest for the Perfect Strawberry* (New York: iUniverse, 2005), 28–29; Wilhelm and Sagen, *History of the Strawberry*, 213.

44. Harold E. Thomas, "*Verticillium* Wilt of Strawberries, Bulletin 530" (Berkeley: University of California Printing Office, 1932).

45. Thomas, "*Verticillium* Wilt of Strawberries"; Wilhelm and Sagen, *History of the Strawberry*, 218.

46. W.G. Keyworth and Margery Bennett, "*Verticillium* Wilt of the Strawberry," *Journal of Horticultural Science* 26, no. 4 (1951): 304–16.

47. Thomas, "*Verticillium* Wilt of Strawberries"; Wilhelm and Sagen, *History of the Strawberry*, 218.

48. Thomas, "*Verticillium* Wilt of Strawberries"; Stephen Wilhelm and Edward C. Koch, "*Verticillium* Wilt Controlled: Chloropicrin Achieves Effective Control of *Verticillium* Wilt in Strawberry Plantings if Properly Applied as Soil Fumigant," *California Agriculture* 10, no. 6 (1956): 3–14, cited in Douglas V. Shaw, Thomas Gordon, Kirk D. Larson, W. Douglas Gubler, John Hansen, and Sharon C. Kirkpatrick, "Strawberry Breeding Improves Genetic Resistance to *Verticillium* Wilt," *California Agriculture* 64, no. 1 (2010): 37–41.

49. Baum, *Quest for the Perfect Strawberry*, 29–32.

50. Pegg, "The Impact of *Verticillium* Diseases in Agriculture," 184.

51. Wilhelm and Sagen, *History of the Strawberry*, 200.

52. Klosterman et al., "Diversity, Pathogenicity, and Management of *Verticillium* Species," 39–62; Wilhelm and Sagen, *History of the Strawberry*, 205. Later it was discovered that lettuce is also a host plant and can tolerate high levels of *Verticillium*. Pegg, "The Impact of *Verticillium* Diseases in Agriculture," 191. This is significant, since many strawberry growers rotate with lettuce growers for reasons that will become clear in chapter 5. Recent research, however, suggests that lettuce is not so tolerant of *Verticillium*, with wilt becoming much more apparent in California lettuce since the mid-1990s. See Zahi K. Atallah, Ryan J. Hayes, and Krishna V. Subbarao, "Fifteen Years of *Verticillium* Wilt of Lettuce in America's Salad Bowl: A Tale of Immigration, Subjugation, and Abatement," *Plant Disease* 95, no. 7 (2011): 784–92.

53. Wilhelm and Sagen, *History of the Strawberry*, 43.

54. Baum, *Quest for the Perfect Strawberry*, 30–31. Still, given the longevity of sclerotia in the soil, one might wonder just how effective this was.

55. Wilhelm and Sagen, *History of the Strawberry*, 213.

56. Quoted in Wilhelm and Sagen, *History of the Strawberry*, 219.

57. Darrow, *The Strawberry*, 233.

58. Wilhelm and Sagen, *History of the Strawberry*, 221.

59. Thomas, "*Verticillium* Wilt of Strawberries."

60. Koike et al., "Vegetable Diseases Caused by Soilborne Pathogens." Much later they would learn that high nitrate levels increase the severity of *Verticillium*, while decreasing that of *Fusarium*.

61. Paulus, "Fungal Diseases of Strawberry," 887.

62. Wilhelm and Koch, "*Verticillium* Wilt Controlled," 3–14.

63. Darrow, *The Strawberry*, 229.

64. Paulus, "Fungal Diseases of Strawberry," 885–89.

65. Baum, *Quest for the Perfect Strawberry*, 29–30. With furrow irrigation, the water left in the furrows evaporates, leaving the salt on the ground.

66. Baum, *Quest for the Perfect Strawberry*, 30

67. Stephen Wilhelm, R.S. Storkan, and John M. Wilhelm, "Preplant Soil Fumigation with Methyl Bromide-Chloropicrin Mixtures for Control of Soil-Borne

Diseases of Strawberries: A Summary of Fifteen Years of Development," *Agriculture and Environment* 1 (1974): 227–36.

68. Darrow, *The Strawberry*, 232.

69. Adam M. Romero, "'From Oil Well to Farm': Industrial Waste, Shell Oil, and the Petrochemical Turn (1927–1947)," *Agricultural History* 90, no. 1 (2016): 71.

70. Stephen Wilhelm and Albert O. Paulus, "How Soil Fumigation Benefits the California Strawberry Industry," *Plant Disease* 64, no. 3 (1980): 264–70.

71. Pegg, "The Impact of *Verticillium* Diseases in Agriculture," 186.

72. As explained to me by a soil scientist, when methyl bromide kills the microbes, it leaves goo. Bacteria and fungi then move in and transform this dead matter into mineral forms.

73. Agro-ecologists tend to think that adding soil amendments is akin to taking probiotics and generally not very effective. Adding back a select group of microbes is not the same as creating a diverse environment through cover cropping, composting, and biodiverse cropping, in which microbes not identified might be performing functions unknown or recognized. Plus many introduced microbes do not get established. See Raaijmakers and Mazzola, "Soil Immune Responses," 1392–93. See also chapter 8.

74. M. Mazzola, J. Muramoto, and C. Shennan, "Transformation of Soil Microbial Community Structure in Response to Anaerobic Soil Disinfestation for Soilborne Disease Control in Strawberry," paper presented at the American Phytopathological Society Annual Meeting, Providence, Rhode Island, August 2–8, 2012. For further discussions of disturbance and colonization see Stephen R. Gliessman, Eric Engles, and Robin Krieger, *Agroecology: Ecological Processes in Sustainable Agriculture* (Boca Raton, FL: CRC, 1998); J. Philip Grime, *Plant Strategies, Vegetation Processes, and Ecosystem Properties* (Chichester, England: John Wiley and Sons, 2006).

75. For extended discussions of these two fungi see Cook and Baker, *The Nature and Practice of Biological Control of Plant Pathogens*.

76. This was suggested to me by a soil ecologist.

77. Kaur et al., "Emerging Phytopathogen *Macrophomina phaseolina*," 138.

78. Steven T. Koike, Renee S. Arias, Cliff S. Hogan, Frank N. Martin, and Thomas R. Gordon, "Status of *Macrophomina phaseolina* on Strawberry in California and Preliminary Characterization of the Pathogen," *International Journal of Fruit Science* 16, no. sup1 (2016): 148–59.

79. R. F. Cerkauskas, O. D. Dhingra, and J. B. Sinclair, "Effect of Herbicides on Competitive Saprophytic Colonization by *Macrophomina phaseolina* of Soybean Stems," *Transactions of the British Mycological Society* 79, no. 2 (1982): 201–5.

80. I am told that such phenotype-based naming has actually thwarted understanding of the pathogen since some variants cause yellowing (chlorosis) and wilt while others just cause wilt.

81. Koike et al., "Recent Developments on Strawberry Plant Collapse Problems in California Caused by *Fusarium* and *Macrophomina*," 76–83.

82. R. Kaur, J. Kaur, and Rama S. Singh, "Nonpathogenic *Fusarium* as a Biological Control Agent," *Plant Pathology Journal* 9, no. 3 (2011): 79–91. For further dis-

cussion of *Fusarium* see Cook and Baker, *The Nature and Practice of Biological Control of Plant Pathogens*; Van Bruggen and Finckh, "Plant Diseases and Management Approaches in Organic Farming Systems," 25–54.

83. Steven T. Koike and Thomas R. Gordon, "Management of *Fusarium* Wilt of Strawberry," *Crop Protection* 73 (2015): 67–72.

84. Koike et al., "Status of *Macrophomina phaseolina* on Strawberry in California and Preliminary Characterization of the Pathogen," 148–59.

85. Chapters 5 and 7 will provide more detail on the geography of strawberry production in California.

86. Koike et al., "Recent Developments on Strawberry Plant Collapse Problems in California Caused by *Fusarium* and *Macrophomina*," 76–83.

87. Koike et al., "Recent Developments on Strawberry Plant Collapse Problems in California Caused by *Fusarium* and *Macrophomina*," 76–83.

88. Koike et al., "Recent Developments on Strawberry Plant Collapse Problems in California Caused by *Fusarium* and *Macrophomina*," 76–83. Another theory is that *Verticillium* likes heavy, wet soil, which was in short supply during the drought years. Koike et al., "Vegetable Diseases Caused by Soilborne Pathogens." It is also the case that some *Verticillium*-resistant varieties had been released. Shaw et al., "Strawberry Breeding Improves Genetic Resistance to *Verticillium* Wilt," 37–41.

89. "Buffer-zone incentives" refers to regulatory protocols that allow growers smaller buffer zones if they adopt practices such as bed fumigation, which both reduces chemical load and the potential for off-gassing.

90. Koike et al., "Recent Developments on Strawberry Plant Collapse Problems in California Caused by *Fusarium* and *Macrophomina*," 76–83.

91. P. M. Henry, S. C. Kirkpatrick, C. M. Islas, A. M. Pastrana, J. A. Yoshisato, S. T. Koike, O. Daugovish, and T. R. Gordon, "The Population of *Fusarium oxysporum f. sp. fragariae*, Cause of *Fusarium* Wilt of Strawberry, in California," *Plant Disease* 101, no. 4 (2017): 550–56; Koike and Gordon, "Management of *Fusarium* Wilt of Strawberry," 67–72.

92. G. E. Short, T. D. Wyllie, and P. R. Bristow, "Survival of *Macrophomina phaseolina* in Soil and in Residue of Soybean," *Phytopathology* 70, no. 1 (1980): 13–17.

93. Koike et al., "Status of *Macrophomina phaseolina* on Strawberry in California and Preliminary Characterization of the Pathogen," 148–59; Koike and Gordon, "Management of *Fusarium* Wilt of Strawberry," 67–72. I learned at a strawberry field day that *Verticillium* is also worsened by extreme heat.

94. Tsing, *The Mushroom at the End of the World*, 143. Tsing discusses that some fungi are given to exchanging genes in nonreproductive encounters that can defy species boundaries. For bacteria, however, horizontal gene transfer is apparently the norm and microbes are apparently highly plastic. See Hannah Landecker, "Antibiotic Resistance and the Biology of History," *Body and Society* 22, no. 4 (2015): 1–34; McFall-Ngai, "Noticing Microbial Worlds," M51–M72.

95. Gilbert, "Holobiont by Birth," M73–M90; Haraway, "Symbiogenesis, Sympoiesis, and Art Science Activisms for Staying with the Trouble," M25–M50; Margulis, *Symbiotic Planet*; McFall-Ngai, "Noticing Microbial Worlds," M51–M72.

96. These included fungi with tough, resistant spores; the capacity of *Fusarium* to produce mycotoxins, which could have suppressed *Verticillium* for a time; and that fumigation eliminated protective communities such as mycorrhizae and endophytes (fungi that grow on leaves that do not cause disease), allowing the fungi to gain an upper hand. Together we concluded that there are many plausible explanations given that fumigation changes the entire composition of soil, flora, and fauna. Another informant communicated that organisms are actually much more versatile and plastic that previously thought, and can change their physiology to survive in existing conditions even without a mutation or epigenetic effect.

97. Some of these possibilities are beginning to gain attention. In 2018, I went to a strawberry field day at Cal Poly San Luis Obispo and learned of a field trial assessing the relationship between soil microbial communities and tolerance to *Verticillium* and *Macrophomina*. According to the graduate student researcher conducting the trial, such studies have been slow in coming because of both lack of knowledge that these might be important relationships, and insufficient technologies to test the relationships.

98. On undone science see Scott Frickel, Sahra Gibbon, Jeff Howard, Joanna Kempner, Gwen Ottinger, and David J. Hess, "Undone Science: Charting Social Movement and Civil Society Challenges to Research Agenda Setting," *Science, Technology and Human Values* 35, no. 4 (2010): 444–73; Naomi Oreskes and Erik M. Conway, *Merchants of Doubt: How a Handful of Scientists Obscured the Truth on Issues from Tobacco Smoke to Global Warming* (New York: Bloomsbury USA, 2011); Robert Proctor and Londa L. Schiebinger, eds., *Agnotology: The Making and Unmaking of Ignorance* (Palo Alto, CA: Stanford University Press, 2008).

99. Frank Uekötter, "Ignorance Is Strength: Science-Based Agriculture and the Merits of Incomplete Knowledge," in *Managing the Unknown: Essays on Environmental Ignorance*, ed. Frank Uekötter and Uwe Lübken (New York: Berghahn, 2014), 122–39.

100. Relatively recent genomic technologies are allowing scientists to learn more about pathogen evolution. For instance Klosterman et al. sequenced the genomes of *V. dahliae*, *V. albo-atrum*, and *Fusarium oxysporum* and identified a set of proteins that are shared among all three wilt pathogens. Steven J. Klosterman, Krishna V. Subbarao, Seogchan Kang, Paola Veronese, Scott E. Gold, Bart P. H. J. Thomma, Zehua Chen, et al., "Comparative Genomics Yields Insights into Niche Adaptation of Plant Vascular Wilt Pathogens," *PLOS Pathogens* 7, no. 7 (2011): e1002137 (open access e-journal).

101. Landecker, "Antibiotic Resistance and the Biology of History," 1–34.

CHAPTER THREE. CURIOUSLY BRED PLANTS
AND PROPRIETARY INSTITUTIONS

1. Stephen Wilhelm and James E. Sagen, *History of the Strawberry: From Ancient Gardens to Modern Markets* (Berkeley: University of California Press, 1974), xiii.

2. Jack Kloppenburg, *First the Seed: The Political Economy of Plant Biotechnology* (Madison: University of Wisconsin Press, 2005), 201.

3. Organisms with more sets of chromosomes also tend to be bigger than their counterparts.

4. Ploidy is also important for sexual compatibility: octoploids only hybridize easily with other octoploids, although breeders have developed techniques to overcome these limits. R. Bringhurst and Victor Voth, "Hybridization in Strawberries," *California Agriculture* 36, no. 8 (1982): 25.

5. George M. Darrow, *The Strawberry: History, Breeding, and Physiology* (New York: Holt, Rinehart, and Winston, 1966), 336.

6. The relatively nascent field of genomics can only help so much. Genomics enables breeders to identify and locate genetic markers that are responsible for certain traits and perhaps eliminate others from field trials that don't have markers for that trait. They may, for example, send plant material to a lab to identify which seedlings carry day-neutral varieties (varieties that produce regardless of day length). "Instead of using a tweezer to sort through the haystack you're using a pitchfork," one breeder quipped. However, there are only a handful of markers that plant geneticists have identified thus far. Day neutrality is one; as it happens, so is *Fusarium* resistance. Without identified markers, or when there may be several markers that seem to correspond with a given trait, genomic breeders are pretty much left to employ statistical approaches. They might use pedigrees to predict for certain traits, but that is an inexact process.

7. On how "plantiness"—the set of characteristics and capacities specific to certain plants—shapes political landscapes see Jake Fleming, "Toward Vegetal Political Ecology: Kyrgyzstan's Walnut–Fruit Forest and the Politics of Graftability," *Geoforum* 79 (2017): 27.

8. Wilhelm and Sagen, *History of the Strawberry*, 2, 20–21.

9. Wilhelm and Sagen, *History of the Strawberry*, 25.

10. Wilhelm and Sagen, *History of the Strawberry*, 6–7.

11. Jared Diamond, *Guns, Germs, and Steel* (New York: W. W. Norton, 1997), 116.

12. Wilhelm and Sagen, *History of the Strawberry*, 16.

13. Wilhelm and Sagen, *History of the Strawberry*, 44.

14. Wilhelm and Sagen, *History of the Strawberry*, 76, 86.

15. Darrow, *The Strawberry*, 73; Wilhelm and Sagen, *History of the Strawberry*, 201, 128, 48, 103.

16. Darrow, *The Strawberry*, 54, 38; Wilhelm and Sagen, *History of the Strawberry*, 128, 101.

17. Darrow, *The Strawberry*, 97, see also 86.

18. Darrow, *The Strawberry*, 98.

19. Stevenson Whitcomb Fletcher, *The Strawberry in North America: History, Origin, Botany, and Breeding* (New York: Macmillan, 1917), cited in Darrow, *The Strawberry*, 122.

20. Darrow, *The Strawberry*, 149, see also 130.

21. Darrow, *The Strawberry*, 150, 130.

22. Darrow, *The Strawberry*, 131.

23. See James F. Hancock, Arturo Lavín, and J.B. Retamales, "Our Southern Strawberry Heritage: *Fragaria chiloensis* of Chile," *Horticultural Science* 34, no. 5 (1999): 815. They note that most of the native land races have been contaminated by hybridization and transfer of genetic information with native clones.

24. Darrow, *The Strawberry*, 169.

25. Darrow, *The Strawberry*, 150, 66.

26. Wilhelm and Sagen, *History of the Strawberry*, 107, 112.

27. Wilhelm and Sagen, *History of the Strawberry*, 106. For more on *F. californica* see Wilhelm and Sagen, *History of the Strawberry*, 109. For more on the F. *chiloensis* see Hancock, Lavín, and Retamales, "Our Southern Strawberry Heritage," 815.

28. Wilhelm and Sagen, *History of the Strawberry*, 113.

29. [No author], "The Cinderella Strawberry," *Pacific Rural Press* 15, no. 8 (1878): https://cdnc.ucr.edu/cgi-bin/cdnc?a=d&d=PRP18780223.2.4.

30. Wilhelm and Sagen, *History of the Strawberry*, 176.

31. Wilhelm and Sagen, *History of the Strawberry*, 176. For more information on the Parry see [no author], "A New Strawberry," *Pacific Rural Press* 28, no. 7 (1884): https://cdnc.ucr.edu/cgi-bin/cdnc?a=d&d=PRP18840816.2.3;

32. Wilhelm and Sagen, *History of the Strawberry*, 182.

33. Wilhelm and Sagen, *History of the Strawberry*, 176–78.

34. Darrow, *The Strawberry*, 121.

35. Wilhelm and Sagen, *History of the Strawberry*, 187, 188, 91.

36. Wilhelm and Sagen, *History of the Strawberry*, 192–96.

37. Wilhelm and Sagen, *History of the Strawberry*, 197–98.

38. Wilhelm and Sagen, *History of the Strawberry*, 192, 201.

39. Wilhelm and Sagen, *History of the Strawberry*, 199. See chapter 5 for further discussion.

40. John Tierney, "A Patented Berry Has Sellers Licking Their Lips," *New York Times*, October 14, 1991, http://www.nytimes.com/1991/10/14/us/a-patented-berry-has-sellers-licking-their-lips.html; Wilhelm and Sagen, *History of the Strawberry*, 200–201.

41. It is unclear whether they ever patented the Banner. Except for the Driscoll's website, most sources suggest it remained unpatented.

42. Wilhelm and Sagen, *History of the Strawberry*, 184–85, 129.

43. Wilhelm and Sagen, *History of the Strawberry*, 215.

44. Darrow, *The Strawberry*, 228.

45. Wilhelm and Sagen, *History of the Strawberry*, 224.

46. Darrow, *The Strawberry*, 155.

47. Darrow, *The Strawberry*, 228, 156, 228, 158, 225.

48. Wilhelm and Sagen, *History of the Strawberry*, 227–28.

49. Herbert Baum, *Quest for the Perfect Strawberry* (New York: iUniverse, 2005), 6. See chapter 7 of this book for a discussion of the boom and bust cycles.

50. Darrow, *The Strawberry*, 206; Wilhelm and Sagen, *History of the Strawberry*, 224.

51. Darrow, *The Strawberry*, 206; Wilhelm and Sagen, *History of the Strawberry*, 230.

52. Baum, *Quest for the Perfect Strawberry*, 20; Darrow, *The Strawberry*, 197.

53. Baum, *Quest for the Perfect Strawberry*, 5. It should be noted that Baum worked for Naturipe, a competitor of Driscoll's and one of the few strawberry shippers that never ventured into proprietary varietals, relying solely on UC varietals.

54. Wilhelm and Sagen, *History of the Strawberry*, 207.

55. Wilhelm and Sagen, *History of the Strawberry*, 230.

56. Darrow, *The Strawberry*, 158.

57. Baum, *Quest for the Perfect Strawberry*, 21.

58. Darrow, *The Strawberry*, 200.

59. Wilhelm and Sagen, *History of the Strawberry*, 230.

60. Industry interview.

61. Baum, *Quest for the Perfect Strawberry*, 6; Darrow, *The Strawberry*, 193.

62. Darrow, *The Strawberry*, 129. As I will discuss further in chapter 5, strawberries grown in warm climates need to experience a period of cold to stimulate growth following a period of manufactured dormancy.

63. Darrow, *The Strawberry*, 159.

64. Darrow, *The Strawberry*, 228.

65. R. Bringhurst and Victor Voth, "New Strawberry Varieties: Fresno, Torrey, Wiltguard," ed. University of California Division of Agriculture and Natural Resources, 1961. In 2018, I spoke with a UC breeding student who had run experiments with Wiltguard. He corroborated that the cultivar had never been taken up.

66. Darrow, *The Strawberry*, 229.

67. Baum, *Quest for the Perfect Strawberry*, 25; Ann Filmer, "Strawberry Breeding Program Backgrounder: A Historical Timeline," UC Davis Department of Plant Sciences News Blog, May 11, 2016, https://news.plantsciences.ucdavis.edu/2016/05/11/strawberry-breeding-program-backgrounder-a-historical-timeline/.

68. Baum, *Quest for the Perfect Strawberry*, 140.

69. Douglas V. Shaw, Thomas Gordon, Kirk D. Larson, W. Douglas Gubler, John Hansen, and Sharon C. Kirkpatrick, "Strawberry Breeding Improves Genetic Resistance to *Verticillium* Wilt," *California Agriculture* 64, no. 1 (2010): 37–41.

70. Mark Bolda, Doug Shaw, and Tom Gordon, "Strawberries and Caneberries," University of California Division of Agriculture and Natural Resources blog, September 16, 2015, https://ucanr.edu/blogs/blogcore/postdetail.cfm?postnum=18979.

71. Baum, *Quest for the Perfect Strawberry*, 177.

72. [No author], "State Audit Advises Adjustments for UC Davis Strawberry Program," *Daily Democrat* (Woodland, CA), June 9, 2015, https://www.dailydemocrat.com/2015/06/09/state-audit-advises-adjustments-for-uc-davis-strawberry-program/.

73. Kloppenburg, *First the Seed*.

74. Rebecca Lave, Martin Doyle, and Morgan Robertson, "Privatizing Stream Restoration in the US," *Social Studies of Science* 40, no. 5 (2010): 677–703; Rebecca Lave, Philip Mirowski, and Samuel Randalls, "Introduction: STS and Neoliberal Science," *Social Studies of Science* 40, no. 5 (2010): 659–75; Philip Mirowski, "The Modern Commercialization of Science Is a Passel of Ponzi Schemes," *Social Epistemology* 26, nos. 3/4 (2012): 285–310; Henry Giroux, "Neoliberalism, Corporate Culture, and the Promise of Higher Education: The University as a Democratic Public Sphere," *Harvard Educational Review* 72, no. 4 (2002): 425–64.

75. Dan Charles, "Big Bucks from Strawberry Genes Lead to Conflict at UC Davis," NPR, *The Salt: What's on Your Plate*, July 2, 2014, https://www.npr.org/sections/thesalt/2014/07/02/327355935/big-bucks-from-strawberry-genes-lead-to-conflict-at-uc-davis.

76. Dale Kasler, "Long-Standing Marriage Goes Sour for UC Davis, Strawberry Industry," *Sacramento Bee*, August 16, 2014, https://www.sacbee.com/news/business/article2606860.html. That a former California Secretary of Agriculture was involved in the deal has added fuel to the fire of contention around it.

77. Some of these partners have since pulled out—definitely Lassen Canyon and likely Giant.

78. Mark Anderson, "Strawberry Suit Settled with Appointment of New UC Breeder," *Sacramento Business Journal*, February 9, 2015, https://www.bizjournals.com/sacramento/news/2015/02/09/strawberry-suit-settled-with-appointment-of-new-uc.html.

79. Kasler, "Long-Standing Marriage Goes Sour for UC Davis, Strawberry Industry."

80. They are not the only ones who have done this. In addition to the murky reports around the founding of Driscoll's discussed earlier, several informants alleged that the director of Plant Sciences, a relatively new private breeding company, had done something similar. He had worked for Driscoll's as the director of research. Two years after quitting Driscoll's and beginning his own company he was marketing a new variety, a nearly impossible undertaking without access to patented material.

81. Kasler, "Long-Standing Marriage Goes Sour for UC Davis, Strawberry Industry."

82. Dale Kasler, "UC Davis Fires Back in Strawberry Controversy, Sues Growers' Group," *Sacramento Bee*, October 31, 2014, https://www.sacbee.com/news/business/article3493488.html.

83. [No author], "UC Davis Breeding Strawberries for the 21st Century," *Seed World* podcast audio, 2015, http://seedworld.com/uc-davis-breeding-strawberries-21st-century/.

84. Dana Goodyear, "How Driscoll's Reinvented the Strawberry," *New Yorker*, August 21, 2017, https://www.newyorker.com/magazine/2017/08/21/how-driscolls-reinvented-the-strawberry.

85. Goodyear, "How Driscoll's Reinvented the Strawberry."

86. Ann Filmer, "Jury Sides with UC Davis in Strawberry Breeding Trial," UC Davis Department of Plant Sciences News Blog, May 25, 2017, https://news.plantsciences.ucdavis.edu/2017/05/25/jury-sides-with-uc-davis-in-strawberry-breeding-trial/.

87. Dale Kasler, "The Quest to Breed Better Strawberries Landed UC Davis in Court. Here's What Happened," *Sacramento Bee*, September 15, 2017, http://www.sacbee.com/news/local/article173591931.html.

88. Dan Charles, "Breeding Battle Threatens Key Source of California Strawberries," NPR, *The Salt: What's on Your Plate*, July 1, 2014, https://www.npr.org/sections/thesalt/2014/07/01/327256662/breeding-battle-threatens-key-source-of-california-strawberries.

89. Titus Fey Cronise, *The Natural Wealth of California: Comprising Duly History, Geography, Topography, and Scenery; Climate; Agriculture and Commercial Products; Geology, Zoology, and Botany; Mineralogy, Mines, and Mining Processes; Manufactures; Steamship Lines, Railroads, and Commerce; Immigration, a Detailed Description of Each County* (San Francisco: H. H. Hancroft, 1868), cited in Wilhelm and Sagen, *History of the Strawberry*, 171.

90. Wilhelm and Sagen, *History of the Strawberry*, 7.

91. Baum, *Quest for the Perfect Strawberry*, 8, 20.

92. Baum, *Quest for the Perfect Strawberry*, 20–21.

93. Baum, *Quest for the Perfect Strawberry*, 28; Darrow, *The Strawberry*, 215.

94. Darrow, *The Strawberry*, 216–17.

95. Darrow, *The Strawberry*, 164.

96. Darrow, *The Strawberry*, 229.

97. California Department of Pesticide Regulation, "Nonfumigant Strawberry Production Working Group Action Plan," April 2013, http://www.cdpr.ca.gov/docs/pestmgt/strawberry/work_group/action_plan.pdf.

98. Discussion with competing plant breeder, 2017.

99. Typically inbreeding decreases hybrid vigor, but that has to do with productivity, not susceptibility to disease.

100. Darrow, *The Strawberry*, 229.

101. Amanda Hodson and Edwin E. Lewis, "Managing for Soil Health Can Suppress Pests," *California Agriculture* 70, no. 3 (2016): 140.

102. Surinder Kaur, Gurpreet Singh Dhillon, Satinder Kaur Brar, Gary Edward Vallad, Ramesh Chand, and Vijay Bahadur Chauhan, "Emerging Phytopathogen *Macrophomina Phaseolina*: Biology, Economic Importance and Current Diagnostic Trends," *Critical Reviews in Microbiology* 38, no. 2 (2012): 136–51.

103. Darrow, *The Strawberry*, 383.

104. Breeder interview.

105. Darrow, *The Strawberry*, 122, 383.

106. Hancock, Lavín, and Retamales, "Our Southern Strawberry Heritage," 814–16.

107. California Department of Pesticide Regulation, "Nonfumigant Strawberry Production Working Group Action Plan," 8; interviews.

108. Baum, *Quest for the Perfect Strawberry*, 20–25.

109. Dune Lawrence, "How Driscoll's Is Hacking the Strawberry of the Future," *Bloomberg Businessweek*, July 29, 2015, https://www.bloomberg.com/news /features/2015-07-29/how-driscoll-s-is-hacking-the-strawberry-of-the-future.

110. To be fully transparent, I am on the project team to assess grower willingness to adopt pathogen-resistant varietals.

CHAPTER FOUR. CHEMICAL SOLUTIONS
AND REGULATORY PUSHBACK

1. Adam M. Romero, "'From Oil Well to Farm': Industrial Waste, Shell Oil, and the Petrochemical Turn (1927–1947)," *Agricultural History* 90, no. 1 (2016): 86.

2. More specifically, it measures the difference between near-infrared (which vegetation strongly reflects) and red light (which vegetation absorbs).

3. Environmental Working Group, "Dirty Dozen: EWG's 2018 Shopper's Guide to Pesticides in Produce," accessed February 16, 2018, https://www.ewg.org/foodnews /dirty_dozen_list.php.

4. Romero, "'From Oil Well to Farm,'" 70–93; Edmund P. Russell, "'Speaking of Annihilation': Mobilizing for War against Human and Insect Enemies, 1914– 1945," *Journal of American History* 82, no. 4 (1996): 1505–29.

5. Carl B. Marquand, "Contributions to Better Living from Chemical Corps Research," *Journal of Chemical Education* 34, no. 11 (1957): 532. On Americans' seeming acceptance of the conveniences that applied chemistry brought forth see Carolyn Thomas de la Peña, *Empty Pleasures: The Story of Artificial Sweeteners from Saccharin to Splenda* (Chapel Hill: University of North Carolina Press, 2010); Ken Geiser, *Chemicals without Harm: Policies for a Sustainable World* (Cambridge, MA: MIT Press, 2015); Matthew T. Huber, *Lifeblood: Oil, Freedom, and the Forces of Capital* (Minneapolis: University of Minnesota Press, 2013).

6. Metam sodium and metam potassium are also used in fumigation settings, but have for the most part been bit players.

7. John J. Davis, "Miscellaneous Soil Insecticide Tests," *Soil Science* 10, no. 1 (1920): 61; Roy E. Campbell, "The Concentration of Wireworms by Baits before Soil Fumigation with Calcium Cyanide," *Journal of Economic Entomology* 19, no. 4 (1926): 636–42.

8. Roy E. Campbell and M. W. Stone, "The Effect of Sulphur on Wireworms," *Journal of Economic Entomology* 25, no. 5 (1932): 967–70; Anthony Spuler, "Baiting Wireworms," *Journal of Economic Entomology* 18, no. 5 (1925): 703–7; Russell S. Lehman, "Laboratory Experiments with Various Fumigants against the Wireworm *Limonius (Pheletes) Californicus Mann*," *Journal of Economic Entomology* 26, no. 6 (1933): 1042–51.

9. M.W. Stone and Roy E. Campbell, "Chloropicrin as a Soil Insecticide for Wireworms," *Journal of Economic Entomology* 26, no. 1 (1933): 237–43.

10. John Stenhouse, "III. On Chloranil and Bromanil, No. II," *Journal of the Chemical Society* 23 (1870): 6–14.

11. Ruric Creegan Roark, *A Bibliography of Chloropicrin, 1848–1932* (Washington, DC: US Department of Agriculture, 1934), 1–2.

12. Russell, "'Speaking of Annihilation,'" 1512.

13. William Moore, "Volatility of Organic Compounds as an Index of the Toxicity of Their Vapors to Insects," *Journal of Agricultural Research* 10, no. 7 (1917): 365; William Moore, "Fumigation with Chloropicrin," *Journal of Economic Entomology* 11 (1918): 357–62, cited in R.N. Chapman and A.H. Johnson, "Possibilities and Limitations of Chloropicrin as a Fumigant for Cereal Products," *Journal of Agricultural Research* 31, no. 8 (1925): 745–60.

14. G.F. Pegg, "The Impact of *Verticillium* Diseases in Agriculture," *Phytopathologia Mediterranea* 23, nos. 2/3 (1984): 185; A.L. Taylor, "Nematocides and Nematicides: A History," *Nemaptropica* 33 (2003): 225–32; Edmund Russell, *War and Nature: Fighting Humans and Insects with Chemicals from World War I to Silent Spring* (Cambridge, England: Cambridge University Press, 2001), 83.

15. Russell, "'Speaking of Annihilation,'" 1512.

16. [No author], "To Use Chloropicrin on Pineapple Pests," *New York Times*, December 23, 1928, 42.

17. Nathan R. Smith, "The Partial Sterilization of Soil by Chloropicrin," *Soil Science Society of America Journal* 3, no. C (1939): 188.

18. Roark, *A Bibliography of Chloropicrin, 1848–1932*, 2.

19. Stone and Campbell, "Chloropicrin as a Soil Insecticide for Wireworms," 237–43.

20. R.W. Doane, *Common Pests* (Springfield, IL: C.C. Thomas, 1931), 121.

21. Taylor, "Nematocides and Nematicides," 27; Russell, *War and Nature*, 83.

22. Romero, "'From Oil Well to Farm,'" 81.

23. Bryon M. Vanderbilt, "Chlorination of Nitromethane," US patent filed February 10, 1938, IMC Chemical Group Inc. assignee, patent version no. US2181411A.

24. Huber, *Lifeblood*; Romero, "'From Oil Well to Farm,'" 70–93.

25. Romero, "'From Oil Well to Farm,'" 71.

26. Romero, "'From Oil Well to Farm,'" 81.

27. Romero, "'From Oil Well to Farm,'" 80–81.

28. Walter Carter, "Soil Treatments with Special Reference to Fumigation with D-D Mixture," *Journal of Economic Entomology* 38, no. 1 (1945): 35–44; Romero, "'From Oil Well to Farm,'" 80–81.

29. Romero, "'From Oil Well to Farm,'" 82.

30. Romero, "'From Oil Well to Farm,'" 82, 83.

31. M.W. Stone, "Dichloropropane-Dichloropropylene, a New Soil Fumigant for Wireworms," *Journal of Economic Entomology* 37, no. 2 (1944): 297–99.

32. Keith Barrons, "Chapter Six, Soil Fumigants," Dow Chemical, unpublished manuscript, author-obtained photocopy, book title unknown; A. G. Newhall, "Disinfestation of Soil by Heat, Flooding and Fumigation," *Botanical Review* 21, no. 4 (1955): 189–250.

33. Gordon W. Gribble, "The Natural Production of Organobromine Compounds," *Environmental Science and Pollution Research* 7, no. 1 (2000): 37–49.

34. Stefan Helmreich, *Alien Ocean: Anthropological Voyages in Microbial Seas* (Berkeley: University of California Press, 2009).

35. David B. Harper and John T. G. Hamilton, "The Global Cycles of the Naturally-Occurring Monohalomethanes," in *Natural Production of Organohalogen Compounds*, ed. G. Gribble (Berlin and Heidelberg, Germany: Springer, 2003), 17–41.

36. Robert C. Rhew, Benjamin R. Miller, and Ray F. Weiss, "Natural Methyl Bromide and Methyl Chloride Emissions from Coastal Salt Marshes," *Nature* 403, no. 6767 (2000): 292–95.

37. J. Gan, S. R. Yates, H. D. Ohr, and J. J. Sims, "Production of Methyl Bromide by Terrestrial Higher Plants," *Geophysical Research Letters* 25, no. 19 (1998): 3595–98.

38. Evan Hepler-Smith, "Molecular Bureaucracy: Toxicological Information and Environmental Protection," *Environmental History* 24, no. 3 (forthcoming 2019).

39. Tom Cheetham, "Pathological Alterations in Embryos of the Codling Moth (Lepidoptera: Tortricidae) Induced by Methyl Bromide," *Annals of the Entomological Society of America* 83, no. 1 (1990): 59–67. One account published in 1996 stated that the mode of action for methyl bromide on target organisms had been much studied but never clearly elucidated. O. C. MacDonald and C. Reichmuth, "Effects on Target Organisms," in *The Methyl Bromide Issue*, ed. C. H. Bell, N. Price, and B. Chakrabarti (Chichester, England: John Wiley and Sons, 1996), 150; Agency for Toxic Substances and Disease Registry, "Medical Management Guidelines for Methyl Bromide," n.d., https://www.atsdr.cdc.gov/MMG/MMG.asp?id=818&tid=160.

40. D. B. Mackie, "Methyl Bromide—Its Expectancy as a Fumigant," *Journal of Economic Entomology* 31, no. 1 (1938): 70–79.

41. Mackie, "Methyl Bromide—Its Expectancy as a Fumigant," 70–79.

42. Mackie, "Methyl Bromide—Its Expectancy as a Fumigant," 70–79.

43. D. L. Lindgren, "Vacuum Fumigation," *Journal of Economic Entomology* 29, no. 6 (1936): 1132–37.

44. Frank W. Fisk and Harold H. Shepard, "Laboratory Studies of Methyl Bromide as an Insect Fumigant," *Journal of Economic Entomology* 31, no. 1 (1938): 79–84.

45. Heber C. Donohoe, A. C. Johnson, and J. W. Bulger, "Methyl Bromide Fumigation for Japanese Beetle Control," *Journal of Economic Entomology* 33, no. 2 (1940): 296–302; Harold H. Shepard and Albert W. Buzicky, "Further Studies of Methyl Bromide as an Insect Fumigant," *Journal of Economic Entomology* 32, no. 6 (1939): 854–59; Mackie, "Methyl Bromide—Its Expectancy as a Fumigant," 70–79.

46. Randall Latta, "Methyl Bromide Fumigation for the Delousing of Troops," *Journal of Economic Entomology* 37, no. 1 (1944): 103.

47. F.C. Bishopp, "The Insecticide Situation," *Journal of Economic Entomology* 39, no. 4 (1946): 449–59.

48. H.T. Reynolds and J.W. Huffman, "Methyl Bromide Fumigation for Control of Cyclamen Mite on Strawberries," *Journal of Economic Entomology* 50, no. 4 (1957): 525–26.

49. Taylor, "Nematocides and Nematicides," 225–32; Barrons, "Chapter Six, Soil Fumigants."

50. C.G. Lincoln, H.H. Schwardt, and C.E. Palm, "Methyl Bromide-Dichloroethyl Ether Emulsion as a Soil Fumigant," *Journal of Economic Entomology* 35, no. 2 (1942): 238–39.

51. Newhall, "Disinfestation of Soil by Heat, Flooding and Fumigation," 220.

52. Fisk and Shepard, "Laboratory Studies of Methyl Bromide as an Insect Fumigant," 79.

53. Barrons, "Chapter Six, Soil Fumigants."

54. D.L. Lindgren, "Methyl Iodide as a Fumigant," *Journal of Economic Entomology* 31 (1938): 320.

55. Howard D. Ohr, Nigel M. Grech, and James J. Sims, "Methyl Iodide as a Fumigant," Google Patents, 1998, University of California assignee, patent version no. US5753183A.

56. Barrons, "Chapter Six, Soil Fumigants."

57. This is calculated from agricultural census data. US Bureau of the Census, "1978 Census of Agriculture," National Agricultural Statistics Service, 1981, http://usda .mannlib.cornell.edu/usda/AgCensusImages/1978/01/05/1978-01-05.pdf versus "1950 Census of Agriculture," National Agricultural Statistics Service, 1952, http://usda .mannlib.cornell.edu/usda/AgCensusImages/1950/01/33/1812/34101884v1p33ch1.pdf. Productivity increased from 9,281 pounds per acre to 34,116 pounds per acre.

58. Laura Tourte, Mark Bolda, and Karen Klonsky, "The Evolving Fresh Market Berry Industry in Santa Cruz and Monterey Counties," *California Agriculture* 70, no. 3 (2016): 109.

59. Jill Lindsey Harrison, *Pesticide Drift and the Pursuit of Environmental Justice* (Cambridge, MA: MIT Press, 2011); Christopher R. Henke, *Cultivating Science, Harvesting Power* (Cambridge, MA: MIT Press, 2008). Harrison explains the rationality of prophylactic use: growers will not lose a crop if they use a treatment unnecessarily, but they almost certainly will if they do not treat (70). For similar points see Steven Stoll, *The Fruits of Natural Advantage: Making the Industrial Countryside in California* (Berkeley: University of California Press, 1998). As Stoll writes, pest control became more than a prophylactic—it became another tool growers use to increase profitability, leading to "habits of excess" (117).

60. Brian J. Gareau, "Dangerous Holes in Global Environmental Governance: The Roles of Neoliberal Discourse, Science, and California Agriculture in the Montreal Protocol," *Antipode* 40, no. 1 (2008): 102–30. Some have argued that the "crop protection" industry has worked hard to defend and promulgate its use of products through a variety of regulatory science, marketing, PR, and extension tactics. See Harrison, *Pesticide Drift and the Pursuit of Environmental Justice*, 51–84; Paul

Robbins, *Lawn People: How Grasses, Weeds, and Chemicals Make Us Who We Are* (Philadelphia: Temple University Press, 2007), 72–95. Such promotion has been particularly important given the enormous research and development investments the industry risks if a chemical isn't widely adopted. Yet, as Gareau shows, the strawberry industry itself has also worked hard to defend fumigation because it has become so vital to the industry.

61. Geiser, *Chemicals without Harm*, 17–36.

62. Geiser, *Chemicals without Harm*, 23–24.

63. Geiser, *Chemicals without Harm*, 42; Harrison, *Pesticide Drift and the Pursuit of Environmental Justice*, 92–93.

64. Harrison, *Pesticide Drift and the Pursuit of Environmental Justice*, 104–5. For further discussion on regulatory capture see Naomi Oreskes and Erik M. Conway, *Merchants of Doubt: How a Handful of Scientists Obscured the Truth on Issues from Tobacco Smoke to Global Warming* (New York: Bloomsbury, 2011).

65. Geiser, *Chemicals without Harm*, 41; Soraya Boudia, "Managing Scientific and Political Uncertainty: Environmental Risk Assessment in a Historical Perspective," in *Powerless Science? Science and Politics in a Toxic World*, ed. Soraya Boudia and Nathalie Jas (New York: Berghahn, 2014), 95–112. See also William Boyd, "Genealogies of Risk: Searching for Safety, 1930s–1970s," *Ecology LQ* 39 (2012): 895–989; Scott Frickel and Michelle Edwards, "Untangling Ignorance in Environmental Risk Assessment," in *Powerless Science?*, 215–33. Harrison's book *Pesticide Drift and the Pursuit of Environmental Justice*, especially chapter 4, makes a sustained critique of how the risks in risk assessment are consistently understated.

66. Joshua Dunsby, "Measuring Environmental Health Risks: The Negotiation of a Public Right-to-Know Law," *Science, Technology and Human Values* 29, no. 3 (2004): 269–90; Harrison, *Pesticide Drift and the Pursuit of Environmental Justice*, 88–89.

67. Soraya Boudia and Nathalie Jas, "Introduction: The Greatness and Misery of Science in a Toxic World," in *Powerless Science?*, 1–26; Dunsby, "Measuring Environmental Health Risks," 269–90; Sheldon Krimsky, "Low Dose Toxicology: Narratives from the Science-Transcience Interface," in *Powerless Science?*, 234–53; Jill Lindsey Harrison, "'Accidents' and Invisibilities: Scaled Discourse and the Naturalization of Regulatory Neglect in California's Pesticide Drift Conflict," *Political Geography* 25, no. 5 (2006): 506–29; Michelle Murphy, *Sick Building Syndrome and the Problem of Uncertainty: Environmental Politics, Technoscience, and Women Workers* (Durham, NC: Duke University Press, 2006).

68. Geiser, *Chemicals without Harm*, 43.

69. Suzana Sawyer in Adam M. Romero et al., "Chemical Geographies," *GeoHumanities* 3, no. 1 (2017): 172.

70. Here again Jill Harrison's book *Pesticide Drift and the Pursuit of Environmental Justice* is fundamental.

71. Bernadette Bensaude-Vincent and Jonathan Simon, *Chemistry: The Impure Science* (London: Imperial College Press, 2012); Hepler-Smith, "Molecular Bureaucracy"; Sawyer in Romero et al., "Chemical Geographies," 170–72.

72. Hepler-Smith, "Molecular Bureaucracy."

73. Bensaude-Vincent and Simon, *Chemistry: The Impure Science*, 201–14. On this point see also Max Liboiron, Manuel Tironi, and Nerea Calvillo, "Toxic Politics: Acting in a Permanently Polluted World," *Social Studies of Science* 48, no. 3 (2018): 331–49.

74. For a similar point see Alex M. Nading, "Local Biologies, Leaky Things, and the Chemical Infrastructure of Global Health," *Medical Anthropology* 36, no. 2 (2017): 141–56.

75. Harrison, *Pesticide Drift and the Pursuit of Environmental Justice*, 3–4. Yet, as others have suggested, even in a best-case scenario liberal approaches to environmental protection are limited by particular evidentiary methods and a presumption of possible purity that is simply not available. See Liboiron, Tironi, and Calvillo, "Toxic Politics," 331–49.

76. Harrison, *Pesticide Drift and the Pursuit of Environmental Justice*, 34; Vincent J. Piccirillo, "Methyl Bromide" in *Hayes' Handbook of Pesticide Toxicology*, ed. Robert Krieger and William Krieger (London: Academic Press, 2010), 2:1837–47; Agency for Toxic Substances and Disease Registry, "Toxicological Profile for Bromomethane," 1992, https://www.atsdr.cdc.gov/ToxProfiles/tp27.pdf; Maria Mergel, "Methyl Bromide," in the online database *Toxipedia* (2011), accessed December 27, 2017, http://www.toxipedia.org/display/toxipedia/Methyl+Bromide (URL no longer available).

77. Brian J. Gareau, "The Limited Influence of Global Civil Society: International Environmental Non-Governmental Organisations and the Methyl Bromide Controversy in the Montreal Protocol," *Environmental Politics* 21, no. 1 (2012): 88–107.

78. Gareau suggests that the strawberry industry was able to throw its weight around by making claims that without a ban, international competition would do it in. "Dangerous Holes in Global Environmental Governance," 102–30.

79. E. Melanie DuPuis and Brian J. Gareau, "From Public to Private Global Environmental Governance: Lessons from the Montreal Protocol's Stalled Methyl Bromide Phase-Out," *Environment and Planning A* 41, no. 10 (2009): 1225.

80. DuPuis and Gareau, "From Public to Private Global Environmental Governance," 2305–23; Gareau, "Dangerous Holes in Global Environmental Governance," 102–30; Erin N. Mayfield and Catherine Shelley Norman, "Moving away from Methyl Bromide: Political Economy of Pesticide Transition for California Strawberries since 2004," *Journal of Environmental Management* 106 (2012): 93–101.

81. See the following chapter for a thorough explanation of the clean plant program and propagation process.

82. At least one nursery is using steam instead of fumigants in the screen houses, but that represents a very small percentage of the space used in nursery production. See chapter 8 on other alternatives.

83. California Department of Pesticide Regulation, "Report of the Scientific Review Committee on Methyl Iodide to the Department of Pesticide Regulation," February 5, 2010, http://www.cdpr.ca.gov/docs/risk/mei/peer_review_report.pdf. Also see this book's preface.

84. Julie Guthman and Sandy Brown, "Midas' Not-So-Golden Touch: On the Demise of Methyl Iodide as a Soil Fumigant in California," *Journal of Environmental Policy and Planning* 18, no. 3 (2016): 324–41.

85. Carolyn M. Lewis and Marilyn H. Silva, "Evaluation of Chloropicrin as a Toxic Air Contaminant, Part B: Human Health Effects" (Sacramento: State of California Department of Pesticide Regulation, 2010), accessed April 4, 2014, http://www.cdpr.ca.gov/docs/emon/pubs/tac/part_b_0210.pdf. The report has apparently been removed from the DPR site but can currently be found at https://pdfs.semanticscholar.org/765e/2f7e9d7ccd53eca2d022b02c9c72dc476d25.pdf.

86. For more on this see Julie Guthman and Sandra Brown, "Whose Life Counts: Biopolitics and the 'Bright Line' of Chloropicrin Mitigation in California's Strawberry Industry," *Science, Technology and Human Values* 41, no. 3 (2016): 461–92.

87. Actually it was banned previous to that following evidence of cancer clusters for its use in the 1980s. Dow Chemical, the owner, conducted studies and did some reformulation to allow it to reappear with restrictions. Department of Pesticide Regulation, "Memorandum: Recommendation on Township Cap Exception Requests for 1,3-Dichloropropene," 2014, accessed February 18, 2016, http://www.cdpr.ca.gov/docs/emon/methbrom/telone/rec_on_twnshp_telone.pdf. The report has apparently been removed from the DPR site.

88. Department of Pesticide Regulation, "Memorandum: Recommendation on Township Cap Exception Requests for 1,3-Dichloropropene."

89. [No author], "California DPR Caps Telone Fumigant Use on Jan. 1," *Western Farm Press*, October 6, 2016, http://www.westernfarmpress.com/regulatory/california-dpr-caps-telone-fumigant-use-jan-1.

90. Some growers were very slow to give up methyl bromide, and those who were able to obtain the last remaining inventories paid dearly for it (in dollars), under the assumption that higher yields would make up for the cost. Those who obtained the last remaining inventories also appear to have had privileged relationships with Tri-Cal, the main fumigation company, which received the allocations from the EPA.

91. This is based on an analysis my team conducted on fumigant use between 2004 and 2013. The source of the data was California's Pesticide Use Reporting System. A visualization can be found in Julie Guthman, "Land Access and Costs May Drive Strawberry Growers' Increased Use of Fumigation," *California Agriculture* 71, no. 3 (2017): 184–91.

92. That nursery growers were staunch critics of the enhanced mitigation measures for chloropicrin shows their investment in this chemical as well as in methyl bromide.

93. Harrison, "'Accidents' and Invisibilities," 506–29; also discussed in Harrison, *Pesticide Drift and the Pursuit of Environmental Justice*.

94. Lily Dayton, "Dangerous Drift: Students, Strawberries and Harmful Fumigants Collide in California's Agricultural Belts" (Davis: California Institute for Rural Studies, 2015), http://www.cirsinc.org/rural-california-report/entry/dangerous-drift-students-strawberries-and-harmful-fumigants-collide-in-california-s-agricultural-belts.

95. Harrison, *Pesticide Drift and the Pursuit of Environmental Justice*, 29–30.

96. H. Johnson, A. Holland, A. Paulus, and S. Wilhelm, "Soil Fumigation Found Essential for Maximum Strawberry Yields in Southern California," *Califor-

nia Agriculture 16, no. 10 (1962): 4–6; Stephen Wilhelm and Albert O. Paulus, "How Soil Fumigation Benefits the California Strawberry Industry," *Plant Disease* 64, no. 3 (1980): 264–70.

97. D. Wang, S. R. Yates, F. F. Ernst, J. Gan, and W. A. Jury, "Reducing Methyl Bromide Emission with a High Barrier Plastic Film and Reduced Dosage," *Environmental Science and Technology* 31, no. 12 (1997): 3686–91.

98. L. Klein, "Methyl Bromide as a Soil Fumigant," in *The Methyl Bromide Issue*, 222.

99. Hannah Landecker, "Antibiotic Resistance and the Biology of History," *Body and Society* 22, no. 4 (2015): 5.

100. Here I'm directly referencing a new literature in chemical geographies and anthropology that takes chemicals and toxins as ubiquitous, no longer (if ever) subject to control by liberal regulatory mechanisms, and, as such, human activity that has become permanently "sedimented into the planet's terrestrial record." Andrew Barry, "Manifesto for a Chemical Geography," lecture, University College of London, 2017, http://www.academia.edu/32374031/Manifesto_for_a_Chemical_Geography_2017_. See also Liboiron, Tironi, and Calvillo, "Toxic Politics," 332; Romero et al., "Chemical Geographies," 158–77; Nicholas Shapiro and Eben Kirksey, "Chemo-Ethnography: An Introduction," *Cultural Anthropology* 32, no. 4 (2017): 481–93; Soraya Boudia, Angela N. H. Creager, Scott Frickel, Emmanuel Henry, Nathalie Jas, Carsten Reinhardt, and Jody A. Roberts. "Residues: Rethinking Chemical Environments," *Engaging Science, Technology and Society* 4 (2018): 172.

101. Boudia and Jas, "Introduction: The Greatness and Misery of Science in a Toxic World," 1–26; Boyd, "Genealogies of Risk," 895–989; Guthman and Brown, "Whose Life Counts," 461–92.

102. Alison Gemmill, Robert B. Gunier, Asa Bradman, Brenda Eskenazi, and Kim G. Harley, "Residential Proximity to Methyl Bromide Use and Birth Outcomes in an Agricultural Population in California," *Environmental Health Perspectives* 121, no. 6 (2013): 737.

103. A. J. Idrovo, L. H. Sanìn, D. Cole, J. Chavarro, H. Cáceres, J. Narváez, and M. I. Restrepo, "Time to First Pregnancy among Women Working in Agricultural Production," *International Archives of Occupational and Environmental Health* 78, no. 6 (2005): 493–500.

104. California Department of Pesticide Regulation, "Nonfumigant Strawberry Production Working Group Action Plan," April 2013, http://www.cdpr.ca.gov/docs/pestmgt/strawberry/work_group/action_plan.pdf.

105. Harrison, *Pesticide Drift and the Pursuit of Environmental Justice*, 56–57.

CHAPTER FIVE. SOILED ADVANTAGES
AND HIGHLY VALUED LAND

1. David Harvey, *Limits to Capital* (Chicago: University of Chicago Press, 1982), 334, 347, 368, emphasis in original.

2. Miriam Wells, *Strawberry Fields: Politics, Class, and Work in California Agriculture* (Ithaca, NY: Cornell University Press, 1996), 29.

3. Many of these developments are detailed in Wells, *Strawberry Fields*, chapter 2.

4. California Strawberry Commission, "Growing California Strawberries," http://www.calstrawberry.com/Growing-California-Strawberries. This page was last accessed December 27, 2017, and has since been taken down. The California Strawberry Commission routinely changes its website and does not archive older pages.

5. Steven Stoll, *The Fruits of Natural Advantage: Making the Industrial Countryside in California* (Berkeley: University of California Press, 1998), 2.

6. For foundational writings on this point see Ted Benton, "Marxism and Natural Limits: An Ecological Critique and Reconstruction," *New Left Review* 178 (1989): 51–86; David Harvey, *Justice, Nature, and the Geography of Difference* (Cambridge, MA: Blackwell, 1996); Neil Smith, *Uneven Development: Nature, Capital, and the Production of Space* (Oxford: Basil Blackwell, 1984).

7. Wells, *Strawberry Fields*, 29.

8. Stephen Wilhelm and James E. Sagen, *History of the Strawberry: From Ancient Gardens to Modern Markets* (Berkeley: University of California Press, 1974), 103.

9. Wells, *Strawberry Fields*, 29; PAST Consultants, "Historic Context Statement for Agricultural Resources in the North County Planning Area, Monterey County," (Petaluma, CA: PAST Consultants, 2010), http://ohp.parks.ca.gov/pages/1054/files/nomontereyco.pdf, 45–46.

10. Wells, *Strawberry Fields*, 29.

11. Stoll, *The Fruits of Natural Advantage*, xv.

12. Marcos López, "In Hidden View: How Water Became a Catalyst for Indigenous Farmworker Resistance in Baja California, Mexico," in *The Politics of Fresh Water: Access, Conflict, and Identity*, ed. C. Ashcraft and T. Mayer (New York: Routledge, 2017), 195.

13. Malcolm Margolin, *The Ohlone Way: Indian Life in the San Francisco–Monterey Bay Area* (Berkeley: Heyday Books, 1978), 8.

14. Wells, *Strawberry Fields*, 103.

15. Wilhelm and Sagen, *History of the Strawberry*, 171.

16. Wells, *Strawberry Fields*, 31.

17. Wells, *Strawberry Fields*, 31.

18. Laura-Anne Minkoff-Zern, Nancy Peluso, Jennifer Sowerwine, and Christy Getz, "Race and Regulation: Asian Immigrants in California Agriculture," in *Cultivating Food Justice: Race, Class, and Sustainability*, ed. Alison Alkon and Julian Agyeman (Cambridge, MA: MIT Press, 2011), 65–86.

19. The advantages are lower transportation costs. Theoretically, land with such locational advantages should cost more. Harvey, *Limits to Capital*, 333–37.

20. And therefore there is a dearth of literature on the history of coastal agricultural land, making this story of bit of a challenge to put together. In contrast, the large landholdings of the Central Valley have seen an abundance of scholarly treatment. See for example Ellen Leibman, *California Farmland: A History of Large*

Agricultural Land Holdings (Totowa, NJ: Rowman and Allanheld, 1983); Paul W. Gates, "Public Land Disposal in California," *Agricultural History* 49 (1975): 158–78; Tamara Venit Shelton, *A Squatter's Republic: Land and the Politics of Monopoly in California, 1850–1900* (Berkeley: University of California Press, 2013).

21. California residents acquired twenty-five major land grants ranging from four thousand to three hundred thousand acres, a pattern of land distribution that continued in the Mexican period. C. F. Nuckton, R. I. Rochin, and A. F. Scheuring, "California Agriculture: The Human Story," in *A Guidebook to California Agriculture*, ed. Anne Foley Scheuring (Berkeley: University of California Press, 1983), 10.

22. Leibman, *California Farmland*, 26.

23. Leibman, *California Farmland*; Nuckton, Rochin, and Scheuring, "California Agriculture," 9–38.

24. Kent Seavey, "A Short History of Salinas, California," Monterey County Historical Society, accessed December 27, 2017, http://mchsmuseum.com/salinasbrief.html.

25. Oxnard Downtowners, "General Historical Overview until 1898," accessed December 27, 2017, http://oxnarddowntowners.org/downtown-history.html.

26. Tina Daunt, "McGraths Everywhere: Founders: In 1876, the First One Arrived, Setting up a Family Dynasty on a Patchwork of Farms. There Are Now 500 Descendants," *Los Angeles Times*, May 6, 1991, http://articles.latimes.com/1991-05-06/local/me-848_1_charles-mcgrath; McGrath Family Farm, "Farm History," accessed December 27, 2017, http://www.mcgrathfamilyfarm.com/farm-history/.

27. PAST Consultants, "Historic Context Statement for Agricultural Resources in the North County Planning Area, Monterey County," 42.

28. Margaret FitzSimmons, "The New Industrial Agriculture," *Economic Geography* 62, no. 4 (1986): 334–53; Wells, *Strawberry Fields*, 104–5.

29. Shirley Contreras, "A Firsthand History of Santa Maria 50 Years after It All Began," *Santa Maria Times*, August 15, 2010, https://santamariatimes.com/lifestyles/columnist/shirley_contreras/a-firsthand-history-of-santa-maria-years-after-it-all/article_d2d4eb60-a823-11df-b747-001cc4c03286.html.

30. Santa Maria Valley Historical Society Museum, "A Brief History," accessed December 27, 2017, http://santamariahistory.com/history.html.

31. George L. Henderson, *California and the Fictions of Capital* (New York: Oxford University Press, 1998), 4–7; Leibman, *California Farmland*. For a general overview and novel analysis of California's agricultural development see Richard Walker, *The Conquest of Bread: 150 Years of Agribusiness in California* (New York: New Press, 2004).

32. FitzSimmons, "The New Industrial Agriculture," 337–38; Torsten A. Magnuson, "History of the Beet Sugar Industry in California," *Annual Publication of the Historical Society of Southern California* 11, no. 1 (1918): 68–79; Oxnard Downtowners, "General Historical Overview until 1898"; Santa Maria Valley Historical Society Museum, "A Brief History."

33. Federal Writers' Project, "Labor in California Sugar Beet Crop" (1938), http://content.cdlib.org/view?docId=hb88700929;NAAN=13030&doc.view=frames&chunk.id=div00115&toc.depth=1&toc.id=div00115&brand=calisphere.

34. PAST Consultants, "Historic Context Statement for Agricultural Resources in the North County Planning Area, Monterey County," 86.

35. Oxnard Downtowners, "General Historical Overview until 1898."

36. Ross Eric Gibson, "Agricultural Legacy of Serbo-Croatian Community," n.d., https://history.santacruzpl.org/omeka/items/show/134320#?c=0&m=0&s=0&cv=0; Wells, *Strawberry Fields*, 104.

37. FitzSimmons, "The New Industrial Agriculture," 338.

38. William H. Friedland, Amy E. Barton, and Robert J. Thomas, *Manufacturing Green Gold* (Cambridge, England: Cambridge University Press, 1981), 45–49.

39. Wells, *Strawberry Fields*, 110–14.

40. Nuckton, Rochin, and Scheuring, "California Agriculture," 9–38.

41. FitzSimmons, "The New Industrial Agriculture," 334–53.

42. Daunt, "McGraths Everywhere"; Los Angeles Food Policy Council, "Interview with Phil McGrath, Owner of McGrath Family Farms," December 13, 2013, accessed December 27, 2017, http://goodfoodla.org/2013/12/16/interview-with-phil-mcgrath-owner-of-mcgrath-family-farms/, page no longer available.

43. United States Bureau of the Census, "2012 Census of Agriculture," National Agricultural Statistics Service, 2014, https://www.agcensus.usda.gov/Publications/2012/Full_Report/Volume_1,_Chapter_2_County_Level/California/cav1.pdf. Without fruit trees, which tend to be grown by farmers who own the land, this data would suggest that almost all row-crop land is leased.

44. PAST Consultants, "Historic Context Statement for Agricultural Resources in the North County Planning Area, Monterey County," 86.

45. G. F. Pegg, "The Impact of *Verticillium* Diseases in Agriculture," *Phytopathologia Mediterranea* 23, nos. 2/3 (1984): 184.

46. Adam M. Romero, "'From Oil Well to Farm': Industrial Waste, Shell Oil, and the Petrochemical Turn (1927–1947)," *Agricultural History* 90, no. 1 (2016): 71.

47. George M. Darrow, *The Strawberry: History, Breeding, and Physiology* (New York: Holt, Rinehart, and Winston, 1966), 232.

48. Frank Uekötter, "Ignorance Is Strength: Science-Based Agriculture and the Merits of Incomplete Knowledge," in *Managing the Unknown: Essays on Environmental Ignorance*, ed. Frank Uekötter and Uwe Lübken (New York: Berghahn, 2014), 122–39.

49. Darrow, *The Strawberry*, 232. Although plantings are staggered by region, most of the crop is planted in fall or late fall for a late winter or spring harvest. Ventura (Oxnard) and Santa Barbara (Santa Maria) Counties see some "summer plantings" as well, for harvest in the fall. These plants are kept in cold storage for a while after propagation (as discussed further in the chapter).

50. The use of different kinds of plastic tarps helped to extend the seasons as well, by tricking plants into certain conditions. For instance, growers in the foggier cool areas tend to use black plastic, which absorbs heat, while growers planting summer berries in places like Santa Maria and Oxnard will use white plastic to help keep the ground cool.

51. Wells, *Strawberry Fields*, 180–85.

52. DPR requires growers to report every time they use an agricultural chemical.

53. DPR director, personal communication November 25, 2015. The "label" refers to restrictions by USEPA or California DPR on what chemicals can be used on what crops, in what amounts, and with what mitigation measures.

54. Steven J. Klosterman, Zahi K. Atallah, Gary E. Vallad, and Krishna V. Subbarao, "Diversity, Pathogenicity, and Management of *Verticillium* Species," *Annual Review of Phytopathology* 47 (2009): 39–62.

55. Wells, *Strawberry Fields*, 185.

56. The meristem process is described in detail at Foundation Plant Services, "Introduction to the Strawberry Clean Plant Program," University of California, College of Agriculture and Environmental Sciences, http://fps.ucdavis.edu/strawberry .cfm; Larry Strand, *Integrated Pest Management for Strawberries* (Oakland: University of California Division of Agriculture and Natural Resources, 2008), 36–40. The description that follows this note in the text is also derived from my interviews with nursery people, which sometimes left me more confused than not.

57. Strand, *Integrated Pest Management for Strawberries*, 36–40. I use the "mother" and "daughter" language because it is consistent in all writings and discussions of the industry. No doubt it could be critiqued.

58. Strand, *Integrated Pest Management for Strawberries*, 36.

59. University of California varieties are used all over the world for strawberry production. These plants are shipped in test tubes, in advance of propagation. Foundation Plant Services, "Introduction to the Strawberry Clean Plant Program."

60. I note here that the industry is more stringent on disease buffer zones than chemical buffer zones.

61. Plant growers pinch (remove) blossoms to promote runner production, while fruit growers pinch runners to promote fruit production.

62. Apparently some of the newer varieties don't require as much chill.

63. Herbert Baum, *Quest for the Perfect Strawberry* (New York: iUniverse, 2005), 32–32; Wilhelm and Sagen, *History of the Strawberry*, 226.

64. Julie Guthman, *Agrarian Dreams: The Paradox of Organic Farming in California*, 2nd ed. (Berkeley: University of California Press, 2014), 67–68. For an extended discussion of how this works see chapter 11 of Harvey, *Limits to Capital*.

65. Laura Tourte, Mark Bolda, and Karen Klonsky, "The Evolving Fresh Market Berry Industry in Santa Cruz and Monterey Counties," *California Agriculture* 70, no. 3 (2016): 107–15.

66. Paul Rogers, "Farmland Measure Divides Monterey County," *San Jose Mercury News*, October 31, 1996, no longer available online. Interviews my research assistant conducted in Ventura County revealed similar opposition.

67. Gladstone Land Corporation, "Company Overview," accessed October 30, 2017, http://ir.gladstoneland.com/company-overview.

68. Gladstone Land Corporation, "Gladstone Land Corporation Announces Farmland Acquisition in California," news release, July 24, 2014, http://ir.gladstoneland

.com/news-releases/news-release-details/gladstone-land-corporation-announces-farmland-acquisition-6?releaseid=861798.

69. University of California Cooperative Extension, "Agriculture and Natural Resources Ventura County: General Soil Map," Division of Agriculture and Natural Resources, University of California, accessed December 27, 2017, http://ceventura.ucanr.edu/Com_Ag/Soils/The_environamental_characteristics_of_Ventura_County_and_its_soils_/General_Soil_Map/; E.J. Carpenter and Stanley W. Cosby, "Soil Survey of the Salinas Area, California," United States Department of Agriculture, Bureau of Chemistry and Soils, 1925, https://www.nrcs.usda.gov/Internet/FSE_MANUSCRIPTS/california/salinasareaCA1925/salinasarea CA1925.pdf.

70. Jason W. Moore, *Capitalism in the Web of Life: Ecology and the Accumulation of Capital* (New York: Verso, 2015), 273. Moore also notes that the fixes become more toxic so that rising public health costs also produce negative value—what others have called externalities. Yet, as I have argued elsewhere, these externalities are not as visible when the health costs are associated with "disposable" populations. Julie Guthman, "Lives versus Livelihoods? Deepening the Regulatory Debates on Soil Fumigants in California's Strawberry Industry," *Antipode* 49, no. 1 (2017): 86–105.

71. For a specific discussion see Guthman, *Agrarian Dreams*, 10–12. For renditions of the agrarian imaginary see Wendell Berry, *The Unsettling of America: Culture and Agriculture*, 2nd ed. (San Francisco: Sierra Club Books, 1986); Marty Strange, *Family Farming: A New Economic Vision* (Lincoln: University of Nebraska Press / Institute for Food and Development Policy, 1988).

72. Agricultural historian Colin Duncan has argued that the separation of ownership and management was critical to maintaining soil fertility in England's period of high agrarian capitalism because leases would stipulate extensive rotations whereas owner-operators were too inclined to respond to markets. Colin A.M. Duncan, *The Centrality of Agriculture: Between Humankind and the Rest of Nature* (Montreal: McGill-Queen's University Press, 1996). And rural sociologists have shown that tenant farmers interested in more sustainable methods often encounter landlord skepticism, also suggesting that nothing inherent about landownership leads to better stewardship. Michael S. Carolan, "Barriers to the Adoption of Sustainable Agriculture on Rented Land: An Examination of Contesting Social Fields," *Rural Sociology* 70, no. 3 (2005): 387–413; Douglas H. Constance, J. Sanford Rikoon, and Jian C. Ma, "Landlord Involvement in Environmental Decision-Making on Rented Missouri Cropland: Pesticide Use and Water Quality Issues," *Rural Sociology* 61, no. 4 (1996): 577–605.

73. Christine L. Carroll, Colin A. Carter, Rachael E. Goodhue, and C.-Y. Cynthia Lin Lawell, "The Economics of Decision-Making for Crop Disease Control" (Davis: University of California at Davis, 2017), http://www.des.ucdavis.edu/faculty/Lin/Vwilt_short_long_paper.pdf. Note that their method does not allow them to differentiate owners from renters. They used data from the California Pesticide Use Reporting System to view turnover and used very short-term (one year)

growers as proxies for renters. An alternative explanation, not explored in their models, is that these short-term renters are very-low-resource growers who can ill afford preventive treatments.

74. For many, these would be utterly contrasting approaches to soil knowledge and care. See for instance Uekötter, "Ignorance Is Strength," 122–39.

75. The subtitle "opposing materialities" alludes to the two different notions of materialism that seem to be at work here, with one referring to the tendencies and dynamics of capitalism and the other referring to the tangible, earthly material world. See chapter 1 for a discussion.

76. Harvey, *Limits to Capital*, 336. Those with access to high quality land stand to gain excess profits in perpetuity by the natural advantages they enjoy.

77. FitzSimmons, "The New Industrial Agriculture," 345.

78. Harvey, *Limits to Capital*, 333–37; Richard Walker, "Urban Ground Rent: Building a New Conceptual Framework," *Antipode* 6, no. 1 (1974): 51–59.

79. Economic Research Service, "Table 2. Deflated US Strawberry Prices and Values, 1970–2012," United States Departure of Agriculture, updated June 2013, http://usda.mannlib.cornell.edu/MannUsda/viewDocumentInfo.do?documentID=1381 (table 2 must be downloaded).

80. Mark Bolda, Laura Tourte, Jeremy Murdock, and Daniel Sumner, "Sample Costs to Produce and Harvest Strawberries, Central Coast Region," UC Cooperative Extension and Agricultural Issues Center, 2016, https://coststudyfiles.ucdavis.edu/uploads/cs_public/e7/6d/e76dceb8-f0f5-4b60-bcb8-76b88d57e272/strawberrycentralcoast-2016-final2-5-1-2017.pdf.

81. Compare Santa Barbara County Agricultural Commissioner, "Agricultural Production Report," Weights and Measures, County of Santa Barbara, 2016, http://cosb.countyofsb.org/uploadedFiles/agcomm/crops/2016.pdf, versus the 2001 report, http://cosb.countyofsb.org/uploadedFiles/agcomm/crops/2001.pdf. The Santa Maria / Guadalupe region spans Santa Barbara and San Luis Obispo Counties. Previous to the contemporary period, the region was used primarily for a bridge season. After 2000 more growers began to locate there as their primary site of business.

82. Crop advisor interview, 2017. Here Ryan Galt's research is apt, as he shows how land with poor soils requires more pesticide applications and is often not as productive. Ryan E. Galt, *Food Systems in an Unequal World: Pesticides, Vegetables, and Agrarian Capitalism in Costa Rica* (Tucson: University of Arizona Press, 2014).

83. Joseph Cerna, "San Francisco Sets All-Time Heat Record Downtown at 106 Degrees during State's Hottest Recorded Summer," *Los Angeles Times*, September 1, 2017, http://www.latimes.com/local/lanow/la-me-ln-california-heat-wave-weekend-20170901-story.html.

84. Paul Rogers, "California Heat Wave: How Much Is from Climate Change?," *San Jose Mercury News*, September 1, 2017, https://www.mercurynews.com/2017/09/01/california-heat-wave-how-much-is-from-climate-change/.

85. López, "In Hidden View," 189–204.

86. Carroll et al., "The Economics of Decision-Making for Crop Disease Control," 6; Christine L. Carroll, Colin A. Carter, Rachael E. Goodhue, C.-Y. Cynthia Lin Lawell, and Krishna V. Subbarao, "The Economics of Managing *Verticillium* Wilt, an Imported Disease in California Lettuce," *California Agriculture* 71, no. 3 (2017): 178–83.

CHAPTER SIX. SCARCE LABOR AND DISPOSABLE BODIES

1. Miriam Wells, *Strawberry Fields: Politics, Class, and Work in California Agriculture* (Ithaca, NY: Cornell University Press, 1996), 19.

2. Don Mitchell, *"They Saved the Crops": Labor, Landscape, and the Struggle over Industrial Farming in Bracero-Era California* (Athens: University of Georgia Press, 2012), 73.

3. For a detailed description of the strawberry harvest see Wells, *Strawberry Fields*, 166–71.

4. A 2016 cost study assumed $11.50 per hour as the average wage. Mark Bolda, Laura Tourte, Jeremy Murdock, and Daniel Sumner, "Sample Costs to Produce and Harvest Strawberries, Central Coast Region," UC Cooperative Extension and Agricultural Issues Center, 2016, https://coststudyfiles.ucdavis.edu/uploads/cs_public/e7/6d/e76dceb8-f0f5-4b60-bcb8-76b88d57e272/strawberrycentralcoast-2016-final2-5-1-2017.pdf.

5. Sarah Bronwen Horton, *They Leave Their Kidneys in the Fields: Illness, Injury, and Illegality among US Farmworkers* (Oakland: University of California Press, 2016).

6. Seth Holmes, *Fresh Fruit, Broken Bodies: Migrant Farmworkers in the United States* (Berkeley: University of California Press, 2013); Wells, *Strawberry Fields*.

7. Jason De Léon, *The Land of Open Graves: Living and Dying on the Migrant Trail* (Berkeley: University of California Press, 2015).

8. Wells, *Strawberry Fields*.

9. James C. Scott, *Weapons of the Weak* (New Haven, CT: Yale University Press, 1985).

10. For histories of how wages were set so low see Cletus Daniel, *Bitter Harvest: A History of California Farm Workers, 1870–1941* (Ithaca, NY: Cornell University Press, 1981); Mitchell, *"They Saved the Crops."*

11. Wells, *Strawberry Fields*, 55.

12. Tomás Almaguer, *Racial Fault Lines: The Historical Origins of White Supremacy in California* (Berkeley: University of California Press, 1994); Daniel, *Bitter Harvest*; Lawrence J. Jelinek, *Harvest Empire: A History of California Agriculture* (San Francisco: Boyd and Fraser, 1979); Richard Walker, *The Conquest of Bread: 150 Years of Agribusiness in California* (New York: New Press, 2004).

13. For comparison, see for example Jamie Peck, "The Right to Work, and the Right at Work," *Economic Geography* 92, no. 1 (2016): 4–30.

14. On the notion of the prevailing wage see Mitchell, *"They Saved the Crops,"* 87–92.

15. Wells, *Strawberry Fields*, 55–90.

16. Frank Bardacke, *Trampling out the Vintage: Cesar Chavez and the Two Souls of the United Farm Workers* (New York: Verso, 2012), chapter 32; Wells, *Strawberry Fields*, 90–96.

17. Ann Aurelia López, *Farmworker's Journey* (Berkeley: University of California Press, 2007); Linda C. Majka and Theo J. Majka, "Organizing U.S. Farmworkers: A Continuous Struggle," in *Hungry for Profit: The Agribusiness Threat to Farmers, Food, and the Environment*, ed. Frederick H. Buttel, Fred Magdoff, and John Bellamy Foster (New York: Monthly Review Press, 2000), 161–74.

18. Bardacke, *Trampling out the Vintage*; Matt García, *From the Jaws of Victory: The Triumph and Tragedy of Cesar Chavez and the Farm Worker Movement* (Berkeley: University of California Press, 2012). See also Marshall Ganz, *Why David Sometimes Wins: Leadership, Organization, and Strategy in the California Farm Worker Movement* (New York: Oxford University Press, 2009).

19. Farmworker-led strategies pursued elsewhere, such as in the retail- and fast-food-oriented campaigns of the Coalition of Immokalee Workers, have not taken hold in California for reasons that to my knowledge have not been analyzed. For an analysis of CIW's work, though, see Laura-Anne Minkoff-Zern, "Challenging the Agrarian Imaginary: Farmworker-Led Food Movements and the Potential for Farm Labor Justice," *Human Geography* 7, no. 1 (2014): 85–101.

20. Leo Chavez, *Covering Immigration: Popular Images and the Politics of the Nation* (Berkeley: University of California Press, 2001).

21. Joseph Nevins, *Operation Gatekeeper: The Rise of the "Illegal Alien" and the Remaking of the U.S.-Mexico Boundary* (London: Routledge, 2001); Maoyong Fan, Susan Gabbard, Anita Alves Pena, and Jeffrey M. Perloff, "Why Do Fewer Agricultural Workers Migrate Now?," *American Journal of Agricultural Economics* 97, no. 3 (2015): 665–79.

22. Mathew Coleman, "Immigration Geopolitics Beyond the Mexico-US Border," *Antipode* 39, no. 1 (2007): 54–76; De León, *The Land of Open Graves*, 98, 155.

23. On this point see also Holmes, *Fresh Fruit, Broken Bodies*, 21–27.

24. Christina Gathmann, "Effects of Enforcement on Illegal Markets: Evidence from Migrant Smuggling along the Southwestern Border," *Journal of Public Economics* 92, no. 10 (2008): 1926–41; Kristen Parks, Gabriel Lozada, Miguel Mendoza, and Lourdes Garcia Santos, "Strategies for Success: Border Crossing in an Era of Heightened Security," in *Migration from the Mexican Mixteca: A Transnational Community in Oaxaca and California*, ed. Wayne A. Cornelius, David FitzGerald, Jorge Hernandez-Diaz, and Scott Borger (Boulder, CO: Lynne Rienner, 2009), 31–61.

25. Margaret Gray, *Labor and the Locavore: The Making of a Comprehensive Food Ethic* (Berkeley: University of California Press, 2013), 63–64; Jill Lindsey Harrison,

Pesticide Drift and the Pursuit of Environmental Justice (Cambridge, MA: MIT Press, 2011), 28–30; Holmes, *Fresh Fruit, Broken Bodies*, 37.

26. Growers' efforts to maintain a low prevailing wage are discussed throughout Mitchell, *"They Saved the Crops."*

27. Philip Martin, "Farm Labor Shortages: How Real? What Response? Backgrounder No. 9–07" (Washington, DC: Center for Immigration Studies, 2007), http://www.cis.org/articles/2007/back907.html. Mitchell more forcefully argues that there have never been labor shortages—that labor surplus is the defining condition of the agricultural landscape. Mitchell, *"They Saved the Crops,"* 229.

28. Stephan G. Bronars, "A Vanishing Breed: How the Decline in U.S. Farm Laborers over the Last Decade Has Hurt the U.S. Economy and Slowed Production on American Farms," Partnership for a New American Economy, 2015, http://www.renewoureconomy.org/wp-content/uploads/2015/08/PNAE_FarmLabor_August-3-3.pdf.

29. Jeffrey S. Passel and D'Vera Cohn, "Unauthorized Immigrant Totals Rise in 7 States, Fall in 14," Pew Research Center, 2014, http://www.pewhispanic.org/2014/11/18/unauthorized-immigrant-totals-rise-in-7-states-fall-in-14/.

30. See for instance Melissa Block and Marisa Penaloza, "'They're Scared': Immigration Fears Exacerbate Migrant Farmworker Shortage," NPR, *The Salt: What's on Your Plate*, September 27, 2017, https://www.npr.org/sections/thesalt/2017/09/27/552636014/theyre-scared-immigration-fears-exacerbate-migrant-farmworker-shortage.

31. Natalie Kitroeff and Geoffrey Mohan, "Wages Rise on California Farms. Americans Still Don't Want the Job," *Los Angeles Times*, March 17, 2017, https://www.latimes.com/projects/la-fi-farms-immigration/.

32. Bolda et al., "Sample Costs to Produce and Harvest Strawberries, Central Coast Region."

33. Wells, *Strawberry Fields*, 229, 233.

34. Wells, *Strawberry Fields*, 229–43.

35. Wells, *Strawberry Fields*, 235, 37.

36. Wells, *Strawberry Fields*, 235, 268.

37. Wells, *Strawberry Fields*, 268–70; Julie Guthman, "Life Itself under Contract: Rent-Seeking and Biopolitical Devolution through Partnerships in California's Strawberry Industry," *Journal of Peasant Studies* 44, no. 1 (2017): 100–117.

38. Wells, *Strawberry Fields*, 45. These per-acre costs included the cost of labor.

39. Wells, *Strawberry Fields*, 44.

40. Wells, *Strawberry Fields*, 150–51.

41. Wells, *Strawberry Fields*, 159; See also Almaguer, *Racial Fault Lines*; and on the construction of worker vulnerability as a form of labor control see Robert J. Thomas, *Citizenship, Gender, and Work* (Berkeley: University of California Press, 1985).

42. Wells, *Strawberry Fields*, 165.

43. On this point see also Gray, *Labor and the Locavore*, 53–61.

44. Wells, *Strawberry Fields*, 146.

45. Wells, *Strawberry Fields*, 158.

46. Wells, *Strawberry Fields*, 173.

47. Wells, *Strawberry Fields*, 19.

48. Mark Bolda, Laura Tourte, Karen Klonsky, and R. L. De Moura, "Sample Costs to Produce Strawberries, Central Coast Region," UC Cooperative Extension, 2010, https://coststudyfiles.ucdavis.edu/uploads/cs_public/1f/29/1f290e6f-c512-4f08-9695-517da36d10f8/strawberrycc2010.pdf; Bolda et al., "Sample Costs to Produce and Harvest Strawberries, Central Coast Region." By 2016, harvest labor costs represented about 60 percent of total costs. Laura Tourte, Mark Bolda, and Karen Klonsky, "The Evolving Fresh Market Berry Industry in Santa Cruz and Monterey Counties," *California Agriculture* 70, no. 3 (2016): 111.

49. Precise numbers are hard to come by, since so many growers rotate with vegetable growers. Some rotate plots of land, but others go in and out of production. Plus these are tracked by permit number, and permit numbers change based on who in the company obtains the permits.

50. Wells, *Strawberry Fields*, 176.

51. In fairly typical doublespeak, growers also testified in meetings about fumigant regulation—that the loss of fumigants would destroy the jobs that the nursery business brought to those regions.

52. Strangely, I was not able to locate any recent research to support or refute this contention. Most of the research on farm labor contracting dates back to 1993, during its heyday.

53. Dan Wheat, "Use of H-2A Workers to Increase Despite Wage Hike," *Capital Press*, January 8, 2015, http://www.capitalpress.com/Nation_World/Nation/20150108/use-of-h-2a-workers-to-increase-despite-wage-hike.

54. Geoffrey Mohan, "As California's Labor Shortage Grows, Farmers Race to Replace Workers with Robots," *Los Angeles Times*, July 21, 2017, https://www.latimes.com/projects/la-fi-farm-mechanization/.

55. On how these sorts of practices can intensify paternalistic labor relations see Gray, *Labor and the Locavore*, 53–61.

56. As further corroboration of this point, at a 2018 strawberry field day I observed farm advisors suggesting that if growers are experiencing labor shortages they should look at the conditions of their plants and ensure they're receiving the optimal amount of pesticide treatment.

57. Mitchell, *"They Saved the Crops,"* 70–73.

58. Quoted in Jason W. Moore, *Capitalism in the Web of Life: Ecology and the Accumulation of Capital* (New York: Verso, 2015), 224.

59. Mitchell, *"They Saved the Crops,"* 71.

60. Vinay Gidwani and Rajyashree N. Reddy, "The Afterlives of 'Waste': Notes from India for a Minor History of Capitalist Surplus," *Antipode* 43, no. 5 (2011): 1625–58; Michael McIntyre and Heidi J. Nast, "Bio(necro)polis: Marx, Surplus Populations, and the Spatial Dialectics of Reproduction and 'Race,'" *Antipode* 43, no. 5 (2011): 1465–88; Michelle Yates, "The Human-as-Waste, the Labor Theory of Value and Disposability in Contemporary Capitalism," *Antipode* 43, no. 5 (2011): 1679–95.

61. McIntyre and Nast, "Bio(necro)polis," 472.

62. Melissa W. Wright, *Disposable Women and Other Myths of Global Capitalism* (New York and London: Routledge, 2006), 49.

63. Horton, *They Leave Their Kidneys in the Fields*, 40.

64. Holmes, *Fresh Fruit, Broken Bodies*.

65. Horton, *They Leave Their Kidneys in the Fields*, 40, 108. On similar points see Megan A. Carney, *The Unending Hunger: Tracing Women and Food Insecurity across Borders* (Berkeley: University of California Press, 2015).

66. Harrison, *Pesticide Drift and the Pursuit of Environmental Justice*; Linda Nash, "The Fruits of Ill-Health: Pesticides and Workers' Bodies in Postwar California," *Osiris* 19 (2004): 203–19; Dvera I. Saxton, "Strawberry Fields as Extreme Environments: The Ecobiopolitics of Farmworker Health," *Medical Anthropology* 34, no. 2 (2015): 166–83.

67. Horton, *They Leave Their Kidneys in the Fields*, 40, 166.

68. Holmes, *Fresh Fruit, Broken Bodies*, chapter 5, 111–54.

69. Harrison, *Pesticide Drift and the Pursuit of Environmental Justice*, 29–30.

70. Nash, "The Fruits of Ill-Health," 203–19.

71. McIntyre and Nast, "Bio(necro)polis," 1465–88.

72. Thomas A. Arcury, Sara A. Quandt, Altha J. Cravey, Rebecca C. Elmore, and Gregory B. Russell, "Farmworker Reports of Pesticide Safety and Sanitation in the Work Environment," *American Journal of Industrial Medicine* 39, no. 5 (2001): 487–98; Shedra Amy Snipes, Beti Thompson, Kathleen O'connor, Bettina Shell-Duncan, Denae King, Angelica P. Herrera, and Bridgette Navarro, "'Pesticides Protect the Fruit, but Not the People': Using Community-Based Ethnography to Understand Farmworker Pesticide-Exposure Risks," *American Journal of Public Health* 99, no. S3 (2009): S616–S21. On farmworkers neglecting their health more generally to perform as good workers see Holmes, *Fresh Fruit, Broken Bodies*.

73. Philip Martin and Daniel Costa, "Farmworker Wages in California: Large Gap between Full-Time Equivalent and Actual Earnings," *Working Economics Blog*, December 27, 2017, http://www.epi.org/blog/farmworker-wages-in-california-large-gap-between-full-time-equivalent-and-actual-earnings/. This same article thus suggests that the seasonal work may be contributing to the labor shortage.

74. Mitchell, *"They Saved the Crops,"* 201.

75. United States Department of Agriculture, "Noncitrus Fruits and Nuts 2016 Summary," National Agricultural Statistics Service, 2017, formerly at http://usda.mannlib.cornell.edu/usda/current/NoncFruiNu/NoncFruiNu-06-27-2017.pdf (website being revamped at the time of this printing).

CHAPTER SEVEN. PRECARIOUS REPAIRS AND GROWING PATHOLOGIES

1. Ryan Galt saw something similar in his study of vegetable production in Costa Rica. There, growers are faced with the demands of export companies to have low

chemical residues on their produce. So the more highly capitalized farms that sell to exporters were using fewer chemicals than farms growing for domestic markets. As with strawberry production, access to capital alone did not makes this possible. These growers also had access to land above the cloud line, where the air is dry and pest virulence lower. Ryan E. Galt, *Food Systems in an Unequal World: Pesticides, Vegetables, and Agrarian Capitalism in Costa Rica* (Tucson: University of Arizona Press, 2014).

2. Although the salience of the Driscoll's story may possibly reflect too much hindsight, sources from which I draw also wrote in hindsight of the company's rise to industry dominance.

3. Here I allude to the processes discussed in chapter 1 that Goodman et al. called appropriationism and substitutionism. David Goodman, Bernardo Sorj, and John Wilkinson, *From Farming to Biotechnology* (Oxford: Basil Blackwell, 1987).

4. Christopher R. Henke, *Cultivating Science, Harvesting Power* (Cambridge, MA: MIT Press, 2008), 36.

5. Stephen Wilhelm and James E. Sagen, *History of the Strawberry: From Ancient Gardens to Modern Markets* (Berkeley: University of California Press, 1974), 165.

6. Wilhelm and Sagen, *History of the Strawberry*, 175, 169.

7. Wilhelm and Sagen, *History of the Strawberry*, 175; Kazuko Nakane, *Nothing Left in My Hands: The Issei of a Rural California Town, 1900–1942* (Berkeley, CA: Heyday Books, 2008), 8.

8. PAST Consultants, "Historic Context Statement for Agricultural Resources in the North County Planning Area, Monterey County" (Petaluma, CA: PAST Consultants, 2010), http://ohp.parks.ca.gov/pages/1054/files/nomontereyco.pdf, 45–46.

9. PAST Consultants, "Historic Context Statement for Agricultural Resources in the North County Planning Area, Monterey County," 86, 85.

10. Wilhelm and Sagen, *History of the Strawberry*, 176.

11. Nakane, *Nothing Left in My Hands*, 11, quoted in PAST Consultants, "Historic Context Statement for Agricultural Resources in the North County Planning Area, Monterey County," 85.

12. Wilhelm and Sagen, *History of the Strawberry*, 195; Driscoll's, "Our Story: 1872–Now," accessed January 3, 2018, https://www.driscolls.com/about/heritage.

13. Wilhelm and Sagen, *History of the Strawberry*, 200.

14. Wilhelm and Sagen, *History of the Strawberry*, 200.

15. H. Vincent Moses, "'The Orange-Grower Is Not a Farmer': G. Harold Powell, Riverside Orchardists, and the Coming of Industrial Agriculture, 1893–1930," *California History* 74, no. 1 (1995): 22–37.

16. Miriam Wells, *Strawberry Fields: Politics, Class, and Work in California Agriculture* (Ithaca, NY: Cornell University Press, 1996), 111–12.

17. Jane W. Borg and Kathy McKenzie Nichols, "Nihon Bunka/Japanese Culture: One Hundred Years in the Pajaro Valley," n.d., accessed January 3, 2018, https://history.santacruzpl.org/omeka/items/show/134512.

18. Borg and Nichols, "Nihon Bunka/Japanese Culture."

19. Wilhelm and Sagen, *History of the Strawberry*, 206–7, 210.

20. Borg and Nichols, "Nihon Bunka/Japanese Culture"; Wilhelm and Sagen, *History of the Strawberry*, 213; Naturipe Farms, "Timeline," accessed January 3, 2018, https://www.naturipefarms.com/timeline/.

21. Wilhelm and Sagen, *History of the Strawberry*, 203; Borg and Nichols, "Nihon Bunka/Japanese Culture."

22. Borg and Nichols, "Nihon Bunka/Japanese Culture."

23. Borg and Nichols, "Nihon Bunka/Japanese Culture"; Wilhelm and Sagen, *History of the Strawberry*, 204.

24. Wilhelm and Sagen, *History of the Strawberry*, 180, 217.

25. Herbert Baum, *Quest for the Perfect Strawberry* (New York: iUniverse, 2005), 6.

26. Beatrice M. Bain and Sidney Hoos, "The California Strawberry Industry: Changing Economic and Marketing Relationships," *California Agriculture* 17, no. 11 (1963): 4. Technology to overcome the problem of perishability for fresh foods is thoroughly examined in Susanne Freidberg, *Fresh: A Perishable History* (Cambridge, MA: Harvard University Press, 2009).

27. George M. Darrow, *The Strawberry: History, Breeding, and Physiology* (New York: Holt, Rinehart, and Winston, 1966), 207.

28. Wilhelm and Sagen, *History of the Strawberry*, 212.

29. Darrow, *The Strawberry*, 207.

30. Baum, *Quest for the Perfect Strawberry*, 7–8; Driscoll's, "Our Story: 1872–Now."

31. Baum, *Quest for the Perfect Strawberry*, 161.

32. Baum, *Quest for the Perfect Strawberry*, 8.

33. Bain and Hoos, "The California Strawberry Industry," 4–6; Wells, *Strawberry Fields*, 33–34; Shane Hamilton, "The Economies and Conveniences of Modern-Day Living: Frozen Foods and Mass Marketing, 1945–1965," *Business History Review* 77, no. 1 (2003): 33–60.

34. Bain and Hoos, "The California Strawberry Industry," 5.

35. Darrow, *The Strawberry*, 207.

36. Bain and Hoos, "The California Strawberry Industry," 4.

37. Here is a case where practice is not in keeping with theories that businesses surrounding the farm shun risk. Plant propagation is subject to some of the same threats as fruit production in terms of land and labor. In addition, nursery growers must predict fruit grower demand for particular cultivars four years out. They also must meet fruit growers' timing demands for planting dates while dealing with unpredictable conditions, for instance having to wait until snow melts in the spring to plant and weather cools in the fall to harvest. Disease is the most significant risk, however, and can wipe out not only a crop but a reputation if you're the nursery grower who sells diseased stock. Nurseries have even faced lawsuits because of diseased plants. Some of these risks are presumably included in the price of a plant.

38. Lassen Canyon Nursery, "About Us," accessed January 3, 2018, https://www.lassencanyonnursery.com/about-us/; also interview.

39. It was the Sakuma brothers' labor practices in Washington state that made Driscoll's the target of a well-publicized secondary boycott in the 2010s, mentioned in the book's introduction. The boycott ended in 2016, when an agreement was made with Familias Unidas por la Justicia, a union representing harvest workers at Sakuma Farm. See Fair World Project, "The Hidden History Made at Sakuma Brothers Farms," accessed January 3, 2018, http://fairworldproject.org/voices-of-fair-trade/the-hidden-history-made-at-sakuma-brothers-farms/.

40. Wells, *Strawberry Fields*, 113.

41. Driscoll's, "Our Story: 1872–Now."

42. Dune Lawrence, "How Driscoll's Is Hacking the Strawberry of the Future," *Bloomberg Businessweek*, July 29, 2015, https://www.bloomberg.com/news/features/2015-07-29/how-driscoll-s-is-hacking-the-strawberry-of-the-future.

43. Wells, *Strawberry Fields*, 113.

44. Baum, *Quest for the Perfect Strawberry*, 14.

45. Baum, *Quest for the Perfect Strawberry*, 15; Wells, *Strawberry Fields*, 43.

46. Stacy Nichole Roberts, "How We Have Forgotten: Chemical Strawberries and Their Archived Alternatives in the Nineteenth and Twentieth Centuries" (MA thesis, North Carolina State University, 2010), 87.

47. Wells, *Strawberry Fields*, 43.

48. Baum, *Quest for the Perfect Strawberry*, 14; Stephen Wilhelm, R. S. Storkan, and John M. Wilhelm, "Preplant Soil Fumigation with Methyl Bromide-Chloropicrin Mixtures for Control of Soil-Borne Diseases of Strawberries: A Summary of Fifteen Years of Development," *Agriculture and Environment* 1 (1974): 229.

49. Baum, *Quest for the Perfect Strawberry*, 3, 43; Bain and Hoos, "The California Strawberry Industry," 5; Roberts, "How We Have Forgotten," 89.

50. Baum, *Quest for the Perfect Strawberry*, 43.

51. Baum, *Quest for the Perfect Strawberry*, 29–31; California Department of Pesticide Regulation "Nonfumigant Strawberry Production Working Group Action Plan," April 2013, http://www.cdpr.ca.gov/docs/pestmgt/strawberry/work_group/action_plan.pdf.

52. Baum, *Quest for the Perfect Strawberry*, 20.

53. Wells, *Strawberry Fields*, 114–15.

54. Roberts, "How We Have Forgotten," 98.

55. Baum, *Quest for the Perfect Strawberry*, 38–39; Roberts, "How We Have Forgotten," 95–96. On the production of fresh food see Freidberg, *Fresh*.

56. Baum, *Quest for the Perfect Strawberry*, 41; Wells, *Strawberry Fields*, 44.

57. Baum, *Quest for the Perfect Strawberry*, 41, 42, 49.

58. Baum, *Quest for the Perfect Strawberry*, 50, 142, 159.

59. Wells, *Strawberry Fields*, 51, 34.

60. Laura Tourte, Mark Bolda, and Karen Klonsky, "The Evolving Fresh Market Berry Industry in Santa Cruz and Monterey Counties," *California Agriculture* 70, no. 3 (2016): 109.

61. WellPict, "About Us," accessed December 21, 2018, http://www.wellpict.com/about-us/.

62. California Giant Berry Farms, "Our Farms," accessed January 3, 2018, https://www.calgiant.com/our-company.

63. Naturipe Berry Growers, "Who We Are," accessed January 3, 2018, http://www.naturipeberrygrowers.com/who-we-are/; also interviews.

64. Hannah Madans, "Irvine Businessman A.G. Kawamura Appointed to Trump's Agriculture Group," *Orange County Register*, August 22, 2016, https://www.ocregister.com/2016/08/22/irvine-businessman-ag-kawamura-appointed-to-trumps-agriculture-group/.

65. Cal Poly Strawberry Center, "Our Team," accessed January 31, 2018, https://strawberry.calpoly.edu/our-team.

66. Maura Maxwell, "Planasa Acquires US's NorCal Nursery," Fruitnet.com, June 1, 2017, http://www.fruitnet.com/americafruit/article/172374/planasa-acquires-uss-norcal-nursery.

67. Fumigation does involve obvious economies of scope. There are huge data management and coordination issues of having to track various caps and acreage limits, with hundreds of growers wanting to fumigate within a brief temporal window.

68. Wells, *Strawberry Fields*, 197–98.

69. Henke, *Cultivating Science, Harvesting Power*.

70. Wells also placed significant emphasis on grower difference, particularly as it pertained to labor control, one of the central concerns of the book. For her, grower difference mapped onto the particularities of place and ethnicity. The Salinas and the Pajaro Valleys, along with an in-between area of small valleys and hamlets that she dubbed the Monterey Hills, constituted three micro-regions that she found worthy of juxtaposition. She characterized the Salinas Valley micro-region as having the largest-scale, most well-financed farms (average acres of seventy-one in 1985). Although most farmers were owner operated, these were also the ones most likely to be absentee owned or manager operated. And except for one large manager-operated Japanese-owned farm, these were most often what she called Anglo growers. In contrast, farms were markedly smaller in the more marginal Monterey Hills micro-region, with an average of twenty-seven acres and the majority of farms less than fifteen acres. Of the growers here, 75 percent were of Mexican origin, and came to be farmers in the way described earlier; according to her analysis, they lacked capital, farming experience, facility in English, and social connections to the local farming system. The Pajaro Valley region was somewhere in the middle, with an average farm size of forty acres, albeit the most wide ranging. The plurality of growers there were of Japanese origin, but whites farmed there, too, as well as a handful of those of Mexican origin. With one exception all farms were owner operated, and most were solidly financed—not because of access to corporate money but because of years of experience and accumulated creditworthiness. Growers there were also most socially connected and made the highest per-acre investments—this not surprising for the birthplace of the industry.

71. Mark Bolda, Laura Tourte, Jeremy Murdock, and Daniel Sumner, "Sample Costs to Produce and Harvest Strawberries, Central Coast Region," UC Coopera-

tive Extension and Agricultural Issues Center, 2016, https://coststudyfiles.ucdavis.edu/uploads/cs_public/e7/6d/e76dceb8-f0f5-4b60-bcb8-76b88d57e272/strawberrycentralcoast-2016-final2-5-1-2017.pdf.

72. California Strawberry Commission, "Growing the American Dream," July 4, 2014, accessed December 27, 2017, http://www.calstrawberry.com/Portals/2/Reports/Community%20Reports/Growing%20the%20American%20Dream.pdf?ver=2018-01-12-075427-827.

73. In late 2018 it was announced that Miles Reiter was to resume his earlier post as the CEO of Driscoll's. Koger, Chris. "Miles Reiter Assumes Driscoll's CEO Post Again," *The Packer*, December 3, 2018, https://www.thepacker.com/article/miles-reiter-assumes-driscolls-ceo-post-again.

74. Eric Schlosser, "In the Strawberry Fields," *Atlantic Monthly* 276, no. 5 (1995): 80–108.

75. For that matter, buyers can mediate access to knowledge, too. Knowledge or lack thereof about practices such as ground preparation, plant spacing, weeding, fertilizer and pesticide application, and timing are especially important in affecting the robustness of the plant and its ability to withstand soil pathogens. Wells, *Strawberry Fields*, 106. On similar points see Galt, *Food Systems in an Unequal World*.

76. Baum, *Quest for the Perfect Strawberry*, 25; Dale Kasler, "The Quest to Breed Better Strawberries Landed UC Davis in Court. Here's What Happened," *Sacramento Bee*, September 15, 2017, http://www.sacbee.com/news/local/article173591931.html.

77. Steve Hinchliffe, Nick Bingham, John Allen, and Simon Carter, *Pathological Lives: Disease, Space and Biopolitics* (Chichester, England: John Wiley and Sons, 2016).

78. Thus far, Driscoll's is by no means forbidding its growers from using fumigants, nor are they providing financial assistance to wean growers off of them. All they provide is the incentive of an organic price premium should growers choose that direction. The problem for low-resource growers in particular is that it is very difficult to forgo fumigation when plants are stressed and unable to fight off pathogens. And it is even tougher to experiment with nonchemical alternatives without some economic cushion. So these growers are in the worst position to heed industry prescriptions to get off fumigants.

79. Driscoll's, "Labor Standards," accessed February 14, 2018, https://www.driscolls.com/about/labor-standards. There's much to say about such schemes when the initiating organization is itself not subject to the standards. See the ample literature on audits and third-party certification for sweatshops, for example: Jill Esbenshade, *Monitoring Sweatshops: Workers, Consumers, and the Global Apparel Industry* (Philadelphia: Temple University Press, 2009); César A Rodríguez-Garavito, "Global Governance and Labor Rights: Codes of Conduct and Anti-Sweatshop Struggles in Global Apparel Factories in Mexico and Guatemala," *Politics and Society* 33, no. 2 (2005): 203–33. For a piece on how such standards apply to farmworkers see Sandy Brown and Christy Getz, "Privatizing Farm Worker Justice: Regulating Labor through Voluntary Certification and Labeling," *Geoforum* 39, no. 3 (2008): 1184–96.

80. Tourte, Bolda, and Klonsky, "The Evolving Fresh Market Berry Industry in Santa Cruz and Monterey Counties," 108. This reflected enormous increases in tons per acre, from nine to thirty in Santa Cruz County and twelve to thirty-seven in Monterey.

81. Tourte, Bolda, and Klonsky, "The Evolving Fresh Market Berry Industry in Santa Cruz and Monterey Counties," 112–13.

82. Colin A. Carter, James A. Chalfant, Rachael E. Goodhue, Frank M. Han, and Massimiliano DeSantis, "The Methyl Bromide Ban: Economic Impacts on the California Strawberry Industry," *Review of Agricultural Economics* 27, no. 2 (2005): 181–97; Rachael E. Goodhue, Steven A. Fennimore, and Husein A. Ajwa, "The Economic Importance of Methyl Bromide: Does the California Strawberry Industry Qualify for a Critical Use Exemption from the Methyl Bromide Ban?," *Review of Agricultural Economics* 27, no. 2 (2005): 198–211.

83. California Strawberry Commission, "2018 Acreage Survey," 2018, http://www .calstrawberry.com/Portals/2/Reports/Industry%20Reports/Acreage%20Survey /2018%20Acreage%20Survey%20-%20Update.pdf?ver=2018-08-03-100021-043.

84. California Strawberry Commission, "2018 Acreage Survey." The 2017 report, which included 2013 figures, was removed from the website before I obtained the exact URL. Industry spokespeople say that some of the decline in acreage has been compensated by higher productivity from high-yield varieties. Organic Produce Network, "Organic Strawberries in Short Supply," January 11, 2018, http://www .organicproducenetwork.com/article/351/organic-strawberries-in-short-supply.

85. Mike Harris, "Oxnard's Mandalay Berry Farms Closing; 565 Employees Losing Jobs," *Ventura County Star*, June 14, 2016, http://archive.vcstar.com /business/local/oxnards-mandalay-berry-farms-closing-565-employees-losing-jobs-354105ae-4579-7b01-e053-0100007f881f-382982511.html/; Geoffrey Mohan, "Dole, the World's Largest Fresh Fruit and Vegetable Company, Is Stepping Back from Southland," *Los Angeles Times*, August 30, 2017, http://www.latimes.com/business /la-fi-dole-socal-20170830-story.html.

86. Bolda et al., "Sample Costs to Produce and Harvest Strawberries, Central Coast Region."

CHAPTER EIGHT. IMPERFECT ALTERNATIVES
AND TENUOUS FUTURES

1. Jamie Lorimer, "Probiotic Environmentalities: Rewilding with Wolves and Worms," *Theory, Culture and Society* 34, no. 4 (2017): 36.

2. I use "antibiotic" and "probiotic," discussed in what follows, in the broad sense described in Lorimer, "Probiotic Environmentalities," 27–48.

3. Starkly different directions for the future of agriculture have been noted elsewhere. Generally, scholars have identified three distinct paths: a continuation of chemical-intensive industrial food production, which many consider unsustainable; a more decisive shift toward integrated, low-input, and localized agro-ecological

systems, which many consider unrealistic; or an embrace of technologies based in the life sciences, which some consider hopeful and others frightening and dangerous. Warren Belasco, *Meals to Come: A History of the Future of Food* (Berkeley: University of California Press, 2006); Tim Lang and Michael Heasman, *Food Wars: The Global Battle for Mouths, Minds and Markets* (London: Routledge, 2015). This particular taxonomy only roughly maps onto the distinctions I'm making here. In particular, life-science technologies, which many are advocating for the strawberry industry, can be antibiotic and/or probiotic, and even play a role in abiotic systems.

4. Here I refer to scholarship that seems to suggest that plantations are opposite to entangled assemblages, for example Anna Lowenhaupt Tsing, *The Mushroom at the End of the World: On the Possibility of Life in Capitalist Ruins* (Princeton, NJ: Princeton University Press, 2015).

5. Evan Hepler-Smith, "Molecular Bureaucracy: Toxicological Information and Environmental Protection," *Environmental History* 24, no. 3 (forthcoming 2019).

6. California Strawberry Commission, "Investing in a Sustainable Future," January 2013, http://www.calstrawberry.com/Portals/0/Reports/Community%20 Reports/Investing%20in%20a%20Sustainable%20Future.pdf?ver=2017-03-24- 165159-710%3Cspan%20id.

7. I serve as a co-PI on one. This grant is led by UC Davis and is using genomics to develop disease-resistant varieties. As the lead qualitative social scientist, my role is to assess grower willingness to adopt these varieties and suggest policies and practices that would encourage adoption.

8. Facing greater scrutiny and enhanced regulation since *Silent Spring*, the "crop protection" industry has more generally been investing in the development of so-called green chemicals. Ken Geiser, *Chemicals without Harm: Policies for a Sustainable World* (Cambridge, MA: MIT Press, 2015). Some of these are more carefully targeted (rather than broad spectrum) while others dissipate quickly or leave no residues, which, as some have argued, make them safer for consumers rather than for handlers and other workers. Jill Lindsey Harrison, "'Accidents' and Invisibilities: Scaled Discourse and the Naturalization of Regulatory Neglect in California's Pesticide Drift Conflict," *Political Geography* 25, no. 5 (2006): 506–29; Angus Wright, *The Death of Ramón González: The Modern Agricultural Dilemma*, 2nd ed. (Austin: University of Texas Press, 2005).

9. California Department of Pesticide Regulation, "Nonfumigant Strawberry Production Working Group Action Plan," April 2013, https://www.cdpr.ca.gov /docs/pestmgt/strawberry/work_group/action_plan.pdf; Florida Department of Agriculture and Consumer Services, and Florida Department of Health, "Frequently Asked Questions about Dimethyl Disulfide," 2013, https://www.freshfromflorida .com/content/download/3302/20733/DMDS_QnA_3-20-2013.pdf.

10. Isagro-USA, "Dominus," n.d., accessed December 18, 2018, http://www .isagro-usa.com/dominus.html.

11. California Department of Pesticide Regulation, "Nonfumigant Strawberry Production Working Group Action Plan."

12. California Strawberry Commission, "Investing in a Sustainable Future"; Department of Pesticide Regulation, "Nonfumigant Strawberry Production Working Group Action Plan."

13. California Strawberry Commission, "Investing in a Sustainable Future"; Department of Pesticide Regulation, "Nonfumigant Strawberry Production Working Group Action Plan."

14. Amanda Hodson and Edwin E. Lewis, "Managing for Soil Health Can Suppress Pests," *California Agriculture* 70, no. 3 (2016): 137–41.

15. California Strawberry Commission, "Investing in a Sustainable Future"; California Department of Pesticide Regulation, "Nonfumigant Strawberry Production Working Group Action Plan."

16. Hodson and Lewis, "Managing for Soil Health Can Suppress Pests," 137–41.

17. California Strawberry Commission, "Investing in a Sustainable Future"; Department of Pesticide Regulation, "Nonfumigant Strawberry Production Working Group Action Plan."

18. Carol Shennan, Joji Muramoto, M. Mazzola, D. Butler, E. Rosskoph, N. Kokalis-Burelle, K. Momma, Y. Kobara, and J. Lamers, "Anaerobic Soil Disinfestation for Soil Borne Disease Control in Strawberry and Vegetable Systems: Current Knowledge and Future Directions," *Acta Horticulturae* 1137 (2015): 113–20; Joji Muramoto, Carol Shennan, Margherita Zavatta, Graeme Baird, Lucinda Toyama, and Mark Mazzola, "Effect of Anaerobic Soil Disinfestation and Mustard Seed Meal for Control of Charcoal Rot in California Strawberries," *International Journal of Fruit Science* 16, no. sup1 (2016): 59–70; California Strawberry Commission, "Investing in a Sustainable Future"; Department of Pesticide Regulation, "Nonfumigant Strawberry Production Working Group Action Plan."

19. Maria Puig de la Bellacasa, "Making Time for Soil: Technoscientific Futurity and the Pace of Care," *Social Studies of Science* 45, no. 5 (2015): 691–716.

20. California Department of Pesticide Regulation, "Nonfumigant Strawberry Production Working Group Action Plan."

21. Steve Hinchliffe, John Allen, Stephanie Lavau, Nick Bingham, and Simon Carter, "Biosecurity and the Topologies of Infected Life: From Borderlines to Borderlands," *Transactions of the Institute of British Geographers* 38, no. 4 (2013): 531–43; Alex M. Nading, *Mosquito Trails: Ecology, Health, and the Politics of Entanglement* (Berkeley: University of California Press, 2014), 132. Nading critiques "a view of health that has the disentanglement of people, things, vectors, and pathogens as its primary objectives" when health itself depends on entanglements (132).

22. Lorimer, "Probiotic Environmentalities," 28. Many say that the antibiotic approach that dominates both medicine and food production originated with Louis Pasteur (1822–1895), whose discoveries supported a germ theory of disease. With germs isolated as a causal agent of disease, technologies of eradication seemed the obvious next step. Although Pasteur's work on vaccination involved ideas of inoculation rather than killing, he is best known for pasteurization—treatments that kill pathological germs. In agriculture, however, killing as treatment far predates Pasteur.

23. Lorimer, "Probiotic Environmentalities," 27–28.

24. Lorimer, "Probiotic Environmentalities," 33.

25. Nading, *Mosquito Trails*, 132.

26. Hodson and Lewis, "Managing for Soil Health Can Suppress Pests," 137–41.

27. California Department of Pesticide Regulation, "Nonfumigant Strawberry Production Working Group Action Plan."

28. As Lorimer states, "One does not have to look far to detect tendencies towards private property in the probiotic turn." Lorimer, "Probiotic Environmentalities," 38.

29. Jos M. Raaijmakers and Mark Mazzola, "Soil Immune Responses," *Science* 352, no. 6292 (2016): 1392–93. For an analogy in nutrition see Gyorgy Scrinis, *Nutritionism: The Science and Politics of Dietary Advice* (New York: Columbia University Press, 2013).

30. Laura Tourte, Mark Bolda, and Karen Klonsky, "The Evolving Fresh Market Berry Industry in Santa Cruz and Monterey Counties," *California Agriculture* 70, no. 3 (2016): 107–15.

31. Organic Produce Network, "Organic Strawberries in Short Supply," January 11, 2018, http://www.organicproducenetwork.com/article/351/organic-strawberries-in-short-supply.

32. Kayla Webb, "Driscoll's Expands Its Organic Operations to Keep Up with Demand," AndNowUKnow.com, 2018, https://www.andnowuknow.com/whats-store/driscolls-expands-its-organic-operations-keep-demand/kayla-webb/58968?no-trailer=1&utm_source=ANUK&utm_campaign=91bff959c5-EMAIL_CAMPAIGN_2018_07_12_04_07&utm_medium=email&utm_term=0_05a0621836–91bff959c5–16818157. This article, accessed in 2018 from a produce industry news site, is no longer available to the public.

33. Mark Bolda, Laura Tourte, Karen Klonsky, and R. L. De Moura, "Sample Costs to Produce Strawberries, Central Coast Region," UC Cooperative Extension, 2010, https://coststudyfiles.ucdavis.edu/uploads/cs_public/1f/29/1f290e6f-c512-4f08-9695-517da36d10f8/strawberrycc2010.pdf; Mark Bolda, Laura Tourte, Karen Klonsky, Richard L. De Moura, and Kabir P. Tumber, "Sample Costs to Produce Organic Strawberries, Central Coast Region," UC Cooperative Extension, 2014, http://coststudyfiles.ucdavis.edu/uploads/cs_public/94/4b/944b5aad-6660-4dcd-a449-d26361afcae2/strawberry-cc-organic-2014.pdf. Discussions with growers in 2018 suggest that the organic premium is, not surprisingly, eroding from more widespread entry.

34. The primary rationales for transitioning strawberry fields to organic production were pretty much identical to those I learned of for the organic sector as a whole while researching the first edition of *Agrarian Dreams* in the late 1990s. Julie Guthman, *Agrarian Dreams: The Paradox of Organic Farming in California*, 2nd ed. (Berkeley: University of California Press, 2014).

35. This is a gross simplification of how organic standards are set and monitored, and I have written about the nature of this extensively, as have several others. While

individual certifiers once set standards, now the National Organic Standards Board must review all materials and processes, which some say have been coopted by "big organic" to make them less distinct from conventional. See E. Melanie DuPuis and Sean Gillon, "Alternative Modes of Governance: Organic as Civic Engagement," *Agriculture and Human Values* 26 (2009): 43–56; Julie Guthman, "The Trouble with 'Organic Lite' in California: A Rejoinder to the 'Conventionalization' Debate," *Sociologia Ruralis* 44, no. 3 (2004): 301–16; Daniel Jaffee and Philip Howard, "Corporate Cooptation of Organic and Fair Trade Standards," *Agriculture and Human Values* 27 (2010): 387–99; Ben Lilliston and Ronnie Cummins, "Organic Versus 'Organic': The Corruption of a Label," *The Ecologist (U.K. Environmental Magazine)*, July–August 1998, 195–202; Peter M. Rosset and Miguel Altieri, "Agroecology versus Input Substitution: A Fundamental Contradiction of Sustainable Agriculture," *Society and Natural Resources* 10, no. 3 (1997): 283–95.

36. National Organic Standards Board, "Petitioned Material Checklist—Allyl Isothiocyanate," December 16, 2014, https://www.ams.usda.gov/sites/default/files /media/Allyl%20Isothiocyanate%20Proposal.pdf.

37. Caitlin Dewey, "Pioneers of Organic Farming Are Threatening to Leave the Program They Helped Create," *Washington Post*, November 2, 2017, https://www .washingtonpost.com/news/wonk/wp/2017/11/02/pioneers-of-organic-farming-are-threatening-to-leave-the-program-they-helped-create/.

38. Organic regulations stipulate that crops cannot be sold as organic until three years have passed since the application of a disallowed substance. Since growers cannot receive the price premium for organics for these three years, the costs of transitioning are sizable and, thus, discouraging.

39. Organic Produce Network, "Organic Strawberries in Short Supply."

40. A nursery owner estimated that organic starts would cost more than $20,000 per acre (in 2016 dollars).

41. Yet these are precisely the sorts of ideas that get farm advisors in trouble with their clientele audience, who want simple "rules of thumb" even when they come with "significant brutality." Frank Uekötter, "Farming and Not Knowing: Agnotology Meets Environmental History," in *New Natures: Joining Environmental History with Science and Technology Studies*, ed. Dolly Jorgensen, Finn Arne Jorgensen, and Sara Pritchard (Pittsburgh: University of Pittsburgh Press, 2013), 46.

42. In Wells's study, social capital was an additional factor that differentiated growers, precisely in their ability to access information, new technologies, and learning. Miriam Wells, *Strawberry Fields: Politics, Class, and Work in California Agriculture* (Ithaca, NY: Cornell University Press, 1996).

43. Alan P. Rudy, Dawn Coppin, Jason Konefal, Bradley T. Shaw, Toby Van Eyck, Craig Harris, and Lawrence Busch, *Universities in the Age of Corporate Science: The UC Berkeley–Novartis Controversy* (Philadelphia: Temple University Press, 2007).

44. Eben Kirksey, *Emergent Ecologies* (Durham, NC: Duke University Press, 2015), 7, 217; Tsing, *The Mushroom at the End of the World*.

45. Donna Haraway, "When Species Meet: Staying with the Trouble," *Environment and Planning D: Society and Space* 28, no. 1 (2010): 53–55; Kristina Marie

Lyons, "Soil Science, Development, and the 'Elusive Nature' of Colombia's Amazonian Plains," *Journal of Latin American and Caribbean Anthropology* 19, no. 2 (2014): 212–36; Heather Anne Swanson, "The Banality of the Anthropocene," *Cultural Anthropology*, February 22, 2017, https://culanth.org/fieldsights/1074-the-banality-of-the-anthropocene.

46. Jason W. Moore, *Capitalism in the Web of Life: Ecology and the Accumulation of Capital* (New York: Verso, 2015), 273; Heather Swanson et al., "Introduction: Bodies Tumbled into Bodies," in *Arts of Living on a Damaged Planet | Monsters of the Anthropocene*, ed. Anna Tsing, Heather Swanson, Elaine Gan, and Nils Bubandt, (Minneapolis: University of Minnesota, 2017), M1–M12. The notion of "dead land" is attributed to Saskia Sassen, *Expulsions: Brutality and Complexity in the Global Economy* (Cambridge, MA: Harvard University Press, 2014) 149–210.

47. Moore, *Capitalism in the Web of Life*, 273–80.

48. Kirksey, *Emergent Ecologies*, 4.

49. Kirksey, *Emergent Ecologies*, 1, 200–201.

50. Tsing, *The Mushroom at the End of the World*.

51. Kirksey, *Emergent Ecologies*, 209.

52. Lyons, "Soil Science, Development, and the 'Elusive Nature' of Colombia's Amazonian Plains," 231.

53. I use "troubled ecologies" borrowing from Sarah Besky and Alex Blanchette to connote an ecology not yet in ruin, but requiring constant maintenance. Sarah Besky and Alex Blanchette, eds., *How Nature Works* (Santa Fe, NM: School for Advanced Research, forthcoming).

54. To be clear, for Tsing, in *The Mushroom at the End of the World*, patchiness is a way of seeing the world and not a metaphor or model for life after ruin.

55. Nor can these be wholly condemned. As a professor of public policy and old friend reminds me, affordable housing might be California's most critical need.

56. To be sure, a range of scholars drawing on biopolitical analytics have in various ways highlighted how human management of plants, animals, and other organisms, whether in the form of saving endangered species or making cheese, is never about having all life flourish but can often entail neglect, or even eradication, of life that is detrimental to valued life. Christine Biermann and Becky Mansfield, "Biodiversity, Purity, and Death: Conservation Biology as Biopolitics," *Environment and Planning D: Society and Space* 32, no. 2 (2014): 257–73; Aaron Bobrow-Strain, "White Bread Bio-Politics: Purity, Health, and the Triumph of Industrial Baking," *Cultural Geographies* 15, no. 1 (2008): 19–40; Rosemary Collard and Jessica Dempsey, "Capitalist Natures in Five Orientations," *Capitalism Nature Socialism* 28, no. 1 (2017): 78–97; Jamie Lorimer and Clemens Driessen, "Bovine Biopolitics and the Promise of Monsters in the Rewilding of Heck Cattle," *Geoforum* 48 (2013): 249–59; Heather Paxson, "Post-Pasteurian Cultures: The Microbiopolitics of Raw-Milk Cheese in the United States," *Cultural Anthropology* 23, no. 1 (2008): 15–47. The overarching point is that making life live inevitably involves sorting what lives are valuable. Kirksey (*Emergent Ecologies*), too, concedes that being part of heterogeneous political communities requires killing—what he refers to as killing existential threats to species we love.

57. Lorimer, "Probiotic Environmentalities," 38. Also Elizabeth A. Povinelli, *Geontologies: A Requiem to Late Liberalism* (Durham, NC: Duke University Press, 2016), 12, in response to Tsing, *The Mushroom at the End of the World*.

58. On the biopolitics of plantation economies see David Nally, "The Biopolitics of Food Provisioning," *Transactions of the Institute of British Geographers* 36, no. 1 (2011): 37–53. On the biopolitics of labor see those writing on disposability who have noted that racially marked populations may be valued for their labor but are otherwise not counted as important: Collard and Dempsey, "Capitalist Natures in Five Orientations," 78–97; Tania Murray Li, "To Make Live or Let Die? Rural Dispossession and the Protection of Surplus Populations," *Antipode* 41, no. S1 (2010): 66–93; Michelle Yates, "The Human-as-Waste, the Labor Theory of Value and Disposability in Contemporary Capitalism," *Antipode* 43, no. 5 (2011): 1679–95.

59. On the violence that ruination may entail for human lives still attached to capitalist natures see Cynthia Morinville and Nicole Van Lier, "'Dead' Land and the End of Capitalism? Interrogating the Politics of Toxic Environments and Capitalist Accumulation," *Capitalism Nature Socialism* (in review).

60. Sarah Besky, *The Darjeeling Distinction: Labor and Justice on Fair-Trade Tea Plantations in India* (Berkeley: University of California Press, 2013).

61. Elspeth Probyn, *Eating the Ocean* (Durham, NC: Duke University Press, 2016).

CONCLUSION: THE PROBLEM WITH THE SOLUTION

1. Hannah Landecker, "Antibiotic Resistance and the Biology of History," *Body and Society* 22, no. 4 (2015): 1.

2. They could have made conditions more comfortable for workers by raising beds or breeding for taller plants. George M. Darrow, *The Strawberry: History, Breeding, and Physiology* (New York: Holt, Rinehart, and Winston, 1966), 383.

3. Landecker, "Antibiotic Resistance and the Biology of History," 13, 14.

4. Swanton Berry Farm, "History," accessed March 8, 2018, http://www.swantonberryfarm.com/history.

5. Christopher R. Henke, *Cultivating Science, Harvesting Power* (Cambridge, MA: MIT Press, 2008), 144.

6. On the importance of interdisciplinarity see Margaret McFall-Ngai, "Noticing Microbial Worlds: The Postmodern Synthesis in Biology," in *Arts of Living on a Damaged Planet | Monsters of the Anthropocene*, ed. Anna Tsing, Heather Swanson, Elaine Gan, and Nils Bubandt (Minneapolis: University of Minnesota, 2017), M65.

REFERENCES

Abrahamsson, Sebastian, and Filippo Bertoni. "Compost Politics: Experimenting with Togetherness in Vermicomposting." *Environmental Humanities* 4, no. 1 (2014): 125–48.

Agency for Toxic Substances and Disease Registry. "Medical Management Guidelines for Methyl Bromide." N.d. Accessed July 13, 2018. https://www.atsdr.cdc.gov/MMG/MMG.asp?id=818&tid=160.

———. "Toxicological Profile for Bromomethane." 1992. https://www.atsdr.cdc.gov/ToxProfiles/tp27.pdf.

Ahmad, Sameerah. "Strawberries and Solidarity: Farmworkers Build Unity around Driscoll's Berries Boycott." *In These Times*, February 4, 2017, http://inthesetimes.com/working/entry/19927/farmworkers_build_unity_around_driscolls_berries_boycott.

Alkon, Alison, and Julie Guthman. *The New Food Activism: Opposition, Cooperation, and Collective Action*. Oakland: University of California Press, 2017.

Alkon, Alison Hope. "Food Justice and the Challenge to Neoliberalism." *Gastronomica: The Journal of Food and Culture* 14, no. 2 (2014): 27–40.

Allen, John, and Stephanie Lavau. "'Just-in-Time' Disease: Biosecurity, Poultry and Power." *Journal of Cultural Economy* 8, no. 3 (2015): 342–60.

Allen, Patricia, Margaret FitzSimmons, Michael Goodman, and Keith Warner. "Shifting Plates in the Agrifood Landscape: The Tectonics of Alternative Agrifood Initiatives in California." *Journal of Rural Studies* 19, no. 1 (2003): 61–75.

Almaguer, Tomás. *Racial Fault Lines: The Historical Origins of White Supremacy in California*. Berkeley: University of California Press, 1994.

Anderson, Ben, and Colin McFarlane. "Assemblage and Geography." *Area* 43, no. 2 (2011): 124–27.

Anderson, James B. "The Fungus *Armillaria bulbosa* Is among the Largest and Oldest Living Organisms." *Nature* 356 (1992): 428–31.

Anderson, Mark. "Strawberry Suit Settled with Appointment of New UC Breeder." *Sacramento Business Journal*, February 9, 2015, https://www.bizjournals.com

/sacramento/news/2015/02/09/strawberry-suit-settled-with-appointment-of-new-uc.html.

Appel, Hannah, Arthur Mason, and Michael Watts. "Introduction: Oil Talk." In *Subterranean Estates: Life Worlds of Oil and Gas,* edited by Hannah Appel, Arthur Mason, and Michael Watts, 1–26. Ithaca, NY: Cornell University Press, 2015.

Arcury, Thomas A., Sara A. Quandt, Altha J. Cravey, Rebecca C. Elmore, and Gregory B. Russell. "Farmworker Reports of Pesticide Safety and Sanitation in the Work Environment." *American Journal of Industrial Medicine* 39, no. 5 (2001): 487–98.

Ashworth, L. J., and G. Zimmerman. "*Verticillium* Wilt of the Pistachio Nut Tree: Occurrence in California and Control by Fumigation." *Phytopathology* 66 (1976): 1449–51.

Atallah, Zahi K., Ryan J. Hayes, and Krishna V. Subbarao. "Fifteen Years of *Verticillium* Wilt of Lettuce in America's Salad Bowl: A Tale of Immigration, Subjugation, and Abatement." *Plant Disease* 95, no. 7 (2011): 784–92.

Bain, Beatrice M., and Sidney Hoos. "The California Strawberry Industry: Changing Economic and Marketing Relationships." *California Agriculture* 17, no. 11 (1963): 4–6.

Barad, Karen. *Meeting the Universe Halfway: Quantum Physics and the Entanglement of Matter and Meaning.* Durham, NC: Duke University Press, 2007.

Barbour, Madison, and Julie Guthman. "(En)gendering Exposure: Pregnant Farmworkers and the Inadequacy of Pesticide Notification." *Journal of Political Ecology* 25, no. 1 (2018): 332–49.

Bardacke, Frank. *Trampling out the Vintage: Cesar Chavez and the Two Souls of the United Farm Workers.* New York: Verso, 2012.

Barrons, Keith. "Chapter Six, Soil Fumigants." Dow Chemical, unpublished manuscript, author-obtained photocopy, book title unknown.

Barry, Andrew. "Manifesto for a Chemical Geography." Lecture, University College of London, 2017. http://www.academia.edu/32374031/Manifesto_for_a_Chemical_Geography_2017_.

Baum, Herbert. *Quest for the Perfect Strawberry.* New York: iUniverse, 2005.

Belasco, Warren. *Meals to Come: A History of the Future of Food.* Berkeley: University of California Press, 2006.

Benko, Keli. "People Need To Know! Notification and the Regulation of Pesticide Use near Public Schools in California." In review, *Environmental and Planning E: Nature and Space.*

Bensaude-Vincent, Bernadette, and Jonathan Simon. *Chemistry: The Impure Science.* London: Imperial College Press, 2012.

Benton, Ted. "Marxism and Natural Limits: An Ecological Critique and Reconstruction." *New Left Review* 178 (1989): 51–86.

Berlanger, I., and M. L. Powelson. "*Verticillium* Wilt." American Phytopathological Society website. Accessed November 21, 2000. http://www.apsnet.org/edcenter/intropp/Lessons/fungi/ascomycetes/Pages/VerticilliumWilt.aspx.

Berry, Wendell. *The Unsettling of America: Culture and Agriculture.* 2nd ed. San Francisco: Sierra Club Books, 1986.

Besky, Sarah. *The Darjeeling Distinction: Labor and Justice on Fair-Trade Tea Plantations in India*. Berkeley: University of California Press, 2013.

Besky, Sarah, and Alex Blanchette, eds. *How Nature Works*. Santa Fe, NM: School for Advanced Research, forthcoming.

Biehler, Dawn, and John-Henry Pitas. "From Mosquitoes in the System to Transformative Entanglements with Ecologies of Injustice." Unpublished paper presented at the annual meeting of the American Association of Geographers, Chicago, 2015.

Biermann, Christine, and Becky Mansfield. "Biodiversity, Purity, and Death: Conservation Biology as Biopolitics." *Environment and Planning D: Society and Space* 32, no. 2 (2014): 257–73.

Bishopp, F. C. "The Insecticide Situation." *Journal of Economic Entomology* 39, no. 4 (1946): 449–59.

Blanchette, Alex. "Living Dust, Toxic Health, and the Labor of Antibiotic Resistance on Factory Farms." *Medical Anthropology Quarterly,* forthcoming.

Block, Melissa, and Marisa Penaloza. "'They're Scared': Immigration Fears Exacerbate Migrant Farmworker Shortage." NPR, *The Salt: What's on Your Plate*, September 27, 2017, https://www.npr.org/sections/thesalt/2017/09/27/552636014/theyre-scared-immigration-fears-exacerbate-migrant-farmworker-shortage.

Bobrow-Strain, Aaron. "White Bread Bio-Politics: Purity, Health, and the Triumph of Industrial Baking." *Cultural Geographies* 15, no. 1 (2008): 19–40.

Bolda, Mark, Doug Shaw, and Tom Gordon. "Strawberries and Caneberries." University of California Division of Agriculture and Natural Resources, September 16, 2015. https://ucanr.edu/blogs/blogcore/postdetail.cfm?postnum=18979.

Bolda, Mark, Laura Tourte, Karen Klonsky, and R. L. De Moura. "Sample Costs to Produce Strawberries, Central Coast Region." UC Cooperative Extension, 2010. https://coststudyfiles.ucdavis.edu/uploads/cs_public/1f/29/1f290e6f-c512-4f08-9695-517da36d10f8/strawberrycc2010.pdf.

Bolda, Mark, Laura Tourte, Karen Klonsky, Richard L. De Moura, and Kabir P. Tumber. "Sample Costs to Produce Organic Strawberries, Central Coast Region." UC Cooperative Extension, 2014. http://coststudyfiles.ucdavis.edu/uploads/cs_public/94/4b/944b5aad-6660-4dcd-a449-d26361afcae2/strawberry-cc-organic-2014.pdf.

Bolda, Mark, Laura Tourte, Jeremy Murdock, and Daniel Sumner. "Sample Costs to Produce and Harvest Strawberries, Central Coast Region." UC Cooperative Extension and Agricultural Issues Center, 2016. https://coststudyfiles.ucdavis.edu/uploads/cs_public/e7/6d/e76dceb8-f0f5-4b60-bcb8-76b88d57e272/strawberrycentralcoast-2016-final2-5-1-2017.pdf.

Borer, Elizabeth T., Anna-Liisa Laine, and Eric W. Seabloom. "A Multiscale Approach to Plant Disease Using the Metacommunity Concept." *Annual Review of Phytopathology* 54 (2016): 397–418.

Borg, Jane W., and Kathy McKenzie Nichols. "Nihon Bunka/Japanese Culture: One Hundred Years in the Pajaro Valley." N.d. https://history.santacruzpl.org/omeka/items/show/134512.

Boudia, Soraya. "Managing Scientific and Political Uncertainty: Environmental Risk Assessment in a Historical Perspective." In *Powerless Science? Science and Politics in a Toxic World*, edited by Soraya Boudia and Nathalie Jas, 95–112. New York: Berghahn, 2014.

Boudia, Soraya, Angela N. H. Creager, Scott Frickel, Emmanuel Henry, Nathalie Jas, Carsten Reinhardt, and Jody A. Roberts. "Residues: Rethinking Chemical Environments." *Engaging Science, Technology and Society* 4 (2018): 165–78.

Boudia, Soraya, and Nathalie Jas, eds. "Introduction: The Greatness and Misery of Science in a Toxic World." In *Powerless Science? Science and Politics in a Toxic World*, 1–26. New York: Berghahn, 2014.

Boyd, William. "Genealogies of Risk: Searching for Safety, 1930s–1970s." *Ecology LQ* 39 (2012): 895–989.

Boyd, William, Scott Prudham, and Rachel Schurman. "Industrial Dynamics and the Problem of Nature." *Society and Natural Resources* 14, no. 7 (2001): 555–70.

Brasier, Clive. "Plant Pathology: The Rise of the Hybrid Fungi." *Nature* 405, no. 6783 (2000): 134–35.

Braun, Bruce. "Environmental Issues: Global Natures in the Space of Assemblage." *Progress in Human Geography* 30, no. 5 (2006): 644–54.

———. "Thinking the City through SARS: Bodies, Topologies, Politics." In *Networked Disease: Emerging Infections in the Global City*, edited by S. Harris Ali and Roger Keil, 250–66. Oxford: Wiley-Blackwell, 2008.

Bringhurst, R., and Victor Voth. "Hybridization in Strawberries." *California Agriculture* 36, no. 8 (1982): 25.

———. "New Strawberry Varieties: Fresno, Torrey, Wiltguard." Edited by University of California Division of Agriculture and Natural Resources, 1961.

Bronars, Stephan G. "A Vanishing Breed: How the Decline in U.S. Farm Laborers over the Last Decade Has Hurt the U.S. Economy and Slowed Production on American Farms." Partnership for a New American Economy, 2015. http://www.renewoureconomy.org/wp-content/uploads/2015/08/PNAE_FarmLabor_August-3-3.pdf.

Brown, Sandy, and Christy Getz. "Privatizing Farm Worker Justice: Regulating Labor through Voluntary Certification and Labeling." *Geoforum* 39, no. 3 (2008): 1184–96.

Bubandt, Nils. "Haunted Geologies: Spirits, Stones, and the Necropolitics of the Anthropocene." In *Arts of Living on a Damaged Planet | Monsters of the Anthropocene*, edited by Anna Tsing, Heather Swanson, Elaine Gan, and Nils Bubandt, G120–G141. Minneapolis: University of Minnesota Press, 2017.

Cal Poly Strawberry Center. "Our Team." Accessed January 31, 2018. https://strawberry.calpoly.edu/our-team.

California Department of Pesticide Regulation. "Memorandum: Recommendation on Township Cap Exception Requests for 1,3-Dichloropropene." 2014. Accessed February 18, 2016. http://www.cdpr.ca.gov/docs/emon/methbrom/telone/rec_on_twnshp_telone.pdf. No longer available online.

———. "Nonfumigant Strawberry Production Working Group Action Plan." April 2013. https://www.cdpr.ca.gov/docs/pestmgt/strawberry/work_group/action_plan.pdf.

———. "Our Director and Chief Deputy Director." Accessed February 16, 2018. http://www.cdpr.ca.gov/dprbios.htm.

———. "Report of the Scientific Review Committee on Methyl Iodide to the Department of Pesticide Regulation." February 5, 2010. http://www.cdpr.ca.gov/docs/risk/mei/peer_review_report.pdf.

California Giant Berry Farms. "Our History." Accessed January 3, 2018. https://www.calgiant.com/our-company.

California Strawberry Commission. "2018 Acreage Survey." http://www.calstrawberry.com/Portals/2/Reports/Industry%20Reports/Acreage%20Survey/2018%20Acreage%20Survey%20-%20Update.pdf?ver=2018-08-03-100021-043.

———. "California Strawberry Farming." January 2018. Accessed February 16, 2018. http://www.calstrawberry.com/Portals/2/Reports/Industry%20Reports/Industry%20Fact%20Sheets/California%20Strawberry%20Farming%20Fact%20Sheet%202018.pdf.

———. "Growing California Strawberries." Accessed December 27, 2017. http://www.calstrawberry.com/Growing-California-Strawberries. No longer available online.

———. "Growing the American Dream." July 4, 2014. Accessed December 27, 2017. http://www.calstrawberry.com/Portals/2/Reports/Community%20Reports/Growing%20the%20American%20Dream.pdf?ver=2018-01-12-075427-827.

———. "Investing in a Sustainable Future." January 2013. http://www.calstrawberry.com/Portals/0/Reports/Community%20Reports/Investing%20in%20a%20Sustainable%20Future.pdf?ver=2017-03-24-165159-710%3Cspan%20id.

Campbell, Roy E. "The Concentration of Wireworms by Baits before Soil Fumigation with Calcium Cyanide." *Journal of Economic Entomology* 19, no. 4 (1926): 636–42.

Campbell, Roy E., and M. W. Stone. "The Effect of Sulphur on Wireworms." *Journal of Economic Entomology* 25, no. 5 (1932): 967–70.

Carney, Megan A. *The Unending Hunger: Tracing Women and Food Insecurity across Borders.* Berkeley: University of California Press, 2015.

Carolan, Michael S. "Barriers to the Adoption of Sustainable Agriculture on Rented Land: An Examination of Contesting Social Fields." *Rural Sociology* 70, no. 3 (2005): 387–413.

Carpenter, E. J., and Stanley W. Cosby. "Soil Survey of the Salinas Area, California." United States Department of Agriculture, Bureau of Chemistry and Soils, 1925. https://www.nrcs.usda.gov/Internet/FSE_MANUSCRIPTS/california/salinasareaCA1925/salinasareaCA1925.pdf.

Carroll, Christine L., Colin A. Carter, Rachael E. Goodhue, and C.-Y. Cynthia Lin Lawell. "The Economics of Decision-Making for Crop Disease Control." University of California, Davis, 2017. http://www.des.ucdavis.edu/faculty/Lin/Vwilt_short_long_paper.pdf.

Carroll, Christine L., Colin A. Carter, Rachael E. Goodhue, C.-Y. Cynthia Lin Lawell, and Krishna V. Subbarao. "The Economics of Managing *Verticillium* Wilt, an Imported Disease in California Lettuce." *California Agriculture* 71, no. 3 (2017): 178–83.

Carter, Colin A., James A. Chalfant, Rachael E. Goodhue, Frank M. Han, and Massimiliano DeSantis. "The Methyl Bromide Ban: Economic Impacts on the California Strawberry Industry." *Review of Agricultural Economics* 27, no. 2 (2005): 181–97.

Carter, Eric D. *Enemy in the Blood: Malaria, Environment, and Development in Argentina.* Tuscaloosa: University of Alabama Press, 2012.

Carter, Walter. "Soil Treatments with Special Reference to Fumigation with D-D Mixture." *Journal of Economic Entomology* 38, no. 1 (1945): 35–44.

Cerkauskas, R. F., O. D. Dhingra, and J. B. Sinclair. "Effect of Herbicides on Competitive Saprophytic Colonization by *Macrophomina phaseolina* of Soybean Stems." *Transactions of the British Mycological Society* 79, no. 2 (1982): 201–5.

Cerna, Joseph. "San Francisco Sets All-Time Heat Record Downtown at 106 Degrees During State's Hottest Recorded Summer." *Los Angeles Times*, September 1, 2017, http://www.latimes.com/local/lanow/la-me-ln-california-heat-wave-weekend-20170901-story.html.

Chapman, R. N., and A. H. Johnson. "Possibilities and Limitations of Chloropicrin as a Fumigant for Cereal Products." *Journal of Agricultural Research* 31, no. 8 (1925): 745–60.

Charles, Dan. "Big Bucks from Strawberry Genes Lead to Conflict at UC Davis." NPR, *The Salt: What's on Your Plate*, July 2, 2014, https://www.npr.org/sections/thesalt/2014/07/02/327355935/big-bucks-from-strawberry-genes-lead-to-conflict-at-uc-davis.

———. "Breeding Battle Threatens Key Source of California Strawberries." NPR, *The Salt: What's on Your Plate*, July 1, 2014, https://www.npr.org/sections/thesalt/2014/07/01/327256662/breeding-battle-threatens-key-source-of-california-strawberries.

Chavez, Leo. *Covering Immigration: Popular Images and the Politics of the Nation.* Berkeley: University of California Press, 2001.

Cheetham, Tom. "Pathological Alterations in Embryos of the Codling Moth (Lepidoptera: Tortricidae) Induced by Methyl Bromide." *Annals of the Entomological Society of America* 83, no. 1 (1990): 59–67.

Clapp, Jennifer, Peter Newell, and Zoe W. Brent. "The Global Political Economy of Climate Change, Agriculture and Food Systems." *Journal of Peasant Studies* 45, no. 1 (2018): 80–88.

Cochrane, Willard W. *The Development of American Agriculture: A Historical Analysis.* Minneapolis: University of Minnesota Press, 1979.

Coleman, Mathew. "Immigration Geopolitics beyond the Mexico-US Border." *Antipode* 39, no. 1 (2007): 54–76.

Collard, Rosemary, and Jessica Dempsey. "Capitalist Natures in Five Orientations." *Capitalism Nature Socialism* 28, no. 1 (2017): 78–97.

Constance, Douglas H., J. Sanford Rikoon, and Jian C. Ma. "Landlord Involvement in Environmental Decision-Making on Rented Missouri Cropland: Pesticide Use and Water Quality Issues." *Rural Sociology* 61, no. 4 (1996): 577–605.

Contreras, Shirley. "A Firsthand History of Santa Maria 50 Years after It All Began." *Santa Maria Times*, August 15, 2010, https://santamariatimes.com/lifestyles /columnist/shirley_contreras/a-firsthand-history-of-santa-maria-years-after-it-all /article_d2d4eb60-a823-11df-b747-001cc4c03286.html.

Cook, R. James, and Kenneth Frank Baker. *The Nature and Practice of Biological Control of Plant Pathogens.* St. Paul: American Phytopathological Society, 1983.

Cronise, Titus Fey. *The Natural Wealth of California: Comprising Duly History, Geography, Topography, and Scenery; Climate; Agriculture and Commercial Products; Geology, Zoology, and Botany; Mineralogy, Mines, and Mining Processes; Manufactures; Steamship Lines, Railroads, and Commerce; Immigration, a Detailed Description of Each County.* San Francisco: H. H. Hancroft, 1868.

Daily Democrat (Woodland, CA). "State Audit Advises Adjustments for UC Davis Strawberry Program." June 9, 2015, https://www.dailydemocrat.com/2015/06/09 /state-audit-advises-adjustments-for-uc-davis-strawberry-program/.

Daniel, Cletus. *Bitter Harvest: A History of California Farm Workers, 1870–1941.* Ithaca, NY: Cornell University Press, 1981.

Darrow, George M. *The Strawberry: History, Breeding, and Physiology.* New York: Holt, Rinehart, and Winston, 1966.

Daunt, Tina. "McGraths Everywhere: Founders: In 1876, the First One Arrived, Setting up a Family Dynasty on a Patchwork of Farms. There Are Now 500 Descendants." *Los Angeles Times*, May 6, 1991, http://articles.latimes.com/1991-05-06/local/me-848_1_charles-mcgrath.

Davis, John J. "Miscellaneous Soil Insecticide Tests." *Soil Science* 10, no. 1 (1920): 61.

Dayton, Lily. "Dangerous Drift: Students, Strawberries and Harmful Fumigants Collide in California's Agricultural Belts." Davis: California Institute for Rural Studies, 2015. http://www.cirsinc.org/rural-california-report/entry /dangerous-drift-students-strawberries-and-harmful-fumigants-collide-in-california-s-agricultural-belts.

Dean, Warren. *Brazil and the Struggle for Rubber: A Study in Environmental History.* Cambridge, England: Cambridge University Press, 1987.

De la Peña, Carolyn Thomas. *Empty Pleasures: The Story of Artificial Sweeteners from Saccharin to Splenda.* Chapel Hill: University of North Carolina Press, 2010.

De León, Jason. *The Land of Open Graves: Living and Dying on the Migrant Trail.* Berkeley: University of California Press, 2015.

Dewey, Caitlin. "Pioneers of Organic Farming Are Threatening to Leave the Program They Helped Create." *Washington Post*, November 2, 2017, https://www.washingtonpost.com/news/wonk/wp/2017/11/02/pioneers-of-organic-farming-are-threatening-to-leave-the-program-they-helped-create/.

Diamond, Jared. *Guns, Germs, and Steel.* New York: W. W. Norton, 1997.

Dixon, Marion W. "Biosecurity and the Multiplication of Crises in the Egyptian Agri-Food Industry." *Geoforum* 61, no. supplement C (2015): 90–100.

Doane, R. W. *Common Pests.* Springfield, IL: C. C. Thomas, 1931.

Donohoe, Heber C., A. C. Johnson, and J. W. Bulger. "Methyl Bromide Fumigation for Japanese Beetle Control." *Journal of Economic Entomology* 33, no. 2 (1940): 296–302.

Driscoll's. "Labor Standards." Accessed February 14, 2018. https://www.driscolls.com/about/labor-standards.

———. "Our Story: 1872–Now." Accessed January 3, 2018. https://www.driscolls.com/about/heritage.

Duncan, Colin A. M. *The Centrality of Agriculture: Between Humankind and the Rest of Nature.* Montreal: McGill-Queen's University Press, 1996.

Dunsby, Joshua. "Measuring Environmental Health Risks: The Negotiation of a Public Right-to-Know Law." *Science, Technology and Human Values* 29, no. 3 (2004): 269–90.

DuPuis, E. Melanie. "Not in My Body: BGH and the Rise of Organic Milk." *Agriculture and Human Values* 17, no. 3 (2000): 285–95.

DuPuis, E. Melanie, and Brian J. Gareau. "From Public to Private Global Environmental Governance: Lessons from the Montreal Protocol's Stalled Methyl Bromide Phase-Out." *Environment and Planning A* 41, no. 10 (2009): 2305–23.

DuPuis, E. Melanie, and Sean Gillon. "Alternative Modes of Governance: Organic as Civic Engagement." *Agriculture and Human Values* 26 (2009): 43–56.

Economic Research Service. "Table 2. Deflated US Strawberry Prices and Values, 1970–2012." United States Departure of Agriculture. Updated June 2013. Accessed December 27, 2017. http://usda.mannlib.cornell.edu/MannUsda/viewDocumentInfo.do?documentID=1381.

Environmental Working Group. "Dirty Dozen: EWG's 2018 Shopper's Guide to Pesticides in Produce." Accessed February 16, 2018. https://www.ewg.org/foodnews/dirty_dozen_list.php.

Esbenshade, Jill Louise. *Monitoring Sweatshops: Workers, Consumers, and the Global Apparel Industry.* Philadelphia: Temple University Press, 2009.

Fair World Project. "The Hidden History Made at Sakuma Brothers Farms." Accessed January 3, 2018. http://fairworldproject.org/voices-of-fair-trade/the-hidden-history-made-at-sakuma-brothers-farms/.

Fan, Maoyong, Susan Gabbard, Anita Alves Pena, and Jeffrey M. Perloff. "Why Do Fewer Agricultural Workers Migrate Now?" *American Journal of Agricultural Economics* 97, no. 3 (2015): 665–79.

Federal Writers' Project. "Labor in California Sugar Beet Crop." 1938. http://content.cdlib.org/view?docId=hb88700929;NAAN=13030&doc.view=frames&chunk.id=div00115&toc.depth=1&toc.id=div00115&brand=calisphere.

Filmer, Ann. "Jury Sides with UC Davis in Strawberry Breeding Trial." UC Davis Department of Plant Sciences News Blog, May 25, 2017, https://news.plant-sciences.ucdavis.edu/2017/05/25/jury-sides-with-uc-davis-in-strawberry-breeding-trial/.

———. "Strawberry Breeding Program Backgrounder: A Historical Timeline." UC Davis Department of Plant Sciences News Blog, May 11, 2016, https://news

.plantsciences.ucdavis.edu/2016/05/11/strawberry-breeding-program-backgrounder-a-historical-timeline/.

Fine, Ben. "Towards a Political Economy of Food." *Review of International Political Economy* 1, no. 3 (1994): 519–45.

Fisk, Frank W., and Harold H. Shepard. "Laboratory Studies of Methyl Bromide as an Insect Fumigant." *Journal of Economic Entomology* 31, no. 1 (1938): 79–84.

FitzSimmons, Margaret. "The New Industrial Agriculture." *Economic Geography* 62, no. 4 (1986): 334–53.

Fleming, Jake. "Toward Vegetal Political Ecology: Kyrgyzstan's Walnut–Fruit Forest and the Politics of Graftability." *Geoforum* 79 (2017): 26–35.

Fletcher, Stevenson Whitcomb. *The Strawberry in North America: History, Origin, Botany, and Breeding.* New York: Macmillan, 1917.

Florida Department of Agriculture and Consumer Services, and Florida Department of Health. "Frequently Asked Questions about Dimethyl Disulfide." 2013. Accessed February 28, 2018. https://www.freshfromflorida.com/content/download/3302/20733/DMDS_QnA_3-20-2013.pdf.

Foundation Plant Services. "Introduction to the Strawberry Clean Plant Program." University of California, College of Agriculture and Environmental Sciences. Accessed December 27, 2017. http://fps.ucdavis.edu/strawberry.cfm.

Freidberg, Susanne. *Fresh: A Perishable History.* Cambridge, MA: Harvard University Press, 2009.

Frickel, Scott, and Michelle Edwards. "Untangling Ignorance in Environmental Risk Assessment." In *Powerless Science? Science and Politics in a Toxic World*, edited by Soraya Boudia and Nathalie Jas, 215–33. New York: Berghahn, 2014.

Frickel, Scott, Sahra Gibbon, Jeff Howard, Joanna Kempner, Gwen Ottinger, and David J. Hess. "Undone Science: Charting Social Movement and Civil Society Challenges to Research Agenda Setting." *Science, Technology and Human Values* 35, no. 4 (2010): 444–73.

Friedland, William H., Amy E. Barton, and Robert J. Thomas. *Manufacturing Green Gold.* Cambridge, England: Cambridge University Press, 1981.

Froines, John, Susan Kegley, Timothy Malloy, and Sarah Kobylewski. "Risk and Decision: Evaluating Pesticide Approval in California: Review of the Methyl Iodide Registration Process." Sustainable Technology and Policy Program, University of California at Los Angeles, 2013. https://www.pesticideresearch.com/site/wp-content/uploads/2012/05/Risk_and_Decision_Report_2013.pdf.

Galt, Ryan E. *Food Systems in an Unequal World: Pesticides, Vegetables, and Agrarian Capitalism in Costa Rica.* Tucson: University of Arizona Press, 2014.

Gan, J., S.R. Yates, H.D. Ohr, and J.J. Sims. "Production of Methyl Bromide by Terrestrial Higher Plants." *Geophysical Research Letters* 25, no. 19 (1998): 3595–98.

Ganz, Marshall. *Why David Sometimes Wins: Leadership, Organization, and Strategy in the California Farm Worker Movement.* New York: Oxford University Press, 2009.

García, Matt. *From the Jaws of Victory: The Triumph and Tragedy of Cesar Chavez and the Farm Worker Movement.* Berkeley: University of California Press, 2012.

Gareau, Brian J. "Dangerous Holes in Global Environmental Governance: The Roles of Neoliberal Discourse, Science, and California Agriculture in the Montreal Protocol." *Antipode* 40, no. 1 (2008): 102–30.

———. "The Limited Influence of Global Civil Society: International Environmental Non-Governmental Organisations and the Methyl Bromide Controversy in the Montreal Protocol." *Environmental Politics* 21, no. 1 (2012): 88–107.

———. "We Have Never Been Human: Agential Nature, ANT, and Marxist Political Ecology." *Capitalism Nature Socialism* 16, no. 4 (2005): 127–40.

Gates, Paul W. "Public Land Disposal in California." *Agricultural History* 49 (1975): 158–78.

Gathmann, Christina. "Effects of Enforcement on Illegal Markets: Evidence from Migrant Smuggling along the Southwestern Border." *Journal of Public Economics* 92, no. 10 (2008): 1926–41.

Geiser, Ken. *Chemicals without Harm: Policies for a Sustainable World.* Cambridge, MA: MIT Press, 2015.

Gemmill, Alison, Robert B. Gunier, Asa Bradman, Brenda Eskenazi, and Kim G. Harley. "Residential Proximity to Methyl Bromide Use and Birth Outcomes in an Agricultural Population in California." *Environmental Health Perspectives* 121, no. 6 (2013): 737.

Gibson, Ross Eric. "Agricultural Legacy of Serbo-Croatian Community." N.d. https://history.santacruzpl.org/omeka/items/show/134320#?c=0&m=0&s=0&cv=0.

Gidwani, Vinay, and Rajyashree N. Reddy. "The Afterlives of 'Waste': Notes from India for a Minor History of Capitalist Surplus." *Antipode* 43, no. 5 (2011): 1625–58.

Gilbert, Scott F. "Holobiont by Birth: Multilineage Individuals as the Concretion of Cooperative Processes." In *Arts of Living on a Damaged Planet | Monsters of the Anthropocene,* edited by Anna Tsing, Heather Swanson, Elaine Gan, and Nils Bubandt, M73–M90. Minneapolis: University of Minnesota Press, 2017.

Giroux, Henry. "Neoliberalism, Corporate Culture, and the Promise of Higher Education: The University as a Democratic Public Sphere." *Harvard Educational Review* 72, no. 4 (2002): 425–64.

Gladstone Land Corporation. "Company Overview." Accessed October 30, 2017. http://ir.gladstoneland.com/company-overview.

———. "Gladstone Land Corporation Announces Farmland Acquisition in California." News release. July 14, 2014. http://ir.gladstoneland.com/news-releases/news-release-details/gladstone-land-corporation-announces-farmland-acquisition-6?releaseid=861798.

Gliessman, Stephen R., Eric Engles, and Robin Krieger. *Agroecology: Ecological Processes in Sustainable Agriculture.* Boca Raton, FL: CRC, 1998.

Goodhue, Rachael E., Steven A. Fennimore, and Husein A. Ajwa. "The Economic Importance of Methyl Bromide: Does the California Strawberry Industry Qualify for a Critical Use Exemption from the Methyl Bromide Ban?" *Review of Agricultural Economics* 27, no. 2 (2005): 198–211.

Goodman, David, Bernardo Sorj, and John Wilkinson. *From Farming to Biotechnology.* Oxford: Basil Blackwell, 1987.

Goodman, David, and Michael Watts. "Agrarian Questions, Global Appetite, Local Metabolism: Nature, Culture, and Industry in Fin-De-Siècle Food Systems." In *Globalising Food: Agrarian Questions and Global Restructuring,* edited by David Goodman and Michael Watts, 1–33. London and New York: Routledge, 1997.

Goodyear, Dana. "How Driscoll's Reinvented the Strawberry." *New Yorker,* August 21, 2017, https://www.newyorker.com/magazine/2017/08/21/how-driscolls-reinvented-the-strawberry.

Gray, Margaret. *Labor and the Locavore: The Making of a Comprehensive Food Ethic.* Berkeley: University of California Press, 2013.

Gribble, Gordon W. "The Natural Production of Organobromine Compounds." *Environmental Science and Pollution Research* 7, no. 1 (2000): 37–49.

Grime, J. Philip. *Plant Strategies, Vegetation Processes, and Ecosystem Properties.* Chichester, England: John Wiley and Sons, 2006.

Guthman, Julie. *Agrarian Dreams: The Paradox of Organic Farming in California.* 2nd ed. Berkeley: University of California Press, 2014.

———. "Land Access and Costs May Drive Strawberry Growers' Increased Use of Fumigation." *California Agriculture* 71, no. 3 (2017): 184–91.

———. "Life Itself under Contract: Rent-Seeking and Biopolitical Devolution through Partnerships in California's Strawberry Industry." *Journal of Peasant Studies* 44, no. 1 (2017): 100–117.

———. "Lives versus Livelihoods? Deepening the Regulatory Debates on Soil Fumigants in California's Strawberry Industry." *Antipode* 49, no. 1 (2017): 86–105.

———. "The Trouble with 'Organic Lite' in California: A Rejoinder to the 'Conventionalization' Debate." *Sociologia Ruralis* 44, no. 3 (2004): 301–16.

Guthman, Julie, and Sandra Brown. "Whose Life Counts: Biopolitics and the 'Bright Line' of Chloropicrin Mitigation in California's Strawberry Industry." *Science, Technology and Human Values* 41, no. 3 (2016): 461–92.

Guthman, Julie, and Sandy Brown. "I Will Never Eat Another Strawberry Again: The Biopolitics of Consumer-Citizenship in the Fight against Methyl Iodide in California." *Agriculture and Human Values* 33, no. 3 (2016): 575–85.

———. "Midas' Not-So-Golden Touch: On the Demise of Methyl Iodide as a Soil Fumigant in California." *Journal of Environmental Policy and Planning* 18, no. 3 (2016): 324–41.

Hamilton, Shane. "The Economies and Conveniences of Modern-Day Living: Frozen Foods and Mass Marketing, 1945–1965." *Business History Review* 77, no. 1 (2003): 33–60.

Hancock, James F., Arturo Lavín, and J. B. Retamales. "Our Southern Strawberry Heritage: *Fragaria chiloensis* of Chile." *Horticultural Science* 34, no. 5 (1999): 814–16.

Haraway, Donna. "Anthropocene, Capitalocene, Plantationocene, Chthulucene: Making Kin." *Environmental Humanities* 6, no. 1 (2015): 159–65.

————. "Symbiogenesis, Sympoiesis, and Art Science Activisms for Staying with the Trouble." In *Arts of Living on a Damaged Planet | Monsters of the Anthropocene*, edited by Anna Tsing, Heather Swanson, Elaine Gan, and Nils Bubandt, M25–M50. Minneapolis: University of Minnesota Press, 2017.

————. "When Species Meet: Staying with the Trouble." *Environment and Planning D: Society and Space* 28, no. 1 (2010): 53–55.

Haraway, Donna, Noboru Ishikawa, Scott F. Gilbert, Kenneth Olwig, Anna L. Tsing, and Nils Bubandt. "Anthropologists Are Talking—About the Anthropocene." *Ethnos* 81, no. 3 (2016): 535–64.

Harper, David B., and John T. G. Hamilton. "The Global Cycles of the Naturally-Occurring Monohalomethanes." In *Natural Production of Organohalogen Compounds*, edited by G. Gribble, 17–41. Berlin and Heidelberg, Germany: Springer, 2003.

Harris, Mike. "Oxnard's Mandalay Berry Farms Closing; 565 Employees Losing Jobs." *Ventura County Star*, June 14, 2016, http://archive.vcstar.com/business /local/oxnards-mandalay-berry-farms-closing-565-employees-losing-jobs-354105ae-4579- 7b01-e053-0100007f881f-382982511.html/.

Harrison, Jill Lindsey. "'Accidents' and Invisibilities: Scaled Discourse and the Naturalization of Regulatory Neglect in California's Pesticide Drift Conflict." *Political Geography* 25, no. 5 (2006): 506–29.

————. *Pesticide Drift and the Pursuit of Environmental Justice*. Cambridge, MA: MIT Press, 2011.

Harvey, David. *Justice, Nature, and the Geography of Difference*. Cambridge, MA: Blackwell, 1996.

————. *Limits to Capital*. Chicago: University of Chicago Press, 1982.

Helmreich, Stefan. *Alien Ocean: Anthropological Voyages in Microbial Seas*. Berkeley: University of California Press, 2009.

Henderson, George L. *California and the Fictions of Capital*. New York: Oxford University Press, 1998.

Henke, Christopher R. *Cultivating Science, Harvesting Power*. Cambridge, MA: MIT Press, 2008.

Henry, P. M., S. C. Kirkpatrick, C. M. Islas, A. M. Pastrana, J. A. Yoshisato, S. T. Koike, O. Daugovish, and T. R. Gordon. "The Population of *Fusarium oxysporum f. sp. fragariae*, Cause of *Fusarium* Wilt of Strawberry, in California." *Plant Disease* 101, no. 4 (2017): 550–56.

Hepler-Smith, Evan. "Molecular Bureaucracy: Toxicological Information and Environmental Protection." *Environmental History* 24, no. 3 (forthcoming 2019).

Hinchliffe, Steve, John Allen, Stephanie Lavau, Nick Bingham, and Simon Carter. "Biosecurity and the Topologies of Infected Life: From Borderlines to Borderlands." *Transactions of the Institute of British Geographers* 38, no. 4 (2013): 531–43.

Hinchliffe, Steve, Nick Bingham, John Allen, and Simon Carter. *Pathological Lives: Disease, Space and Biopolitics*. Chichester, England: John Wiley and Sons, 2016.

Hodson, Amanda, and Edwin E. Lewis. "Managing for Soil Health Can Suppress Pests." *California Agriculture* 70, no. 3 (2016): 137–41.

Holmes, Seth. *Fresh Fruit, Broken Bodies: Migrant Farmworkers in the United States*. Berkeley: University of California Press, 2013.

Horton, Sarah Bronwen. *They Leave Their Kidneys in the Fields: Illness, Injury, and Illegality among US Farmworkers*. Oakland: University of California Press, 2016.

Huber, Matthew T. *Lifeblood: Oil, Freedom, and the Forces of Capital*. Minneapolis: University of Minnesota Press, 2013.

Idrovo, A.J., L.H. Sanìn, D. Cole, J. Chavarro, H. Cáceres, J. Narváez, and M.I. Restrepo. "Time to First Pregnancy among Women Working in Agricultural Production." *International Archives of Occupational and Environmental Health* 78, no. 6 (2005): 493–500.

Isagro-USA. "Dominus." N.d. Accessed December 18, 2018. http://www.isagro-usa.com/dominus.html.

Jaffee, Daniel, and Philip Howard. "Corporate Cooptation of Organic and Fair Trade Standards." *Agriculture and Human Values* 27 (2010): 387–99.

Jelinek, Lawrence J. *Harvest Empire: A History of California Agriculture*. San Francisco: Boyd and Fraser, 1979.

Jenny, Hans. *The Soil Resource: Origin and Behaviour*. New York: Springer-Verlag, 1980.

Johnson, H., A. Holland, A. Paulus, and S. Wilhelm. "Soil Fumigation Found Essential for Maximum Strawberry Yields in Southern California." *California Agriculture* 16, no. 10 (1962): 4–6.

Johnson, N.C., J.H. Graham, and F.A. Smith. "Functioning of Mycorrhizal Associations along the Mutualism–Parasitism Continuum." *New Phytologist* 135, no. 4 (1997): 575–85.

Kasler, Dale. "Long-Standing Marriage Goes Sour for UC Davis, Strawberry Industry." *Sacramento Bee*, August 16, 2014, https://www.sacbee.com/news/business/article2606860.html.

———. "The Quest to Breed Better Strawberries Landed UC Davis in Court. Here's What Happened." *Sacramento Bee*, September 15, 2017, http://www.sacbee.com/news/local/article173591931.html.

———. "UC Davis Fires Back in Strawberry Controversy, Sues Growers' Group." *Sacramento Bee*, October 31, 2014, https://www.sacbee.com/news/business/article3493488.html.

Kaur, R., J. Kaur, and Rama S. Singh. "Nonpathogenic *Fusarium* as a Biological Control Agent." *Plant Pathology Journal* 9, no. 3 (2011): 79–91.

Kaur, Surinder, Gurpreet Singh Dhillon, Satinder Kaur Brar, Gary Edward Vallad, Ramesh Chand, and Vijay Bahadur Chauhan. "Emerging Phytopathogen *Macrophomina phaseolina*: Biology, Economic Importance and Current Diagnostic Trends." *Critical Reviews in Microbiology* 38, no. 2 (2012): 136–51.

Kautsky, Karl. *The Agrarian Question*. London: Zwan, 1988. Originally published in 1899.

Kegley, Susan E., Stephan Orme, and Lars Neumeister. "Hooked on Poison: Pesticide Use in California, 1991–1998." San Francisco: Pesticide Action Network, 2000. Accessed July 2, 2014. http://www.panna.org/issues/publication/hooked-poison. No longer available online.

Keyworth, W. G., and Margery Bennett. "*Verticillium* Wilt of the Strawberry." *Journal of Horticultural Science* 26, no. 4 (1951): 304–16.

King, Brian. *States of Disease: Political Environments and Human Health.* Oakland: University of California Press, 2017.

Kirksey, Eben. *Emergent Ecologies.* Durham, NC: Duke University Press, 2015.

Kitroeff, Natalie, and Geoffrey Mohan. "Wages Rise on California Farms. Americans Still Don't Want the Job." *Los Angeles Times,* March 17, 2017, https://www.latimes.com/projects/la-fi-farms-immigration/.

Klein, L. "Methyl Bromide as a Soil Fumigant." In *The Methyl Bromide Issue,* edited by C. H. Bell, N. Price, and B. Chakrabarti, 191–235. Chichester, England: John Wiley and Sons, 1996.

Kleinman, Daniel Lee, and Sainath Suryanarayanan. "Dying Bees and the Social Production of Ignorance." *Science, Technology, and Human Values* 38, no. 4 (2013): 492–517.

Kloppenburg, Jack. *First the Seed: The Political Economy of Plant Biotechnology.* Madison: University of Wisconsin Press, 2005.

Klosterman, Steven J., Zahi K. Atallah, Gary E. Vallad, and Krishna V. Subbarao. "Diversity, Pathogenicity, and Management of *Verticillium* Species." *Annual Review of Phytopathology* 47 (2009): 39–62.

Klosterman, Steven J., Krishna V. Subbarao, Seogchan Kang, Paola Veronese, Scott E. Gold, Bart P. H. J. Thomma, Zehua Chen, et al. "Comparative Genomics Yields Insights into Niche Adaptation of Plant Vascular Wilt Pathogens." *PLOS Pathogens* 7, no. 7 (2011): e1002137 (open access e-journal).

Koger, Chris. "Miles Reiter Assumes Driscoll's CEO Post Again." *The Packer,* December 3, 2018, https://www.thepacker.com/article/miles-reiter-assumes-driscolls-ceo-post-again.

Koike, S., K. Subbarao, R. Michael Davis, and T. Turini. "Vegetable Diseases Caused by Soilborne Pathogens." Report 8099. University of California Division of Agriculture and Natural Resources, 2003. https://anrcatalog.ucanr.edu/pdf/8099.pdf.

Koike, Steven T., Renee S. Arias, Cliff S. Hogan, Frank N. Martin, and Thomas R. Gordon. "Status of *Macrophomina phaseolina* on Strawberry in California and Preliminary Characterization of the Pathogen." *International Journal of Fruit Science* 16, no. sup1 (2016): 148–59.

Koike, Steven T., and Thomas R. Gordon. "Management of *Fusarium* Wilt of Strawberry." *Crop Protection* 73 (2015): 67–72.

Koike, Steven T., Thomas R. Gordon, Oleg Daugovish, Husein Ajwa, Mark Bolda, and Krishna Subbarao. "Recent Developments on Strawberry Plant Collapse Problems in California Caused by *Fusarium* and *Macrophomina.*" *International Journal of Fruit Science* 13, nos. 1/2 (2013): 76–83.

Krimsky, Sheldon. "Low Dose Toxicology: Narratives from the Science-Transcience Interface." In *Powerless Science? Science and Politics in a Toxic World,* edited by Soraya Boudia and Nathalie Jas, 234–53. New York: Berghahn, 2014.

Landecker, Hannah. "Antibiotic Resistance and the Biology of History." *Body and Society* 22, no. 4 (2015): 1–34.

Lang, Tim, and Michael Heasman. *Food Wars: The Global Battle for Mouths, Minds and Markets*. London: Routledge, 2015.

Lassen Canyon Nursery. "About Us." Accessed January 3, 2018. https://www.lassencanyonnursery.com/about-us.

Latour, Bruno. *The Pasteurization of France*. Translated by Alan Sheridan and John Law. Cambridge, MA: Harvard University Press, 1993.

Latour, Bruno, and Steve Woolgar. "Laboratory Life: The Social Construction of Scientific Facts." Beverly Hills, CA: Sage, 1979.

Latta, Randall. "Methyl Bromide Fumigation for the Delousing of Troops." *Journal of Economic Entomology* 37, no. 1 (1944): 103.

Lave, Rebecca. "Reassembling the Structural: Political Ecology and Actor-Network Theory." In *The Routledge Handbook of Political Ecology*, edited by Tom Perreault, Gavin Bridge, and James McCarthy, 213–23. Abingdon-on-Thames, England: Routledge, 2015.

Lave, Rebecca, Martin Doyle, and Morgan Robertson. "Privatizing Stream Restoration in the U.S." *Social Studies of Science* 40, no. 5 (2010): 677–703.

Lave, Rebecca, Philip Mirowski, and Samuel Randalls. "Introduction: STS and Neoliberal Science." *Social Studies of Science* 40, no. 5 (2010): 659–75.

Lawrence, Dune. "How Driscoll's Is Hacking the Strawberry of the Future." *Bloomberg Businessweek*, July 29, 2015, https://www.bloomberg.com/news/features/2015-07-29/how-driscoll-s-is-hacking-the-strawberry-of-the-future.

Lehman, Russell S. "Laboratory Experiments with Various Fumigants against the Wireworm *Limonius* (*Pheletes*) *californicus Mann*." *Journal of Economic Entomology* 26, no. 6 (1933): 1042–51.

Leibman, Ellen. *California Farmland: A History of Large Agricultural Land Holdings*. Totowa, NJ: Rowman and Allanheld, 1983.

Lewis, Carolyn M., and Marilyn H. Silva. "Evaluation of Chloropicrin as a Toxic Air Contaminant, Part B: Human Health Effects." Sacramento: State of California Department of Pesticide Regulation, 2010. Accessed April 4, 2014. http://www.cdpr.ca.gov/docs/emon/pubs/tac/part_b_0210.pdf. No longer available at the DPR site, but reposted at https://pdfs.semanticscholar.org/765e/2f7e9d7ccd53eca2d022b02c9c72dc476d25.pdf.

Li, Tania Murray. "To Make Live or Let Die? Rural Dispossession and the Protection of Surplus Populations." *Antipode* 41, no. S1 (2010): 66–93.

Liboiron, Max, Manuel Tironi, and Nerea Calvillo. "Toxic Politics: Acting in a Permanently Polluted World." *Social Studies of Science* 48, no. 3 (2018): 331–49.

Lien, Marianne Elisabeth. *Becoming Salmon: Aquaculture and the Domestication of a Fish*. Berkeley: University of California Press, 2015.

Lilliston, Ben, and Ronnie Cummins. "Organic versus 'Organic': The Corruption of a Label." *The Ecologist (U.K. Environmental Magazine)*, July–August 1998, 195–202.

Lincoln, C.G., H.H. Schwardt, and C.E. Palm. "Methyl Bromide-Dichloroethyl Ether Emulsion as a Soil Fumigant." *Journal of Economic Entomology* 35, no. 2 (1942): 238–39.

Lindgren, D. L. "Methyl Iodide as a Fumigant." *Journal of Economic Entomology* 31 (1938): 320.

———. "Vacuum Fumigation." *Journal of Economic Entomology* 29, no. 6 (1936): 1132–37.

López, Ann Aurelia. *Farmworker's Journey*. Berkeley: University of California Press, 2007.

López, Marcos. "In Hidden View: How Water Became a Catalyst for Indigenous Farmworker Resistance in Baja California, Mexico." In *The Politics of Fresh Water: Access, Conflict, and Identity*, edited by C. Ashcraft and T. Mayer, 189–204. New York: Routledge, 2017.

Lorimer, Jamie. "Probiotic Environmentalities: Rewilding with Wolves and Worms." *Theory, Culture and Society* 34, no. 4 (2017): 27–48.

Lorimer, Jamie, and Clemens Driessen. "Bovine Biopolitics and the Promise of Monsters in the Rewilding of Heck Cattle." *Geoforum* 48 (2013): 249–59.

Los Angeles Food Policy Council. "Interview with Phil McGrath, Owner of McGrath Family Farms." December 13, 2013. Accessed December 27, 2017. http:// goodfoodla.org/2013/12/16/interview-with-phil-mcgrath-owner-of-mcgrath-family-farms/. No longer available online.

Lyons, Kristina Marie. "Soil Science, Development, and the 'Elusive Nature' of Colombia's Amazonian Plains." *Journal of Latin American and Caribbean Anthropology* 19, no. 2 (2014): 212–36.

MacDonald, O. C., and C. Reichmuth. "Effects on Target Organisms." In *The Methyl Bromide Issue*, edited by C. H. Bell, N. Price, and B. Chakrabarti, 149–89. Chichester, England: John Wiley and Sons, 1996.

Mackie, D. B. "Methyl Bromide—Its Expectancy as a Fumigant." *Journal of Economic Entomology* 31, no. 1 (1938): 70–79.

Madans, Hannah. "Irvine Businessman A. G. Kawamura Appointed to Trump's Agriculture Group." *Orange County Register*, August 22, 2016, https://www .ocregister.com/2016/08/22/irvine-businessman-ag-kawamura-appointed-to-trumps-agriculture-group/.

Magnuson, Torsten A. "History of the Beet Sugar Industry in California." *Annual Publication of the Historical Society of Southern California* 11, no. 1 (1918): 68–79.

Majka, Linda C., and Theo J. Majka. "Organizing U.S. Farmworkers: A Continuous Struggle." In *Hungry for Profit: The Agribusiness Threat to Farmers, Food, and the Environment*, edited by Frederick H. Buttel, Fred Magdoff, and John Bellamy Foster, 161–74. New York: Monthly Review Press, 2000.

Mann, Susan A. *Agrarian Capitalism in Theory and Practice*. Chapel Hill: University of North Carolina Press, 1989.

Mann, Susan A., and James M. Dickinson. "Obstacles to the Development of a Capitalist Agriculture." *Journal of Peasant Studies* 5, no. 4 (1978): 466–81.

Mansfield, Becky. "Fish, Factory Trawlers, and Imitation Crab: The Nature of Quality in the Seafood Industry." *Journal of Rural Studies* 19, no. 1 (2003): 9–21.

Margolin, Malcolm. *The Ohlone Way: Indian Life in the San Francisco–Monterey Bay Area*. Berkeley: Heyday Books, 1978.

Margulis, Lynn. *Symbiotic Planet: A New Look at Evolution*. London: Phoenix, 1999.

Marquand, Carl B. "Contributions to Better Living from Chemical Corps Research." *Journal of Chemical Education* 34, no. 11 (1957): 532.

Martin, Philip. "Farm Labor Shortages: How Real? What Response? Backgrounder No. 9–07." Washington, DC: Center for Immigration Studies, 2007. http://www.cis.org/articles/2007/back907.html.

Martin, Philip, and Daniel Costa. "Farmworker Wages in California: Large Gap between Full-Time Equivalent and Actual Earnings." *Working Economics Blog*, December 27, 2017, http://www.epi.org/blog/farmworker-wages-in-california-large-gap-between-full-time-equivalent-and-actual-earnings/.

Maxwell, Maura. "Planasa Acquires US's Norcal Nursery." Fruitnet.com, June 1, 2017, http://www.fruitnet.com/americafruit/article/172374/planasa-acquires-uss-norcal-nursery.

Mayfield, Erin N., and Catherine Shelley Norman. "Moving away from Methyl Bromide: Political Economy of Pesticide Transition for California Strawberries since 2004." *Journal of Environmental Management* 106 (2012): 93–101.

Mazzola, M., J. Muramoto, and C. Shennan. "Transformation of Soil Microbial Community Structure in Response to Anaerobic Soil Disinfestation for Soilborne Disease Control in Strawberry." Paper presented at the American Phytopathological Society Annual Meeting, Providence, Rhode Island, August 2–8, 2012.

Mazzola, Mark. "Mechanisms of Natural Soil Suppressiveness to Soilborne Diseases." *Antonie van Leeuwenhoek* 81, nos. 1–4 (2002): 557–64.

McFall-Ngai, Margaret. "Noticing Microbial Worlds: The Postmodern Synthesis in Biology." In *Arts of Living on a Damaged Planet | Monsters of the Anthropocene*, edited by Anna Tsing, Heather Swanson, Elaine Gan, and Nils Bubandt, M51–M72. Minneapolis: University of Minnesota Press, 2017.

McGrath Family Farm. "Farm History." Accessed December 27, 2017. http://www.mcgrathfamilyfarm.com/farm-history/.

McIntyre, Michael, and Heidi J Nast. "Bio(necro)polis: Marx, Surplus Populations, and the Spatial Dialectics of Reproduction and 'Race.'" *Antipode* 43, no. 5 (2011): 1465–88.

Mergel, Maria. "Methyl Bromide." In the online database *Toxipedia* (2011). Accessed December 27, 2017. http://www.toxipedia.org/display/toxipedia/Methyl+Bromide. No longer available online.

Minkoff-Zern, Laura-Anne. "Challenging the Agrarian Imaginary: Farmworker-Led Food Movements and the Potential for Farm Labor Justice." *Human Geography* 7, no. 1 (2014): 85–101.

Minkoff-Zern, Laura-Anne, Nancy Peluso, Jennifer Sowerwine, and Christy Getz. "Race and Regulation: Asian Immigrants in California Agriculture." In *Cultivating Food Justice: Race, Class, and Sustainability*, edited by Alison Alkon and Julian Agyeman, 65–86. Cambridge, MA: MIT Press, 2011.

Mirowski, Philip. "The Modern Commercialization of Science Is a Passel of Ponzi Schemes." *Social Epistemology* 26, nos. 3/4 (2012): 285–310.

———. *Science-Mart: Privatizing American Science.* Cambridge, MA: Harvard University Press, 2011.

Mitchell, Don. *"They Saved the Crops": Labor, Landscape, and the Struggle over Industrial Farming in Bracero-Era California.* Athens: University of Georgia Press, 2012.

Mitchell, Timothy. *Rule of Experts: Egypt, Techno-Politics, Modernity.* Berkeley: University of California Press, 2002.

Mohan, Geoffrey. "As California's Labor Shortage Grows, Farmers Race to Replace Workers with Robots." *Los Angeles Times,* July 21, 2017, https://www.latimes.com/projects/la-fi-farm-mechanization.

———. "Dole, the World's Largest Fresh Fruit and Vegetable Company, Is Stepping Back from Southland." *Los Angeles Times,* August 30, 2017, http://www.latimes.com/business/la-fi-dole-socal-20170830-story.html.

Moore, Jason W. *Capitalism in the Web of Life: Ecology and the Accumulation of Capital.* New York: Verso, 2015.

Moore, William. "Fumigation with Chloropicrin." *Journal of Economic Entomology* 11 (1918): 357–62.

———. "Volatility of Organic Compounds as an Index of the Toxicity of Their Vapors to Insects." *Journal of Agricultural Research* 10, no. 7 (1917): 365.

Morinville, Cynthia, and Nicole Van Lier. "'Dead' Land and the End of Capitalism? Interrogating the Politics of Toxic Environments and Capitalist Accumulation." *Capitalism Nature Socialism* (in review).

Moses, H. Vincent. "'The Orange-Grower Is Not a Farmer': G. Harold Powell, Riverside Orchardists, and the Coming of Industrial Agriculture, 1893–1930." *California History* 74, no. 1 (1995): 22–37.

Muramoto, Joji, Carol Shennan, Margherita Zavatta, Graeme Baird, Lucinda Toyama, and Mark Mazzola. "Effect of Anaerobic Soil Disinfestation and Mustard Seed Meal for Control of Charcoal Rot in California Strawberries." *International Journal of Fruit Science* 16, no. sup1 (2016): 59–70.

Murphy, Michelle. *Sick Building Syndrome and the Problem of Uncertainty: Environmental Politics, Technoscience, and Women Workers.* Durham, NC: Duke University Press, 2006.

Nading, Alex M. "Local Biologies, Leaky Things, and the Chemical Infrastructure of Global Health." *Medical Anthropology* 36, no. 2 (2017): 141–56.

———. *Mosquito Trails: Ecology, Health, and the Politics of Entanglement.* Berkeley: University of California Press, 2014.

Nakane, Kazuko. *Nothing Left in My Hands: The Issei of a Rural California Town, 1900–1942.* Berkeley: Heyday Books, 2008.

Nally, David. "The Biopolitics of Food Provisioning." *Transactions of the Institute of British Geographers* 36, no. 1 (2011): 37–53.

Nash, Linda. "The Fruits of Ill-Health: Pesticides and Workers' Bodies in Postwar California." *Osiris* 19 (2004): 203–19.

National Organic Standards Board. "Petitioned Material Checklist—Allyl Isothiocyanate." December 16, 2014. https://www.ams.usda.gov/sites/default/files/media/Allyl%20Isothiocyanate%20Proposal.pdf.

Naturipe Berry Growers. "Who We Are." Accessed January 3, 2018. http://www
.naturipeberrygrowers.com/who-we-are/.

Naturipe Farms. "Timeline." Accessed January 3, 2018. https://www.naturipefarms
.com/timeline/.

Neely, Abigail, and Thokozile Nguse. "Entanglements, Intra-Actions, and Diffrac-
tion." In *The Routledge Handbook of Political Ecology*, edited by Tom Perreault,
Gavin Bridge and James McCarthy, 140–49. Abingdon-on-Thames, England:
Routledge, 2015.

Nevins, Joseph. *Operation Gatekeeper: The Rise of the "Illegal Alien" and the Remaking of the U.S.-Mexico Boundary.* London: Routledge, 2001.

Newhall, A. G. "Disinfestation of Soil by Heat, Flooding and Fumigation." *Botanical Review* 21, no. 4 (1955): 189–250.

New York Times. "To Use Chloropicrin on Pineapple Pests." December 23,
1928, 42.

Nichols, Ron. "Soil Health Campaign Turns Two: Seeks to Unlock Benefits on and off
the Farm." Natural Resources Conservation Service, USDA. Accessed July 3, 2018.
https://www.nrcs.usda.gov/wps/portal/nrcs/detail/sd/home/?cid=stelprdb1261962.

Nuckton, C. F., R. I. Rochin, and A. F. Scheuring. "California Agriculture: The
Human Story." In *A Guidebook to California Agriculture*, edited by Anne Foley
Scheuring, 9–38. Berkeley: University of California Press, 1983.

Ogden, Laura A., Billy Hall, and Kimiko Tanita. "Animals, Plants, People, and
Things: A Review of Multispecies Ethnography." *Environment and Society* 4, no.
1 (2013): 5–24.

Ohr, Howard D., Nigel M. Grech, and James J Sims. "Methyl Iodide as a Fumigant."
Google Patents, 1998. University of California assignee, patent version no.
US5753183A.

Oreskes, Naomi, and Erik M. Conway. *Merchants of Doubt: How a Handful of Scientists Obscured the Truth on Issues from Tobacco Smoke to Global Warming.* New
York: Bloomsbury, 2011.

Organic Produce Network. "Organic Strawberries in Short Supply." January 11,
2018. http://www.organicproducenetwork.com/article/351/organic-strawberries-
in-short-supply.

Oxnard Downtowners. "General Historical Overview until 1898." Accessed December 27, 2017. http://oxnarddowntowners.org/downtown-history.html.

Pacific Rural Press. "The Cinderella Strawberry." *Pacific Rural Press* 15, no. 8 (1878).
https://cdnc.ucr.edu/cgi-bin/cdnc?a=d&d=PRP18780223.2.4.

———. "A New Strawberry." *Pacific Rural Press* 28, no. 7 (1884). https://cdnc.ucr.
edu/cgi-bin/cdnc?a=d&d=PRP18840816.2.3.

Parker, Ingrid M., and Gregory S. Gilbert. "The Evolutionary Ecology of Novel
Plant-Pathogen Interactions." *Annual Review of Ecology Evolution and Systematics* 35 (2004): 675–700.

Parks, Kristen, Gabriel Lozada, Miguel Mendoza, and Lourdes Garcia Santos.
"Strategies for Success: Border Crossing in an Era of Heightened Security." In
Migration from the Mexican Mixteca: A Transnational Community in Oaxaca and

California, edited by Wayne A. Cornelius, David FitzGerald, Jorge Hernandez-Diaz, and Scott Borger, 31–61. Boulder, CO: Lynne Rienner, 2009.

Passel, Jeffrey S., and D'Vera Cohn. "Unauthorized Immigrant Totals Rise in 7 States, Fall in 14." Pew Research Center, 2014. http://www.pewhispanic.org/2014/11/18/unauthorized-immigrant-totals-rise-in-7-states-fall-in-14/.

PAST Consultants. "Historic Context Statement for Agricultural Resources in the North County Planning Area, Monterey County." Petaluma, CA: PAST Consultants, 2010. http://ohp.parks.ca.gov/pages/1054/files/nomontereyco.pdf.

Paulus, Albert O. "Fungal Diseases of Strawberry." *Horticultural Science* 25, no. 8 (1990): 885–89.

Paxson, Heather. "Post-Pasteurian Cultures: The Microbiopolitics of Raw-Milk Cheese in the United States." *Cultural Anthropology* 23, no. 1 (2008): 15–47.

Peck, Jamie. "The Right to Work, and the Right at Work." *Economic Geography* 92, no. 1 (2016): 4–30.

Pegg, G. F. "The Impact of *Verticillium* Diseases in Agriculture." *Phytopathologia Mediterranea* 23, nos. 2/3 (1984): 176–92.

Perkins, Harold A. "Ecologies of Actor-Networks and (Non)social Labor within the Urban Political Economies of Nature." *Geoforum* 38, no. 6 (2007): 1152–62.

Piccirillo, Vincent J. "Methyl Bromide." In *Hayes' Handbook of Pesticide Toxicology*, edited by Robert Krieger and William Krieger, 2:1837–47. London: Academic Press, 2010.

Povinelli, Elizabeth A. *Geontologies: A Requiem to Late Liberalism*. Durham, NC: Duke University Press, 2016.

Probyn, Elspeth. *Eating the Ocean*. Durham, NC: Duke University Press, 2016.

Proctor, Robert, and Londa L. Schiebinger, eds. *Agnotology: The Making and Unmaking of Ignorance*. Palo Alto, CA: Stanford University Press, 2008.

Prudham, W. Scott. *Knock on Wood: Nature as Commodity in Douglas-Fir Country*. New York: Routledge, 2005.

Puig de la Bellacasa, Maria. "Making Time for Soil: Technoscientific Futurity and the Pace of Care." *Social Studies of Science* 45, no. 5 (2015): 691–716.

Raaijmakers, Jos M., and Mark Mazzola. "Soil Immune Responses." *Science* 352, no. 6292 (2016): 1392–93.

Ranganathan, Malini. "Storm Drains as Assemblages: The Political Ecology of Flood Risk in Post-Colonial Bangalore." *Antipode* 47, no. 5 (2015): 1300–1320.

Reynolds, H. T., and J. W. Huffman. "Methyl Bromide Fumigation for Control of Cyclamen Mite on Strawberries." *Journal of Economic Entomology* 50, no. 4 (1957): 525–26.

Rhew, Robert C., Benjamin R. Miller, and Ray F. Weiss. "Natural Methyl Bromide and Methyl Chloride Emissions from Coastal Salt Marshes." *Nature* 403, no. 6767 (2000): 292–95.

Ricardo, David. *Principles of Political Economy and Taxation*. London: G. Bell, 1891.

Roark, Ruric Creegan. *A Bibliography of Chloropicrin, 1848–1932*. Washington, DC: US Department of Agriculture, 1934.

Robbins, Paul. *Lawn People: How Grasses, Weeds, and Chemicals Make Us Who We Are*. Philadelphia: Temple University Press, 2007.

Robbins, Paul, and Brian Marks. "Assemblage Geographies." In *The Sage Handbook of Social Geographies*, edited by S. Smith, R. Pain, S. Marston and J. P. Jones, 176–94. Beverly Hills, CA: Sage, 2010.

Roberts, Stacy Nichole. "How We Have Forgotten: Chemical Strawberries and Their Archived Alternatives in the Nineteenth and Twentieth Centuries." MA Thesis, North Carolina State University, 2010.

Rodríguez-Garavito, César A. "Global Governance and Labor Rights: Codes of Conduct and Anti-Sweatshop Struggles in Global Apparel Factories in Mexico and Guatemala." *Politics and Society* 33, no. 2 (2005): 203–33.

Rogers, Paul. "California Heat Wave: How Much Is from Climate Change?" *San Jose Mercury News*, September 1, 2017, https://www.mercurynews.com/2017/09/01/california-heat-wave-how-much-is-from-climate-change/.

———. "Farmland Measure Divides Monterey County." *San Jose Mercury News*, October 31, 1996. No longer available online.

Romero, Adam M. "'From Oil Well to Farm': Industrial Waste, Shell Oil, and the Petrochemical Turn (1927–1947)." *Agricultural History* 90, no. 1 (2016): 70–93.

Romero, Adam M., Julie Guthman, Ryan E. Galt, Matt Huber, Becky Mansfield, and Suzana Sawyer. "Chemical Geographies." *GeoHumanities* 3, no. 1 (2017): 158–77.

Rosset, Peter M., and Miguel Altieri. "Agroecology versus Input Substitution: A Fundamental Contradiction of Sustainable Agriculture." *Society and Natural Resources* 10, no. 3 (1997): 283–95.

Rudy, Alan P., Dawn Coppin, Jason Konefal, Bradley T. Shaw, Toby Van Eyck, Craig Harris, and Lawrence Busch. *Universities in the Age of Corporate Science: The UC Berkeley–Novartis Controversy*. Philadelphia: Temple University Press, 2007.

Rural Migration News. "California: Sales, Strawberries." *Rural Migration News* 15, no. 4 (2009): https://migration.ucdavis.edu/rmn/more.php?id=1491.

Russell, Edmund. *War and Nature: Fighting Humans and Insects with Chemicals from World War I to Silent Spring*. Cambridge, England: Cambridge University Press, 2001.

Russell, Edmund P. "'Speaking of Annihilation': Mobilizing for War against Human and Insect Enemies, 1914–1945." *Journal of American History* 82, no. 4 (1996): 1505–29.

Santa Barbara County Agricultural Commissioner. "Agricultural Production Report." Weights and Measures, County of Santa Barbara, 2001. http://cosb.countyofsb.org/uploadedFiles/agcomm/crops/2001.pdf.

———. "Agricultural Production Report." Weights and Measures, County of Santa Barbara, 2016. http://cosb.countyofsb.org/uploadedFiles/agcomm/crops/2016.pdf.

Santa Maria Valley Historical Society Museum. "A Brief History." Accessed December 27, 2017. http://santamariahistory.com/history.html.

Sassen, Saskia. *Expulsions: Brutality and Complexity in the Global Economy*. Cambridge, MA: Harvard University Press, 2014.

Sawyer, Richard C. *To Make a Spotless Orange: Biological Control in California*. Lafayette, IN: Purdue University Press, 2002.

Saxton, Dvera I. "Strawberry Fields as Extreme Environments: The Ecobiopolitics of Farmworker Health." *Medical Anthropology* 34, no. 2 (2015): 166–83.

Sayer, Andrew. *Method in Social Science*. 2nd ed. London: Routledge, 1992.

Sayes, Edwin. "Actor-Network Theory and Methodology: Just What Does It Mean to Say That Nonhumans Have Agency?" *Social Studies of Science* 44, no. 1 (2014): 134–49.

Schlosser, Eric. "In the Strawberry Fields." *Atlantic Monthly* 276, no. 5 (1995): 80–108.

Schurman, Rachel, and William A. Munro. *Fighting for the Future of Food: Activists versus Agribusiness in the Struggle over Biotechnology*. Minneapolis: University of Minnesota Press, 2010.

Scott, James C. *Seeing Like a State*. New Haven, CT: Yale University Press, 1998.

———. *Weapons of the Weak*. New Haven, CT: Yale University Press, 1985.

Scrinis, Gyorgy. *Nutritionism: The Science and Politics of Dietary Advice*. New York: Columbia University Press, 2013.

Seavey, Kent. "A Short History of Salinas, California." Monterey County Historical Society. Accessed December 27, 2017. http://mchsmuseum.com/salinasbrief.html.

Seed World podcast audio. "UC Davis Breeding Strawberries for the 21st Century." 2015. http://seedworld.com/uc-davis-breeding-strawberries-21st-century/.

Shapin, Steven. *A Social History of Truth: Civility and Science in Seventeenth-Century England*. Chicago: University of Chicago Press, 1994.

Shapiro, Nicholas, and Eben Kirksey. "Chemo-Ethnography: An Introduction." *Cultural Anthropology* 32, no. 4 (2017): 481–93.

Shaw, Douglas V., Thomas Gordon, Kirk D. Larson, W. Douglas Gubler, John Hansen, and Sharon C. Kirkpatrick. "Strawberry Breeding Improves Genetic Resistance to *Verticillium* Wilt." *California Agriculture* 64, no. 1 (2010): 37–41.

Shelton, Tamara Venit. *A Squatter's Republic: Land and the Politics of Monopoly in California, 1850–1900*. Berkeley: University of California Press, 2013.

Shennan, Carol, Joji Muramoto, M. Mazzola, D. Butler, E. Rosskoph, N. Kokalis-Burelle, K. Momma, Y. Kobara, and J. Lamers. "Anaerobic Soil Disinfestation for Soil Borne Disease Control in Strawberry and Vegetable Systems: Current Knowledge and Future Directions." *Acta Horticulturae* 1137 (2015): 113–20.

Shepard, Harold H., and Albert W. Buzicky. "Further Studies of Methyl Bromide as an Insect Fumigant." *Journal of Economic Entomology* 32, no. 6 (1939): 854–59.

Short, G. E., T. D. Wyllie, and P. R. Bristow. "Survival of *Macrophomina phaseolina* in Soil and in Residue of Soybean." *Phytopathology* 70, no. 1 (1980): 13–17.

Smith, Nathan R. "The Partial Sterilization of Soil by Chloropicrin." *Soil Science Society of America Journal* 3, no. C (1939): 188.

Smith, Neil. *Uneven Development: Nature, Capital, and the Production of Space*. Oxford: Basil Blackwell, 1984.

Snipes, Shedra Amy, Beti Thompson, Kathleen O'connor, Bettina Shell-Duncan, Denae King, Angelica P. Herrera, and Bridgette Navarro. "'Pesticides Protect the Fruit, But Not the People': Using Community-Based Ethnography to Understand Farmworker Pesticide-Exposure Risks." *American Journal of Public Health* 99, no. S3 (2009): S616–S21.

Soluri, John. *Banana Cultures: Agriculture, Consumption, and Environmental Change in Honduras and the United States.* Austin: University of Texas Press, 2005.

Spivak, Gayatri Chakravorty. "Can the Subaltern Speak?" In *Can the Subaltern Speak? Reflections on the History of an Idea*, edited by Rosalind Morris, 21–78. New York: Columbia University Press, 1988.

Spuler, Anthony. "Baiting Wireworms." *Journal of Economic Entomology* 18, no. 5 (1925): 703–7.

Stamets, Paul. *Mycelium Running: How Mushrooms Can Help Save the World.* Random House Digital, 2005.

Stenhouse, John. "III. On Chloranil and Bromanil, No. II." *Journal of the Chemical Society* 23 (1870): 6–14.

Stoll, Steven. *The Fruits of Natural Advantage: Making the Industrial Countryside in California.* Berkeley: University of California Press, 1998.

Stone, M. W. "Dichloropropane-Dichloropropylene, a New Soil Fumigant for Wireworms." *Journal of Economic Entomology* 37, no. 2 (1944): 297–99.

Stone, M. W., and Roy E. Campbell. "Chloropicrin as a Soil Insecticide for Wireworms." *Journal of Economic Entomology* 26, no. 1 (1933): 237–43.

Strand, Larry. *Integrated Pest Management for Strawberries.* Oakland: University of California Division of Agriculture and Natural Resources, 2008.

Strange, Marty. *Family Farming: A New Economic Vision.* Lincoln: University of Nebraska Press / Institute for Food and Development Policy, 1988.

Swanson, Heather Anne. "The Banality of the Anthropocene." *Cultural Anthropology*, February 22, 2017. https://culanth.org/fieldsights/1074-the-banality-of-the-anthropocene.

Swanson, Heather, Anna Tsing, Elaine Gan, and Nils Bubandt. "Introduction: Bodies Tumbled into Bodies." In *Arts of Living on a Damaged Planet | Monsters of the Anthropocene*, edited by Anna Tsing, Heather Swanson, Elaine Gan, and Nils Bubandt, M1–M12. Minneapolis: University of Minnesota Press, 2017.

Swanton Berry Farm, "History." Accessed March 8, 2018. http://www.swanton berryfarm.com/history.

Taylor, A. L. "Nematocides and Nematicides: A History." *Nemaptropica* 33 (2003): 225–32.

Thomas, Harold E. *"Verticillium Wilt of Strawberries, Bulletin 530."* Berkeley: University of California Printing Office, 1932.

Thomas, Robert J. *Citizenship, Gender, and Work.* Berkeley: University of California Press, 1985.

Tierney, John. "A Patented Berry Has Sellers Licking Their Lips." *New York Times*, October 14, 1991, http://www.nytimes.com/1991/10/14/us/a-patented-berry-has-sellers-licking-their-lips.html.

Tourte, Laura, Mark Bolda, and Karen Klonsky. "The Evolving Fresh Market Berry Industry in Santa Cruz and Monterey Counties." *California Agriculture* 70, no. 3 (2016): 107–15.

Tsing, Anna Lowenhaupt. *The Mushroom at the End of the World: On the Possibility of Life in Capitalist Ruins.* Princeton, NJ: Princeton University Press, 2015.

Uekötter, Frank. "Farming and Not Knowing: Agnotology Meets Environmental History." In *New Natures: Joining Environmental History with Science and Technology Studies,* edited by Dolly Jorgensen, Finn Arne Jorgensen, and Sara Pritchard, 37–50. Pittsburgh: University of Pittsburgh Press, 2013.

———. "Ignorance Is Strength: Science-Based Agriculture and the Merits of Incomplete Knowledge." In *Managing the Unknown: Essays on Environmental Ignorance,* edited by Frank Uekötter and Uwe Lübken, 122–39. New York: Berghahn, 2014.

Uekötter, Frank, and Uwe Lübken. "Introduction: The Social Functions of Ignorance." In *Managing the Unknown: Essays on Environmental Ignorance,* edited by Frank Uekötter and Uwe Lübken, 1–11. New York: Berghahn, 2014.

United States Bureau of the Census. "1950 Census of Agriculture." National Agricultural Statistics Service, 1952. http://usda.mannlib.cornell.edu/usda/AgCensusImages/1950/01/33/1812/34101884v1p33ch1.pdf.

———. "1978 Census of Agriculture." National Agricultural Statistics Service, 1981. http://usda.mannlib.cornell.edu/usda/AgCensusImages/1978/01/05/1978-01-05.pdf.

———. "2012 Census of Agriculture." National Agricultural Statistics Service, 2014. https://www.agcensus.usda.gov/Publications/2012/Full_Report/Volume_1,_Chapter_2_County_Level/California/cav1.pdf.

United States Department of Agriculture. "Noncitrus Fruits and Nuts 2016 Summary." National Agricultural Statistics Service, 2017. Formerly at http://usda.mannlib.cornell.edu/usda/current/NoncFruiNu/NoncFruiNu-06-27-2017.pdf (website being revamped at the time of this printing).

University of California Cooperative Extension. "Agriculture and Natural Resources Ventura County: General Soil Map." Division of Agriculture and Natural Resources, University of California. Accessed December 27, 2017. http://ceventura.ucanr.edu/Com_Ag/Soils/The_environamental_characteristics_of_Ventura_County_and_its_soils_/General_Soil_Map/.

Van Bruggen, A. H. C., and M. R. Finckh. "Plant Diseases and Management Approaches in Organic Farming Systems." *Annual Review of Phytopathology* 54 (2016): 25–54.

Vanderbilt, Bryon M. "Chlorination of Nitromethane." US patent filed February 10, 1938. IMC Chemical Group Inc. assignee, patent version no. US2181411A.

Walker, Richard. *The Conquest of Bread: 150 Years of Agribusiness in California.* New York: New Press, 2004.

———. "Urban Ground Rent: Building a New Conceptual Framework." *Antipode* 6, no. 1 (1974): 51–59.

Wang, D., S. R. Yates, F. F. Ernst, J. Gan, and W. A. Jury. "Reducing Methyl Bromide Emission with a High Barrier Plastic Film and Reduced Dosage." *Environmental Science and Technology* 31, no. 12 (1997): 3686–91.

Watts, Michael J. "Life under Contract: Contract Farming, Agrarian Restructuring, and Flexible Accumulation." In *Living under Contract*, edited by Michael J. Watts and Peter Little, 21–78. Madison: University of Wisconsin Press, 1993.

Webb, Kayla. "Driscoll's Expands Its Organic Operations to Keep up with Demand." AndNowUKnow.com. 2018. No longer available online.

WellPict. "About Us." Accessed December 21, 2018. http://www.wellpict.com /about-us/.

Wells, Miriam. *Strawberry Fields: Politics, Class, and Work in California Agriculture*. Ithaca, NY: Cornell University Press, 1996.

Western Farm Press. "California DPR Caps Telone Fumigant Use on Jan. 1." October 6, 2016, http://www.westernfarmpress.com/regulatory/california-dpr-caps-telone-fumigant-use-jan-1.

Wheat, Dan. "Use of H-2A Workers to Increase Despite Wage Hike." *Capital Press*, January 8, 2015, http://www.capitalpress.com/Nation_World/Nation/20150108 /use-of-h-2a-workers-to-increase-despite-wage-hike.

Wilhelm, Stephen, and Edward C. Koch. "*Verticillium* Wilt Controlled: Chloropicrin Achieves Effective Control of *Verticillium* Wilt in Strawberry Plantings if Properly Applied as Soil Fumigant." *California Agriculture* 10, no. 6 (1956): 3–14.

Wilhelm, Stephen, and Albert O. Paulus. "How Soil Fumigation Benefits the California Strawberry Industry." *Plant Disease* 64, no. 3 (1980): 264–70.

Wilhelm, Stephen, and James E. Sagen. *History of the Strawberry: From Ancient Gardens to Modern Markets*. Berkeley: University of California Press, 1974.

Wilhelm, Stephen, Richard C. Storkan, and John M. Wilhelm. "Preplant Soil Fumigation with Methyl Bromide-Chloropicrin Mixtures for Control of Soil-Borne Diseases of Strawberries: A Summary of Fifteen Years of Development." *Agriculture and Environment* 1 (1974): 227–36.

Wilson, Clevo, and Clem Tisdell. "Why Farmers Continue to Use Pesticides Despite Environmental, Health and Sustainability Costs." *Ecological Economics* 39, no. 3 (2001): 449–62.

Wright, Angus. *The Death of Ramón González: The Modern Agricultural Dilemma*. 2nd ed. Austin: University of Texas Press, 2005.

Wright, Melissa W. *Disposable Women and Other Myths of Global Capitalism*. New York and London: Routledge, 2006.

Yates, Michelle. "The Human-as-Waste, the Labor Theory of Value and Disposability in Contemporary Capitalism." *Antipode* 43, no. 5 (2011): 1679–95.

Yeung, Bernice, Kendall Taggart, and Andrew Donohue. "California's Strawberry Industry Is Hooked on Dangerous Pesticides." *Reveal News* (Center for Investigative Reporting), November 10, 2014. https://www.revealnews.org/article /californias-strawberry-industry-is-hooked-on-dangerous-pesticides/.

INDEX

actor network theory (ANT), 18, 213n49
agnotology (undone science), 22–23
Agrarian Dreams: The Paradox of Organic Farming in California, J. Guthman, 121, 190, 257n34
The Agrarian Question, K. Kautsky, 210n26
Agricultural Experiment Stations, 16, 41, 62. *See also* research (agricultural scientists)
agricultural machinery, disease virulence and, 37, 220n33
agricultural scientists. *See* breeding strawberries; research
agrochemical industry: achieving autonomy from, 195; Dow Chemical, 86–87, 88, 89, 236n87; efforts to reduce regulation, 102, 165; green chemicals developed by, 255n8; political-economic dynamics, 11–16, 24; Roundup, 52; Shell Oil, 83–84; societal concerns about, 203; support for soil fumigants, 78, 89; tactics used by, 233–34n60. *See also* pesticide *entries*; soil fumigant *entries*
agrochemicals. *See* pesticide *entries*; soil fumigant *entries*; *specific chemicals*
agro-ecological methods for controlling pests, 188–89, 190, 191
agro-forestry industry, 15–16
Albion variety, 66
Alien Land Laws, 112–13, 136, 156, 167
Alien Ocean, S. Helmreich, 84–85

ALRA (California Agricultural Labor Relations Act), 133
anaerobic soil disinfestation (ASD), 182–83, 183*fig*, 190–91, 197
Andrew & Williamson, 164
ANT (actor network theory), 18, 213n49
Anthropocene: crisis of, 10; fumigation as contributor to, 100, 201, 209n18, 237n100; necropolitics of, 209n18
antibiotic approaches: giardiasis treatment, 176–78; human health and, 177, 256n22; to replace soil fumigants, 180–84, 193, 198, 202, 254–55n3
antibiotic resistance, 100, 202, 223n94
appropriationism, 13, 249n2
Arysta LifeScience, 2, 95
assemblage, defined, 25
assemblage approach (more-than-human assemblages), 17–20, 213n48; actor network theory, 18, 213n49, 215n58; defined, 11; dengue fever, 19; disrupted by soilless systems, 185–86; dynamism of, 17–18, 213–14n51; iatrogenic harm, 20–24; infrastructure role, 215n62; labor role, 146, 150–51; land values, 105–6, 201; limits of repair, 11, 25, 200–203; livestock diseases, 18–19, 215n63; malaria example, 17–18, 23, 212n46, 212n47; overview, 24–26; plantation agriculture, 255n4; socio-natural assemblages, 17, 21, 24, 215n60; soil viewed as, 34; strawberry pathogens, 19–20, 200

bacteria: *Campylobacter*, 18–19; drug-resistant, 4; horizontal gene transfer, 223n94; as soil microbes, 218–19n15, 222n72; that protect plant roots, 29

Baker, Richard, 65

banana production, 15

Bankhead-Jones Act, 212n42

bank loans, access to capital, 13–14, 88, 169, 204

Banner variety, 61–62, 63–64, 71, 156, 158, 226n41

Bard, Thomas, 111

Barron Ranch, 38

Baum, Herbert, 64, 158, 227n53

Bay Area counties, history of strawberry growing, 155

Bayh-Dole Act of 1980, 22

Becoming Salmon, Lien, 216n70

bed fumigation, 47, 96–98, 126

Benicia variety, 66

Bensaude-Vincent, Bernadette, 91

Berry Bowl, 159

Berry Central, labor exchange, 144

Berry Genetics, 164

bio-fumigation, 182, 183–84, 190

biological controls, research on, 16, 23

biopolitical concerns, 31, 259n56, 260n58

blackberry production, 7, 141

Blanchette, Alex, 215n63

blueberry production, 7

Bracero Program, 133, 135, 136

brassica plants: crop rotation with, 38, 191, 204; fumigant-like properties, 85, 180, 188, 191

Brazil and the Struggle for Rubber, W. Dean, 15

breeding clubs, 164, 169

breeding strawberries: Cal Poly Strawberry Center, 164; in fumigated fields, 72; high-elevation fields, 55*fig*; history, 37–38, 58–59; history in California, 59–62, 62–70, 161; interviews with breeders, 51–52; land values, 105–6; loss of useful traits, 201; marketing orders, 160; overproduction and, 158–59; plants weakened by breeding, 73; political-economic dynamics, 13, 14; varietal alignment with workforce, 138–39. *See*

also Driscoll Strawberry Associates breeding program; hybridization; patenting plants; proprietary behavior; University of California breeding program

breeding strawberries, for attributes other than disease resistance, 44, 56; adaptability to environmental stresses, 66; day-neutrality, 89; durability, 71–72, 162; labor considerations, 138–39, 260n2; long-bearing varieties, 75; season extension, 70–71; varietal differences, 63, 64, 71; yields, 89, 200

breeding strawberries, for pathogen resistance/tolerance: to crown rot, 66, 75; emphasis on, 102–3; to *Fusarium* wilt, 66, 72; to mildew, 74; to red stele root rot, 74; reduced focus on, 72; soil fumigant role, 3–4, 11, 72, 88; as UC priority, 76; to *Verticillium* wilt, 5, 38, 40–41, 44, 62–63, 65–67, 74. *See also* resistant strawberry varieties

breeding technology: adaptability of genome, 53–54; cloning, 54, 117; ease of breeding, 26; genetically engineered seed, 52; genome as commodity, 55; genomics tools, 67, 225n6; hybrid seedling development, 53–54, 54*fig*, 55–56; inbreeding concerns, 74, 229n99; nuclear stock production, 118, 241n56, 241n59. *See also* heterozygosity (variability) of strawberries

breeding technology, privatization of: alliances with nurseries, 165; competition with universities, 9, 55–56, 64, 69–70, 158, 164, 170–71, 194; Driscoll's role, 75–76, 158; financial issues, 67–68, 69–70; history of, 22; Strawberry Institute of California, 64, 75

Brennan, Eric, 188

Bringhurst, Royce, 65, 66, 119

broadcast fumigation (flat fuming), 47–48, 96, 116

Broome, Jenny, 78–79

Brown, Governor Jerry, 6

brown blight of potatoes, 38

Bubandt, Nils, 209n18

buffer zones: around schools, 30, 217n3; defined, 30, 173; DPR regulations, 95, 96, 241n60; land values reduced by, 126

Butte Valley, strawberry production, 120

California: agricultural history, 110–14; climate change impact, 127; natural advantages, 106–10, 123, 125; Proposition 187, 134; soils, 107, 123–24; water politics, 108. *See also* strawberry industry, California; University of California (UC); *specific growing areas*

California Agricultural Labor Relations Act (ALRA), 133

California Berry Cultivars LLC (CBC), 67, 69, 164, 228n76

California Central Farmers Association, 156–57

California Department of Food and Agriculture (CDFA), 118

California Giant, 67, 152, 163

California Packing Corporation, 62–63

California Strawberry Advisory Board (CSAB), 65, 160, 162, 166

California Strawberry Commission (CSC), 28, 65, 66, 67–68, 140: acreage statistics, 254n84; assessments paid to, 168; lawsuit against UC, 68; marketing program, 162–63; methyl bromide alternatives research, 179; on methyl iodide registration, 2; production statistics, 238n4; research funded by, 166. *See also* California Strawberry Advisory Board

Cal Poly San Luis Obispo, 164, 165, 166, 224n97

Cal Poly Strawberry Center, 164, 224n97

Camarosa variety, 66

"Can the Mosquito Speak?" T. Mitchell, 17, 212n46, 212n47

capital, access to, 13–14, 88, 169, 204

capitalist economy: achieving autonomy from, 195; agrarian capitalism, 12, 210–11n26; exploitation of labor, 129, 146–47, 210n22, 210n24; grower roles, 12, 14, 210n24; productivist logics, 25; reliance on fumigants and, 12, 122, 260n59. *See also* land ownership; land values

carbon-dioxide, to extend shelf life, 162

carcinogens: chloropicrin, 94–95; less toxic fumigants, 180; methyl iodide, 1–2; pesticide registration and, 102; Telone, 95

Carolina Superba variety, 62

Cassin Ranch research facility, 78

causative agents. *See* plant pathogens

CBC (California Berry Cultivars LLC), 67, 69, 164, 228n76

CDFA (California Department of Food and Agriculture), 118

CEA (controlled environment agriculture), 184, 186–87

Cedar Point Nursery, 165

Central California Berry Grower Association (CCBGA), 39, 156, 157, 160, 166

Central Coast: charcoal rot, 46; fumigation timing, 115; immigrant growers, 167–68; Latinx growers, 161; marketing cooperatives, 157; organizing institutions, 166; per-acre costs, 139; seasonal considerations, 107, 115, 120; studies of strawberry production, 139, 166–67; varieties grown, 62

Central Valley, xiii*map*; agricultural history, 110, 238n20; Asian workers and sharecroppers, 109, 112–13, 167; *Fusarium* as nonpathogen, 46; land leasing, 113–14; melon worker study, 147; nurseries, 52; temperatures, 108; vegetable crops, 113

CHAMACOS project, 101

Chandler variety, 66, 204

Chavez, Cesar, 133, 134

Chemical Warfare Service, chloropicrin research, 81–82

chestnut blight, topographical theory of virulence, 36

chilling, benefits of, 119–120, 227n62, 241n62

chloropicrin: with 1,3-D, 236n91; in Dowfume, MC-2, 87; drift monitoring, 98–99; fuming ability, 97; history of use, 80–83, 86, 88, 233n57, 236n91; impact of methyl bromide ban on, 96; mitigation measures, 99; nursery use, 192, 236n92; in propargyl bromide, 88;

chloropicrin *(continued)*
    proposed ban, 152; restrictions on,
    94–95; soils amenable to, 124; with
    Telone, 30; toxicity and health effects,
    82, 94–95, 236n85; *Verticillium dahliae*
    killed by, 42–43, 87; World War I use,
    81–82, 81*fig. See also* methyl bromide
    with chloropicrin
Cinderella variety, 60
Clean Air Act, soil fumigant regulation, 126
Clean Plant program, California, 93–94,
    118, 192, 241n56, 241n59
climate: breeding strawberries for, 70, 75; of
    California, 107, 127; chilling benefits,
    119–120, 227n62, 241n62; of marginal
    strawberry-growing regions, 124
climate change: impact on plant pathogens,
    21; industrial farming and, 10, 195, 201;
    land values and, 127
cloning strawberry plants, 54, 117
Coalition of Immokalee Workers, 245n19
Cochrane, Jim, 203–4
consumer demand for strawberries, 7, 162–63
contract growers. *See* growers, strawberry:
    contractual relationships; sharecroppers
controlled environment agriculture (CEA),
    184, 186–87
Costa Rica, vegetable crops, 248–49n1
cotton, *Verticillium* host, 40
critical use exemptions, methyl bromide
    ban, 93–96, 102, 235n78
"crop protection" industry, 233–34n60,
    255n8. *See also* agrochemical industry
Crop Protection Services, 165
crop rotation: with brassicas, 38, 191, 204;
    land costs and, 120, 121–22; on leased
    land, 114; for organic strawberries, 124,
    191; for pest and disease management,
    44, 114, 120, 188–89, 204, 221n52; veg-
    etables with strawberry crops, 116–17,
    123–25, 191–92, 247n49
crop specialization, 212n44
Crown Nursery, 165
crown rot (*Macrophomina phaseolina*):
    climate and, 124; disease cycle, 45–46;
    heat and salinity and, 127; history in
    California strawberries, 45–46, 48–49;
    increased virulence of, 49, 218n10; as

novel pathogen, 30, 33, 49; resistance to,
    66, 75
CSAB (California Strawberry Advisory
    Board), 65, 160, 162, 166
CSC. *See* California Strawberry
    Commission
cultural controls: integrated pest manage-
    ment, 79, 188–89, 241n57; irrigation
    management, 42, 43, 43–44; research
    on, 64–65; tilling, 32, 220n33. *See also*
    crop rotation

Darrow, George, 53, 72, 74
day-neutral varieties, 71, 115, 225n6
dazomet, new regulations on, 96
D-D (soil fumigant), 83–84
DDT, 19, 23, 86, 89, 212n47
De Léon, Jason, 131, 134
dengue fever, 19, 212n47, 215n63
Department of Pesticide Regulation (DPR),
    California: Brian Leahy, 6, 95, 176, 199;
    buffer zone regulations, 30, 95, 96, 173,
    217n3, 241n60; chemical use facilitated
    by, 92; drop-in methyl bromide replace-
    ment approvals, 181; establishment of,
    90; lawsuits against, 2, 94; Mary-Ann
    Warmerdam, 1, 2, 6; methyl bromide
    alternatives research, 179; methyl iodide
    registration, 1–2, 90–91, 207n3; nonfu-
    migant production plan, 6, 179; Nonfu-
    migant Strawberry Production Working
    Group Action Plan, 101; regulations, 90;
    shortcomings of, 90–91; Telone regula-
    tions, 95–96, 236n87
deportation of immigrants: farmworker
    loyalty and, 140; Trump administration
    and, 135; unwillingness to report sick-
    ness/injury and, 97, 135, 148; wage theft
    and, 135
Diamante variety, 66
"Dirty Dozen," 8, 79, 209n11
discarding crops as unsellable, 14
disease assemblages, 18–19
diseases: germ theory of, 256n22; iatrogenic
    harm, 10, 17, 20–24, 25; immunocompe-
    tence role, 29; livestock, 18–19; salmon
    farming, 18–19. *See also* human diseases;
    plant diseases

disease triangle model, 32
distributed agency, assemblage thinking and, 17
DMDS (Paladin), soil fumigant, 180, 181
Dole, as grower-shipper, 163, 175
Dominus, soil fumigant, 180, 181, 190
Donner variety, 63
Dow Chemical, 86–87, 88, 89, 236n87
Dowfume, MC-2, 87
DPR. *See* Department of Pesticide Regulation, California
drip irrigation, 97, 161
Driscoll, Donald, 158
Driscoll, E. F., 63–64
Driscoll, Ned, 158
Driscoll, Richard, 38, 61–62, 155–56, 157
Driscoll-Reiter, 61, 117
Driscoll Strawberry Associates (DSA): boycotts against, 251n39; contractual relationships with, 168, 168–69; establishment, 64, 159; fumigant-reduction role, 102–3, 173, 179, 194; land scouts for, 171–72, 173; lawsuits against, 137; management of, 168, 253n73; nursery ownership, 165; organic strawberry production, 189, 190; partnerships with, 169–170; plant health research department, 78–79; pre-cooling technology, 162; operations in Mexico, 174; relationship with RAC, 168
Driscoll Strawberry Associates, industry dominance, 249n2; competitors, 163; innovative research, 75; large-scale ranch management, 152; monopolization threat, 77; percentage of world market, 7; political-economic dynamics, 13; proprietary behavior, 158, 171–73, 194; strategies used by, 26, 152, 153–54, 158–160
Driscoll Strawberry Associates, labor relations: farmworker welfare standards, 173; investments in social justice and sustainability, 173; labor contractor loyalty, 171; "Labor Standards," 253n79; robotics research, 144; sharecropping, 9
Driscoll Strawberry Associates breeding program: competition with university varieties, 22; ever-bearing (remontant)

class, 5; history of, 56, 64, 70–71, 164; licensing fees, 171, 172; patent infringement, 228n80; for pathogen resistance, 75–76
drought: grower concerns, 6; impact on assemblages, 21; impact on disease, 32, 127, 172, 223n88
Duchesne, Antoine Nicolas, 58
Duncan, Colin, 242n72
Dutch elm disease, topographical theory of virulence, 36

Eating the Ocean, E. Probyn, 196
Eaton, O. O., 157
economic concerns: access to capital, 13–14, 88, 169, 204; berry shortages, 161; land as fictitious capital, 104, 105–6, 121–25, 127; overproduction, 14, 21, 154, 162, 198. *See also* capitalist economy; land values; marketing strawberries; political-economic dynamics
Ennoble, soil fumigant, 181
environmental complexity, 187–88, 259n56
environmental epigenetics, 3, 49, 224n96
Environmental Working Group, "Dirty Dozen," 8, 79, 209n11
EPA. *See* US Environmental Protection Agency
Etter, Albert, 60–61, 62
Ettersburg 80 variety, 61
Eurosemillas, 67, 68
ever-bearing (remontant) class, 4, 70–71, 158
extensification (increased agricultural acreages), role in disease virulence, 36
extension services, cooperative: cutbacks, 194; methyl bromide alternatives research, 179, 180; privatization and, 22; research roles, 16; support for fumigants, 89

Familias Unidas por la Justicia, 251n39
farm advisers: interviews with, 47, 98, 176, 193; IPM recommendations, 178, 258n41; on labor shortages, 247n56; recommendations, 44, 193
From Farming to Biotechnology, D. Goodman, B. Sorj, J. Wilkinson, 249n2

farm-labor contractors, 25, 133, 137, 143–44, 165

farmworkers: contractual "partnerships" with growers, 137; disloyalty of, 141–42; as disposable, 131, 147–49, 260n58; history in California strawberries, 136–140, 150; interviews with, 26–27, 27, 129; long-term relationships with growers, 138, 140; ownership opportunities for, 204; recruiting strategies, 3, 138, 145; sabotage by, 142; surplus in 1980s, 133–34; tenant farmers, 112. *See also* sharecroppers

farmworkers, for harvesting, 129–151, 131*fig*; control over labor, 129; diseased fields and, 129; field conditions that attract, 131–32; impact on worker health, 260n2; importance of, 129; mechanical harvesters, 115; pace of work, 132

farmworkers, health and safety of, 130–31; basic needs, 146–47; border deaths, 131, 134–35, 151; breaks and downtime, 147–48, 149; capitalist economy vs., 146–47; CHAMACOS project, 101; chronic stress, 147–48; disability income support, 148; Driscoll's role, 173; environmental epigenetics effects, 3; ergonomic concerns, 144, 197, 260n2; exposure to pesticides, 95*fig*, 97, 101, 132, 148, 148–49, 197, 200; food movement focus on, 3; freezing conditions, 149; health care system, 148; heat exhaustion, 147–48; intergenerational effects of pesticides, 101, 132, 148; plastic tarp removal and danger to, 98; work pace that endangers, 148–49

farmworkers, political issues: activism in early twentieth century, 133; boycotts against labor practices, 251n39; Bracero Program, 133–34, 135, 136; Driscoll's "Labor Standards," 253n79; Familias Unidas por la Justicia, 251n39; fear of deportation, 97, 135, 140, 147–48, 148; guest-worker program, 133, 135, 141, 144; improving working conditions, 205; labor control concerns, 137–38; noncitizen status, 97, 135, 138, 150–51; plantationocene, 10, 194–95, 196, 209n19,

210n22, 260n59; political-economic dynamics, 12–13, 24, 150–51; revolt against unauthorized workers, 134–35, 150–51; unionization of, 133–34, 245n19, 251n39; United Farm Workers (UFW), 133, 134

farmworkers, race: Chinese, 112; Japanese, 112–13, 113*fig*, 114, 136, 156, 158, 252n70; Mexican and Central American, 137, 141, 154; racism, 147, 260n58

farmworker shortages, 129–151; Driscoll and, 171; grower concerns, 6, 27, 105, 129, 131–32, 139, 140–42, 146–47, 246n27; grower responses to, 140, 143–46, 145–46, 148–49, 150; immigration restrictions and, 131; impact on assemblages, 21; leverage for workers, 132; as myth, 246n27; pesticide practices/ regulation and, 101, 146, 247n51, 247n56; plant propagation business, 142–43; revolt against unauthorized workers and, 134–35; robotics and, 21, 144; signs of, 140–43; statistics, 135, 140; yield-enhancing technologies and, 137

farmworker wages: agrarian capitalism, 210–11n26; to attract domestic workers, 135; Bracero Program, 133–34; capital to pay, 13–14; Driscoll's role in improving, 173; for grape pickers, 135; grower profitability and, 138, 139, 146–47, 149, 150; increasing, 131–32, 145, 151, 205; infected fields and, 129; minimum wage and overtime laws, 136, 149; obligatory labor costs and, 136; piece rates, 136, 139, 149, 200; soil fumigant role, 3; specialty crops, 9; wage theft, 135; workers leaving for better pay, 141–42

Federal Insecticide, Fungicide, and Rodenticide Act (FIFRA), 89

Federal Insecticide Law, 89

Federal Plant Repository (USDA), 74

fertilizers: as argument for fumigation, 79; cost of, 172; crop rotation and, 114; nitrates in water supply, 79; political-economic dynamics, 13; for sandy soils, 107; for soilless systems, 184; unintended consequences of use, 23; *Verticillium* wilt and, 42 field-scale hydroponics. *See* soilless systems

First the Seed: The Political Economy of Plant Biotechnology, J. Kloppenburg, 52, 55, 66

first-year yields, soil fumigant role, 4

fishing industry, food politics, 196, 216n70

fitness penalty, disease virulence and, 36–37

flat fuming (broadcast fumigation), 47–48, 96–97, 116

Florida, strawberry industry, 174

foodborne illness, 18–19, 215n58

food insecurity, 196

food movement, 1–2, 3, 207n5

Food Systems in an Unequal World, R. Galt, 248–49n1

fossil fuels, industrial agriculture dependent on, 10

foundation blocks, 118

Foundation Plant Services, UC, 54, 74, 118, 241n56, 241n59

Fragaria: California beach strawberry (*F. chiloensis ssp. lucida*), 59–60, 62, 75; Chilean (*F. chiloensis*), 57–58, 59, 61, 74, 226n23; *F. californica*, 59–60; *F. lucida perfecta*, 60; *F. ovalis*, 58; *F. vesca*, 57; origin of name, 57; pine or pineapple (*F. ananassa*), 58; Virginia (*F. virginiana*), 57, 58, 59

Fresno variety, 65

fruit: as fragile and perishable, 107, 138, 150, 157, 162, 200, 204; frozen, 158–59, 160, 163, 166; harvesting skill requirements, 138; morphology, 52–53; technologies to extend shelf life, 162

The Fruits of Natural Advantage: Making the Industrial Countryside in California, S. Stoll, 106

fungi: beneficial, 34; pathogenic, 35; saprobic, 34–35. *See also* plant pathogens

Fusarium oxysporum f. sp. Cubense (Panama disease), of bananas, 15, 46

fusarium wilt (*Fusarium oxysporum f. sp. fragariae*): genome sequencing, 224n100; history in California strawberries, 46–48, 48–49; increased virulence of, 49; as novel pathogen, 30, 33, 49, 224n96; resistant strawberry varieties, 66, 72, 225n6; *Verticillium* suppressed by, 224n96

Galt, Ryan, 20

Gareau, Brian, 24

genetics of strawberries: environmental epigenetics, 3; genetically engineered seed, 52; genome as commodity, 55; inbreeding concerns, 74, 229n99; indeterminate nature, 53, 75; microbiomatics studies, 218–19n15; octoploidy, 53, 58, 74, 225n4. *See also* breeding strawberries; heterozygosity of strawberries

genomics, 67, 225n6

germplasm: maintenance of, 59; opportunities provided by, 74; taken from public domain, 68–70, 75–76, 170–71, 193–94; UC ownership of, 68

germplasm banks, 54, 74

Giant growers, 164, 169, 189, 228n77. *See also* California Giant

giardiasis, drug treatment for, 176–77

Gladstone REIT, 122–23, 127

Goldsmith (Z5A) variety, 64

Goldsmith, Earl, 62–63, 64, 66

Gonzales, Rafael, 111

Goodman, David, 13

grain farming, 111

growers, farmworker shortage concerns, 27, 133–36, 148, 246n27; DPR and, 6; economic concerns, 105, 139; infected fields and, 129; nativism and, 131–32; responses, 140, 143–46; signs of shortages, 140–43

growers, interviews with: difficulty in obtaining, 9; farm sizes, 167; on future of strawberry production, 152–53, 175; on pesticide regulation, 30, 104–5, 217n3; questions asked, 26–27; on varietal choice, 71–72

growers, race: Chinese, 112; Hmong, 109, 167; Japanese, 136, 156, 158, 159, 161, 167, 252n70; Latinx, 154, 161–62, 167–68, 252n70; statistics, 252n70

growers, strawberry: acceptance of pathogen-resistant varietals, 230n110, 255n7; access to critical resources, 170–73, 253n75; consolidation of, 163–64; contractual relationships, 152–53, 163, 168–170; custom growers, 168; demands of, 75; exemption from labor laws, 137;

growers, strawberry *(continued)*
  fumigant safety concerns, 102; number
  in California, 25; partnerships, 169, 170,
  172; public-perception concerns, 6, 9,
  10–11, 102; regulatory concerns, 6; social
  capital, 258n42; support for soil fumigants, 79, 89, 102, 104, 132, 153. *See also*
  sharecroppers; independent growers
growers, supposed intransigence of, 10–14;
  iatrogenic harm, 11, 20–24; limits of
  repair, 11, 20–24; more-than-human
  assemblages, 17–20; overview, 10–11;
  political-economic dynamics, 11–16;
  reliance on fumigants, 236n90; as
  source of frailty, 24–25
growers' advocacy organizations, 25
growers' cooperatives: history of, 13, 156,
  161; as institutions of repair, 25;
  Naturipe Berry Growers, 157; Watsonville Berry Cooperative, 159
grower-shippers: consolidation of, 163–64;
  contractual "partnerships" with farmworkers, 137; custom growers, 168;
  failures of, 169–170; farm ownership by,
  152; land leasing by, 112; marketing by, 13
growing locations in California, xiii*map*
Guadalupe Hidalgo, treaty, 111
guest-worker program, 133, 135, 141, 144

H2A visas, 133, 135, 141, 144
Haack, Ernst, 157
Haraway, Donna, 194, 209n19
Harrison, Jill, 97, 148
harvesting: caneberries vs. strawberries,
  141; competitions, 130; enhanced by
  pesticide applications, 146, 247n51,
  247n56; first-year berries, 4; mechanical, 115; mobile picking aids, 146; overview, 130; piece rates, 136, 139, 149, 200;
  quality control, 139; by robots, 21, 144;
  varietal alignment with workforce,
  138–39, 145–46. *See also* farmworker
  entries
hedgerows, as pathogen reservoirs, 186
Henke, Christopher, 16, 205
Hepler, Evan, 91
Herbarius of Apuleius Barbarus, 57
Herman, Louise, 61

heterozygosity (variability) of strawberries: as
  advantageous, 24, 26; of ancient plants,
  24; as attractive to breeders, 56, 63,
  73–74; of seed-grown plants, 53, 225n4
Hill, James Bryant, 110–111
Hinchcliffe, Steve, 18, 172, 215n58
History of the Strawberry, S. Wilhelm and J.
  Sagen, 28, 51, 57
Holmes, Seth, 130–31, 148
Hoppe family, 61
Horton, Sarah, 130, 147–48
Hovey, C.M., 59
Hovey's Seedling variety, 59
human diseases: antibiotic approaches, 177,
  256n22; assemblage thinking and, 17,
  18–19, 23, 212n46, 212n47, 215n58; dengue fever, 19, 212n47, 215n63; giardiasis,
  176–77; malaria, 17, 19, 23, 212n46,
  212n47; topological understanding of, 18
human health and safety: capitalism vs.,
  260n59; distrust of pesticides, 2, 7–8,
  102, 126, 207n3; food insecurity, 196;
  nutritional benefits of strawberries, 7;
  public health costs, 242n70. *See also*
  farmworkers, health and safety of
hybridization: cost of, 55–56; early experiments, 58; genome as commodity, 55–56;
  genomics tools, 67, 225n6; with native
  species, 60–61, 74, 226n23; ploidy and,
  53, 58, 74, 225n4; privatization and, 22;
  procedures, 53–54, 54*fig*; true to type
  varieties/species collections, 62. *See also*
  breeding strawberries *entries*
hybrid vigor, 55, 229n99
Hyde, Henry A., 157

iatrogenic harm, 10, 17, 20–24, 25
immigrants: Chinese, 112; H2A visas, 144;
  Hmong, 109, 167; Japanese, 112–13,
  113*fig*, 114, 136, 156, 158, 167, 252n70;
  loyalty to growers, 138, 140; Mexican
  and Central American, 137, 141, 154,
  252n70. *See also* farmworker *entries*
immigrants, undocumented: deportation
  concerns, 97, 135, 140, 148; labor shortages and, 131, 140
immigration reform: Bracero Program, 133,
  133–34, 135, 136; efforts to fortify border,

land values, 125–28; agro-ecological methods and, 191–92; Butte Valley, 120; climate change and, 127; as crop-selection criterion, 116, 121–22; for disease-free land, 127, 172, 173, 243n82; early twentieth century, 112; financial speculation, 105–6, 175; global investors, 122–23; grower concerns, 6, 105–6, 201; industrial farming and, 10; land as fictitious capital, 104, 105–6, 121–25, 127, 201; land quality and, 125–26, 243n76; limitations imposed by, 171–72, 175; locational advantages, 109, 125, 238n19; more-than-human assemblages, 109; as nursery location criterion, 121; soil-fumigant role, 3, 126, 200; theory of differential rents, 125; urbanization and, 122, 126–27, 128, 151; vegetable crops and, 116, 121–22

Larson, Kirk, 66, 67–69, 72, 76, 164

Larvacide Products Inc. (Innis, Speiden, and Co.), 82

Lassen Canyon Nursery, 67, 159, 165, 228n77

Lassen variety, 63

Latinx, roles, 154, 161–62, 167–68, 252n70

Latour, Bruno, 213n49

lawsuits: methyl iodide registration, 2, 94; against nurseries, 250n37; by or against UC, 68, 69, 76; by plant breeders, 9; sharecropper system challenges, 137

Leahy, Brian, 6, 95, 176, 199

Leese, Jacob, 111

lettuce, *Verticillium* wilt disease, 221n52

lice, pesticides to control, 82, 86

licensing fees, for patented varieties, 22, 168, 171, 172

livestock: history in California, 110–11; human diseases and, 18–19, 215n63; prophylactic use of antibiotics, 100, 202

Loftus, Charlie, 61

Lorimer, Jamie, 187, 194

Los Angeles, agricultural history, 155

Lyons, Kristina, 34, 195

Macrophomina phaseolina. See crown rot

malaria: mosquito control and, 17, 19, 23, 212n46, 212n47

Mansfield, Becky, 3

Manteca, xiii*map*, 119

Margulis, Lynn, 218–19n15

marketing contracts, 168–69

marketing orders, 160

marketing strawberries, 8*fig*: bad press for fumigant use, 2, 7–8, 11; berry shortages, 161; costs paid by growers, 169; CSAB/CSC funding for, 162–63; Driscoll's role, 168, 172; exports, 160; generating demand, 7, 162–63, 197–98; by growers' cooperatives, 13, 156; history of, 155, 156–57, 159–160; institutional innovations, 166; by Latinx growers, 161–62; organic berries, 189; political-economic dynamics, 13–14; private labels for retailers, 164; strategies, 7, 9; technologies to extend shelf life, 162. *See also* overproduction; seasonal considerations

Marx, Karl, 147

McGrath, Dominick, 111

McGrath Family Farms, 113–14

McIntrye, Michael, 147

Melinda variety, 60

Merced variety, 66

meristem culture, 52, 54*fig*, 118, 241n56

metam potassium, 96, 230n6

metam sodium, 96, 230n6

methyl bromide: agrochemical industry support for, 78; CHAMACOS project, 101; drift potential, 98; effectiveness of, 98, 232n39; history of use, 1, 65, 80, 84–87, 88; non-fumigant uses, 85–86; as ozone-depletion chemical, 1, 85, 92–93; plant injury caused by, 100; research on alternatives to, 179, 180–81; safety claims, 85; soils amenable to, 123–24; toxicity and health concerns, 92; TriCal and, 171; vegetable crops and, 116–17; for weed control, 88; World War II use, 86

methyl bromide, ban on use of: critical use exemptions, 93, 94, 95–96, 102, 179, 235n78; disease-resistance breeding and, 66; exemption for nurseries, 93–94, 120–21, 192; grower concerns, 236n90; impact on strawberry industry, 29, 33,

47, 179; lobbying to prevent, 102, 165, 179; Montreal Protocol and, 1, 92–93, 178–79; quarantine pre-shipment exemption, 93–94, 102, 179; replaced by methyl iodide, 1–2, 6, 80, 94, 179; township caps, 95–96, 126

methyl bromide with chloropicrin: disease-resistance breeding and, 72; formulations, 87; history of use, 43, 65; impact on strawberry industry, 44, 88, 222n72, 233n57; methyl bromide ban and, 96, 98; synergism of, 88

methyl iodide: advantages of, 2, 94; below-ground dispersal, 2; environmental epigenetics effects, 3; history of use, 88, 94; methyl bromide replaced by, 1–2, 6, 80, 94, 179; patent on, 85; protests over registration of, 1–2, 204, 207n3; request for registration of, 2; toxicity and health effects, 1–2, 79–80, 80, 94

Mexico: economic restructuring, 134, 141; fortifying US border with, 134, 135, 140, 151, 200; late-fall and winter fruit from, 71, 109, 161; Latinx roles in California, 154, 161–62, 167–68, 252n70; strawberry industry, 174; treaty of Guadalupe Hidalgo, 110; water resources, 127. *See also* farmworker *entries*; immigrant *entries*

microbiomatics, 218–19n15

Microcylcus ulei, rubber plant disease, 15

mildew, resistant varieties, 74

mineralization in soil, 34

Mitchell, Don, 133, 150

Mitchell, Timothy, 17–19, 23, 135

mobile picking aids, 146

Mojave variety, 66

molecular bureaucracy, 91–92, 100

monoculture, 5, 10, 16, 210n22. *See also* industrial farming

Monterey County, xiii*map*; dwindling profits, 126; *Fusarium* identified from, 47; history of strawberry growing, 155, 157; landownership changes, 161; production statistics, 174; Rancho Nacional, 110–11; rezoning agricultural land, 122; seasonal strawberry production, 115; strawberry concentration, 109*fig*, 116*fig*, 254n80

Montreal Protocol on Substances That Deplete the Ozone Layer, 1, 92–93, 178–79

Moore, Jason, 124, 195, 242n70

more-than-human assemblages. *See* assemblage approach

Morrill Act, 212n42

mosquitoes: DDT to control, 19, 23; dengue fever and, 19; malaria and, 17, 23, 212n46, 212n47

Mosquito Trails, 215n62

multiple and disparate objects, 17

Muscodor, soil fumigant, 180

The Mushroom at the End of the World, A. L. Tsing, 218n14, 223n94, 259n54

mycorrhizae, 34

Nading, Alex, 19

NAFTA, 134

Nash, Linda, 148

Nast, Heidi J., 147

National Organic Standards Board (NOSB), 190, 257–58n35

The Natural Wealth of California, 70

Naturipe Berry Growers: competitors, 163; establishment, 157; expansion, 159; former UC employees at, 158; funding for soilborne disease research, 169; incorporation of, 164; Japanese growers, 157, 160; marketing by, 159; university varieties, 227n53

nematodes: chloropicrin to control, 43, 82; D-D to control, 83–84; soil fumigants to control, 1; as strawberry pests, 38–39

neurotoxins, 1–2, 92

Niklor Chemical Company, 208n1

Nonfumigant Strawberry Production Working Group Action Plan, 101

NorCal Nursery, 159, 165

normalized difference vegetation index (NDVI), 79

novel ecologies, 194–95, 201

novel pathogens, 19–20, 97, 194–95, 201; containment technologies and, 187; crown rot, 30, 33, 45, 49; explanations for, 224n96; fields recolonized by, 171; fusarium wilt, 30, 33, 49, 224n96; novel plant collapse, 47; pesticide regulations and, 56

licensing fees, 168, 171, 172; Plant Sciences, 164, 228n80; by private breeders, 22, 64–65, 68–70, 75, 226n41, 228n80; procedures, 54; royalties, 65, 75; by UC, 76–77

pathogens, defined, 30. *See also* plant pathogens

Pathological Lives: Disease, Space and Biopolitics, S. Hinchcliffe, 31–32

Pegg, G. F., 31, 35, 36, 42, 220n33

pesticide drift, 97–99, 132, 148, 200

pesticide or pest treadmill, 20–21

pesticide regulation: Clean Air Act, 126; efforts to limit, 102, 165, 179; for export market, 248–49n1; history of, 89–92; impact on farming practices, 26–27, 96–100; interviews with regulators, 28; land prices and, 126; mitigation measures, 99; molecular bureaucracy, 91–92, 100; nonchemical disinfestation strategies and, 181–82; Nonfumigant Strawberry Production Working Group Action Plan, 101; ontological criticism, 91; organics in response to, 189; pesticide drift and, 97, 148, 200; proposed ban on all fumigants, 80, 101; proposed reorientation of, 100; toxicology studies, 92, 235n75; TriCal fumigation services, 165, 252n67; Trump administration and, 93; as under-enforced, 91; unintended consequences of, 89, 99–100, 165; use reports, 141–42; vegetable crop fumigation, 116–17, 231n53; VOC restrictions, 126–27; winners and losers, 153. *See also* buffer zones; Department of Pesticide Regulations, California; methyl bromide, ban on use of

pesticide regulation, of specific chemicals: chloropicrin, 94–95, 99; dazomet, 96; metam potassium, 96; metam sodium, 96; Telone, 95–96, 126; Telone and chloropicrin, 236n87

pesticide resistance, 19, 20–21

pesticides, 148–49: appropriationism and substitutionism, 13, 249n2; assemblage thinking and, 19–20; cost of, 172, 195; crops with most intensive use of, 8, 209n11; environmental epigenetics

effects, 3; farmworker exposure to, 95*fig*, 97, 98, 101, 132, 148–49, 197; green chemicals, 255n8; iatrogenic harm and limits of repair, 20–24; industrial agriculture dependent on, 10; insect/disease problems exacerbated by, 5, 10, 16; intergenerational effects of, 101, 132, 148; as nonhuman actors, 215n60; outsourcing application of, 165; pressure to continue using, 153; public distrust of, 2, 7–8, 102, 126, 207n3; study of long-term effects, 101; synergistic reactions, 91–92, 100; toxicology studies, 92, 235n75; unintended consequences of use, 19–20, 23; use with poor soils, 243n82. *See also* agrochemical industry; soil fumigant *entries; specific pesticides*

pesticide toxicity: EPA understanding of, 90–91; studies of, 92, 235n75. *See also* farmworkers, health and safety of

Pesticide Use Reporting System (PURS), 90

Petaluma variety, 66

petroleum industry, soil fumigant role, 82–83

Phytophthora: fumigation and, 45; *Phytophthora fragariae*, 39, 41, 117; *P. infestans*, 21

piece rates, for farmworkers, 136, 139, 149, 200

Pineapple Research Institute, University of Hawaii, 83–84

Pine variety, 74

Planasa, NorCal bought by, 165

plantationocene (plantation agriculture), 10, 194–98, 209n19, 210n22, 255n4, 260n58

plant diseases: biodiversity to control, 13; climate change and, 21, 127; disease-free land, value of, 41, 127, 200, 243n82; disease triangle model, 32; drought and, 32, 172; environmental effects that exacerbate, 21; land values reduced by, 123–25; living with, 32, 218n14; monoculture/industrial farming and, 5, 10; nonchemical management alternatives, 102; salinity and, 127; of strawberries, 25, 29–50; topographical understanding

plant diseases *(continued)*
of, 31–33, 35–37, 220n30. *See also* crown
rot; fusarium wilt; *Phytophthora*; soilborne pathogens; *Verticillium* wilt
disease
"plantiness," political landscapes shaped by,
225n7
planting beds: bed fumigation, 97; from
fumigated soil, 139, 200; harvesting
difficulties and, 150
Plant Patent Act of 1930, 22, 62, 156
plant pathogens: defined, 30; encouraged
by fumigation, 45, 47, 56; evolution of,
75, 100, 218n10, 224n100; fitness penalty, 36–37; moisture preferences, 107;
propagation practices that eliminate,
118; propagation practices that spread,
117–18; reservoirs of infection, 37, 97;
soil conditions and, 33–34; topographical theory of virulence, 35–37, 220n30.
See also novel pathogens
plant pathogens, specific: *Fusarium oxysporum f. sp. Cubense*, 15, 46; *Fusarium
oxysporum f. sp. fragariae*, 30, 33, 46–48,
48–49, 223n88; *Macrophomina phaseolina*, 30, 33, 45–46, 48–49, 75, 124, 127,
218n10, 223n88; *Microcylcus ulei*, 15;
Phytophthora fragariae, 39, 41, 117;
Phytophthora infestans, 21; *Trichoderma*,
45. *See also Verticillium dahliae*
Plant Sciences, 164, 228n80
plant sorting machines, 144
Pliny the Elder, on strawberry fragrance, 57
political-economic dynamics: of agrarian
economies, 153; agricultural scientists
and institutions of repair, 11, 15–16;
assemblage thinking and, 24; grower
intransigence and, 11–16, 150; nonhuman factors, 12; "plantiness" role, 225n7;
risk factors, 12, 200; technological
development, 12–14, 16. *See also* industrial farming
Porter, John T., 111, 155
potato diseases: brown blight, 38; wilt,
35–36
pre-cooling berries, 157, 162
prices for berries: branding costs, 168;
grower concerns, 6; organic berries,

258n38; overproduction and, 14, 21, 126,
160, 162, 166; political-economic
dynamics, 13–14; statistics, 126, 151;
sustainably grown berries, 189, 190, 197
probiotic approaches to pest control, 186–
89, 254–55n3; agro-ecological methods,
188–89, 190, 191; with antibiotic
approaches, 193; cost of, 257n28; disadvantages, 198; giardiasis treatment, 178;
human life and, 196; inputs, 187–88
processing industry: frozen fruit, 158–59,
160, 163, 166; institutional innovations,
166; Latinx growers supported by,
161–62; technologies to extend shelf life,
162
Processing Strawberry Advisory Board, 160
production regions in California, xiv*map*.
See also specific regions
profits: dwindling profits, 126; factors, 115;
farmworker wages and, 138, 139, 146–47,
149, 150; statistics, 174; winners and
losers, 152–53, 170–73, 175, 178
propagating strawberry plants: asexual
methods, 54, 56; Clean Plant program,
California, 93–94, 192; cloning, 117;
diseases spread by, 117–18; foundation
blocks, 118, 119; fumigant use, 119, 204;
at high elevations, 52, 119, 120, 121*fig*,
149, 200; historic planting stock, 157;
history of, 159; increase blocks, 118;
interviews with breeders, 51–52; labor
contractors, 144; labor shortages,
142–43, 247n51; land used for, 119; by
meristem, 54*fig*, 118, 241n56; methyl
bromide use, 93–94, 120–21, 192,
247n51; nuclear stock, 118; organic
starts, 192, 258n38; procedures, 51–52,
53–54, 54*fig*, 117–19; risk factors,
250n37; theft of plant material, 56;
tissue culture, 118. *See also* runners,
strawberry plants
propargyl bromide, 88
property, theories of, 125
Proposition 187, California, 134
proprietary behavior, 73–77; by Driscoll's,
158, 171–73, 226n41; by institutions of
repair, 200; political-economic dynamics, 24, 203; of private breeders, 56, 60,

soil fumigants, importance to strawberry
production *(continued)*
enhanced by, 132, 146; history of, 7, 88,
179, 233n57; increasingly intensive
production, 88–89; political-economic
dynamics, 11–16, 24; resurgence of
pathogens and, 10, 56
soil fumigants, specific: 1,3-dichloropropene
(1,3-D), 80, 82–84, 236n91; Dominus, 180,
181, 190; Dowfume, MC-2, 87; Ennoble,
181; Inline, 47, 84; metam potassium, 96,
230n6; metam sodium, 96, 230n6; Muscodor, 180; Paladin (DMDS), 180, 181;
propargyl bromide, 88; Telone, 84, 94,
95–96, 97, 126; Telone and chloropicrin,
30, 47, 87, 236n87. *See also* chloropicrin;
methyl bromide; methyl iodide
soil fumigants, toxicity and health effects:
drop-in methyl bromide replacements,
181; to farmworkers, 79–80, 95*fig*,
97–98, 101, 132, 137, 148–49, 197, 200;
intergenerational effects of, 101, 132,
148; methyl bromide substitutes, 98;
pesticide drift, 97; public distrust of, 2,
126; World War I soldiers, 81*fig*
soil fumigants, unintended consequences
of use: breeding stock weakening,
72–73; dead and diseased soil, 115, 124,
194, 194–95; declining chemical efficacy, 100; emergent dynamics, 195, 196;
human harm, 194–95, 242n70. *See also*
novel pathogens
soil health: emphasis on, 33–34; fumigation
vs., 44–45, 49, 79, 222n72, 243n82;
Verticillium wilt and, 42
soilless systems (field-scale hydroponics),
184–86, 185*fig*; disadvantages, 106, 128,
185–86, 192; farmworker ergonomics,
144, 197; NOSB ban on, 190
soil microbes: ASD and, 182; benefits of,
33–34, 124, 224n97; killed by chloropicrin, 82; killed by soil-fumigant replacements, 181; killed by soil fumigants, 44,
45, 49, 80, 222n72, 224n96; mycorrhizae,
34; probiotic inputs to pest control,
187–88, 191; supplementing soil with,
222n73; symbiotic bacteria, 218–19n15
soils of California, 107, 123–24

Solanum sarrachoides, as *Verticillium
dahliae* host, 40
solarization of soil, 182
Soluri, John, 15
spacing between rows and plants, 139, 146,
161
Spanish land grants and missions, 110–11,
239n21
Specialty Crop Research Initiative, USDA,
179, 255n7
specialty crops, economic issues, 7
Spray Safe coalitions, 101–2
steam treatment, 182, 192
Stewart, Phillip, 75–76
Storkan, Richard C., 208n1
strawberry blight or yellows (xanthosis).
See *Verticillium* wilt disease
Strawberry Fields, M. Wells: on capital
outlays, 138, 201; on farmworker wages,
139; on grower differences, 252n70,
258n42; on harm to farmworkers,
130–31; on labor recruiting and management, 132, 145; on pollitical forces, 133;
timespan covered by, 166–67; on worker
disloyalty, 141–42
*The Strawberry: History, Breeding, and
Physiology*, G. Darrow, 28, 240n49
strawberry industry, California: acreage
statistics, 63–64, 175, 254n84; businesses
dependant on, 25; coevolution with
multiple factors, 3, 11; consequences of
abandoning, 197–98; consolidation, 152,
194; corporate ownership, 152; CUE user,
93, 235n78; current status, 174–75; distrust of social scientists, 27; economic
importance, 7; growing locations, xii-
imap; iatrogenic harm and limits of
repair, 20–24; impact of pesticide restrictions on, 96–100; impending shakeout,
152–54; Latinx as major players, 154,
161–62; monopolization of, 76–77, 150,
156, 165, 166; moving to Mexico, 174;
pesticide reliance, 8, 10, 209n11; political-
economic dynamics, 11–16, 24; politics of
post-plantationocene ruin, 194–98;
production costs, 138, 139, 152, 200;
production regions, xiv*map*; production
statistics, 160–61, 161–62, 174; techno-

logical developments, 161–62; water requirements, 108; as world leader, 161, 166, 197, 238n4

strawberry industry, history in California, 38–40, 106–8; breeding research, 59–62; Central Valley, 110, 238n20; experimental beginnings, 155–58, 199; farm sizes, 167, 252n70; farmworkers, 136–140, 150; fluctuations from 1960s to present, 161–66; institutional development, 154–166; natural advantages, 106–10, 123, 125; pesticide regulation, 6; post WWII, 158

strawberry industry, United States, 59

strawberry industry, worldwide, 7, 24–25, 58–59

The Strawberry in North America: History, Origin, Botany, and Breeding, 58–59

Strawberry Institute of California, 64, 75, 159–160

Strawberry Plant Certification Program, 42, 118

strawberry plants: adaptability of, 53, 55*fig*; fragrance of, 56–57; fruit morphology, 52–53; growth habit, 51; history of cultivation, 57–59; origin of, 29, 57. *See also* fruit; propagating strawberry plants; runners, strawberry plants

substitutionism, 249n2

suburbs. *See* urbanization and suburbanization

sugar beet farming, 111–12

Sunkist Growers, 162

Sustainable Agriculture Research and Education Program (UC), 78–79

sustainable strawberry production: challenges to, 204–5; collaboration with other species, 195, 196, 259n56; future of, 195–98; integrated pest management, 79, 188–89, 241n57; landownership and, 242–43n73, 242n72; political-economic dynamics that impact, 11; soil health and, 11; Swanton Berry Farm, 203–4; UC research on, 78–79. *See also* organic strawberry production; soil fumigants, farming without

Swain's Ranch, 59

Swanton Berry Farm, 203–4

Sweet Briar Company, 28

Sweet Briar Ranch, 61

Sweet Briar variety, 61, 63–64, 117, 155

Tahoe variety, 63

Tanimura and Antle brand, 163

tarps, plastic: for ASD, 183; for bio-fumigation, 182; as containment technology, 187; effectiveness of, 98; to extend season, 240n50; history of use, 161; off-gassing when removed, 98; shredded by wind, 99*fig*; for solarization, 182; for *Verticillium* wilt control, 43*fig*

technology treadmill, 14

Telone, 84; regulation of, 95–96, 126; toxicity and health effects, 94, 97

Telone and chloropicrin: regulation of, 236n87; as soil fumigant, 30, 47, 84, 87; toxicity and health effects, 236n87

tenant farmers, 112

theory of differential rents, 125

Thomas, Harold, 39–40, 41–42, 62–66, 66

tilling, impact on disease, 32, 220n33

Tioga variety, 65

tissue culture procedures, 118

tomatoes, *Verticillium* wilt, 39, 40

topographical understanding of plant disease, 31–33, 35–37, 218n10, 220n30

Torrey variety, 65

totally impermeable film (TIF), 95, 96–97

township caps, 95–96, 126

toxicity: as Anthropocene feature, 209n18; studies of, 92, 235n75

trap crops, 188

TriCal, fumigation services, 165, 171, 208n1, 236n90, 252n67

Trichoderma, fumigation and, 45

troubled ecologies, 195, 259n53

Trump administration: anti-immigrant hysteria, 135; deportation of illegal immigrants, 135; EPA changes, 93; Kawamura as advisor, 164

Tsing, Anna, 195, 218n14, 223n94, 259n54

Turlock, xiii*map*, 119

Uekötter, Frank, 23, 50

unions, for farmworkers, 133–34, 245n19, 251n39

United Farm Workers (UFW), 133, 134

University of California (UC): agnotology concerns, 22–23; Agricultural Experiment Stations, 16, 41, 62; competition from other institutions, 164, 165–66; Division of Biological Control, 23; as institutions of repair, 15–16, 25, 154, 165–66, 199; strawberry industry connections, 16, 67–70, 75–76; Sustainable Agriculture Research and Education Program, 78–79; *Verticillium* research, 25–26, 33, 157. *See also* extension services, cooperative; land-grant institutions; research (agricultural scientists)

University of California, Berkeley, 38, 39, 160University of California, Cal Poly San Luis Obispo, 164, 165, 166, 224n97

University of California breeding program, 62–70; budgetary cutbacks, 164; CBC lawsuit against, 69; CSC lawsuit against, 68; field trials, 76*fig*; Foundation Plant Services, 54, 74, 118, 241n56, 241n59; germplasm in public domain, 68, 75, 171, 193–94; history of, 60, 119; patents held by, 76–77; priorities, 55–56, 75–76; proprietary threats, 22, 64, 68–70, 73–77, 75, 158, 164, 171

University of California Davis: CBC lawsuit against, 69; methyl bromide research, 86–87; Public Strawberry Breeding Program, 69

University of California Riverside: methyl iodide patent, 94; soil fumigant research, 86

University of Hawaii, Pineapple Research Institute, 83–84

university varieties: competition from private breeders, 22, 227n53; CSC and, 65–70; funding to develop, 168; produced by nurseries, 165; usage statistics, 63, 171; worldwide use of, 241n59

urbanization and suburbanization: land values and, 122, 126–27, 128, 151; in lieu of strawberries, 196, 259n54

US Agency for Toxic Substances and Disease Registry, 92

US Department of Agriculture (USDA): Bureau of Entomology, 80; Federal Plant Repository, 74; fumigation research, 81–82, 84; nonchemical disease management alternatives, 188; pesticide registration, 89; Specialty Crop Research Initiative, 179, 255n7

US Environmental Protection Agency (EPA): approval for drop-in methyl bromide replacements, 181; chemical use facilitated by, 92; chloropicrin reregistration studies, 94–95; establishment, 89; label requirements, 90; methyl iodide registration, 95; Ozone Protection Layer Award, 2; revolving-door politics, 90; shortcomings of, 90–91; Trump administration, 93

US National Labor Relations Act (Wagner Act), 133

varieties: alignment with workforce, 138–39, 145–46; ancient, 24, 53, 56, 74, 75; chilling requirements, 227n62; cloning, 54, 117; day-neutral, 71, 115; DNA fingerprinting of, 118; early-season, 159–160; everbearing, 4, 70–71; historic acreage statistics, 63–64; licensing fees, 168, 171, 172; marketing orders, 160; native or wild, 53, 56, 60–61, 74, 75, 226n23; for organic production, 173, 189; overproduction and, 158–59; short-day, 71, 161; from Strawberry Institute of California, 64; timing of planting, 4; true to type collections, 62. *See* breeding strawberries; germplasm; heterozygosity (variability) of strawberries; patenting plants; proprietary behavior; resistant strawberry varieties

varieties, specific: Albion, 66; Banner, 61–62, 63–64, 71, 156, 158, 226n41; Benicia, 66; Camarosa, 66; Carolina Superba, 62; Chandler, 66, 204; Cinderella, 60; Diamante, 66; Donner, 63; Ettersburg 80, 61; Fresno, 65; Hovey's Seedling, 59; Lassen, 63; Melinda, 60; Merced, 66; Mojave, 66; Pajaro, 66; Palomar, 66; Petaluma, 66; Pine, 74; San Andreas, 66, 72; Seascape, 66; Shasta, 63; Sierra, 63; Sweet Briar, 61, 63–64, 117, 155; Tahoe, 63; Tioga, 65;